Springer Biographies

The books published in the Springer Biographies tell of the life and work of scholars, innovators, and pioneers in all fields of learning and throughout the ages. Prominent scientists and philosophers will feature, but so too will lesser known personalities whose significant contributions deserve greater recognition and whose remarkable life stories will stir and motivate readers. Authored by historians and other academic writers, the volumes describe and analyse the main achievements of their subjects in manner accessible to nonspecialists, interweaving these with salient aspects of the protagonists' personal lives. Autobiographies and memoirs also fall into the scope of the series.

Anna Matalová • Eva Matalová

Gregor Mendel - The Scientist

Based on Primary Sources 1822-1884

Springer

Anna Matalová
Mendelianum of the Moravian Museum
Brno, Czech Republic

Eva Matalová
Institute of Animal Physiology and Genetics
Academy of Sciences
Brno, Czech Republic

ISSN 2365-0613 ISSN 2365-0621 (electronic)
Springer Biographies
ISBN 978-3-030-98925-5 ISBN 978-3-030-98923-1 (eBook)
https://doi.org/10.1007/978-3-030-98923-1

This Springer imprint is published by the registered company Springer Nature Switzerland AG
The registered company address is: Gewerbestrasse 11, 6330 Cham, Switzerland

We dedicate this book to Professor Jaroslav Křiženecký (1896–1964), the founder of the Mendelianum of the Moravian Museum and defender of Mendel's work in the era when Mendelian genetics was banished under the Stalinist regime.

Preface

This book aims to provide a comprehensive insight into the life and work of Johann/ Gregor Mendel with focus on his scientific activities. Importantly, the concept is based exclusively on primary sources and historical documents.

Additionally, this book is a unique collection of Mendelian documents to facilitate access to images of original records, people and places related to Mendel as a teacher and researcher. The historical documents from Mendel's life are rare. Some archival sources of the past have been reported as unattainable or lost. Some documents were relocated several times as a consequence of the Second World War, of the change of the political regime in 1948 in what was then Czechoslovakia and in connection with the Prague Velvet Revolution in 1989.

Mendel's original handwritten manuscript of his 1865 lecture Experiments on Plant Hybrids (published in 1866) was considered lost during the German occupation of Czechoslovakia and emerged unexpectedly in the hands of Mendel's relative Clemens Erik Richter in Germany in 1987. During the Second World War, some original documents including Mendel's original manuscripts on Pisum and Hieracium were in possession of the *Naturforschender Verein* (Iltis 1932), a research association of natural scientists. The archival papers were physically kept in the Escompt Bank safe of *Naturforschender Verein* in Brno. When Nazis seized Czechoslovakia, the documents were hidden in secret places. After the Second World War, the belongings of the German *Naturforschender Verein* were supposed to be confiscated by the Czechoslovak state. At that time, some documents appeared in Brno scientific circles. Some emerged shortly before the Velvet Revolution in St. Monica Ruit in Germany. Some have not yet been discovered. In some cases, their possessors are not transparent. A few of Mendel's artefacts were taken to the United States at the outbreak of the Second World War. The Mendelianum of the Moravian Museum in Brno has been systematically tracing Mendel's documents since 1962.

Since 1966, Folia Mendeliana publishes on historical sources pertaining to Mendel's life and work. The edition of Folia Mendeliana started with the support of UNESCO and an international group of scientists. The advisory publication committee consisted of J. W. Boyes (Montreal), N. P. Dubinin (Moscow), L. C.

Dunn (Irvington on Hudson), A. E. Gaissinovitch (Moscow), B. Glass (Stony Brook, New York), J. Heimans (Amsterdam), F. B. Hutt (Ithaca, New York), H. B. D. Kettlewell (Oxford), E. Lauprecht (Mecklenhorst), R. C. Olby (Oxford), B. Sekla (Prague), R. M. Stecher (Cleveland, Ohio), C. Stern (Berkeley, California), H. Stubbe (Gatersleben) and F. Weiling (Bonn). Jan Jelínek, Director of the Moravian Museum, was the publisher, and V. Orel, Head of the Gregor Mendel Department of Genetics, was the editor. As a consequence of the repressions of the Prague Spring 1968, Folia Mendeliana lost their independence. They became a part of the Acta Musei Moraviae. The international publication committee was resolved by the pro-Soviet administration of the Czechoslovak Ministry of Culture.

The first information about Mendel's early years comes from his family: his younger sister Theresia Schindler, born Mendel, her husband Leopold, and their sons Alois and Ferdinand. Mendel's first biographer, Hugo Iltis, was fortunate to obtain permission to search for Mendel's relics in the Augustinian Monastery. He was given that privilege thanks to the support of Professor William Bateson, a respected Cambridge naturalist who visited the Augustinian Monastery in 1904, seeking Mendel's notes from the apicultural experiments. Gustav Niessl von Mayendorf, secretary of the *Naturforschender Verein*, who arranged for publication of Mendel's 1865 lecture on Pisum hybrids, the 1870 paper on Hieracium hybrids and Mendel's meteorological observations, failed to obtain a permission from the Augustinians. The exclusive right given to Iltis yielded the discovery of Mendel's Pisum manuscript among discarded old papers destined for fire, and other artefacts that are mentioned in his monograph *Gregor Johann Mendel: Leben, Werk und Wirkung* published by Julius Springer in 1924. Iltis also obtained the precious collection of originals from Mendel's 1850 examination from the University of Vienna. Some archive materials originally in Vienna now reside at the University of Illinois in Urbana-Champaign Archives. Some artefacts from the Augustinian Monastery travelled to Villanova University in Pennsylvania or to Montréal in Québec.

We quote from Mendel's documents from Correns' publication of *Mendel's letters to Carl von Nägeli* (1905), the German version of Iltis' Mendel monograph (1924) and its English translation (1966). Some original documents were found and first published by Oswald Richter in 1943. The Germans intended to build a Gregor Mendel Institute for Genetics (*Das Gregor Mendel Institut für Genetik*) in Brno. Lack of money at the end of the Second World War prevented realization of their plans. In view of the Gregor Mendel Institute for Genetics, the Mendel artefacts were taken from the Old Brno monastery and from Iltis' *Das Gregor-Mendel-Museum* of the German Society of Science and Arts and inventoried as one collection.

The most valuable documents were published by Jaroslav Kříženecký as a Mendel source book (1965) by the Leopoldina Academy in Leipzig. His collection of Mendelian pictorial documentation was prepared for print by Kříženecký's successor, Vítězslav Orel and his colleagues Ludmila Marvanová and Josef Sajner in *Iconographia Mendeliana* published by Jan Jelínek, Director of the Moravian Museum in 1965. Vítězslav Orel's monograph, *Gregor Mendel the First Geneticist* (1996), Franz Weiling's *Johann Gregor Mendel* (1993-4), the academic journal *Folia Mendeliana* (since 1966), Josef Sajner's documentation from Mendel's teaching in the *Gymnasium* in Znojmo (1967) and Renée Gicklhorn's analysis of documents from Mendel's student years in Vienna (1969, 1973) represent important reliable scholarly sources. Johann Vollman's collection of newspaper reports on Mendel's activity (2016) and Peter Van Dijk's (2018) discovery of newspaper articles about Mendel's experimentation with plants and vegetables are excellent examples of continued discovery of facts about Mendel. Jan Klein and Norman Klein with their monumental Mendel monograph *Solitude of a Humble Genius: Gregor Johann Mendel—Formative Years* (2013) composed an ever-lasting Mendelian symphony.

We publish this source book on the occasion of the bicentennial of Mendel's birth to go to the roots of his accomplishment and to support continuation of Mendelian historical research. Mendelian documents in facsimile, German transcriptions and/or English translations may be a valuable resource of inspiration and study.

Geographical names then and now

Brünn/Brno
Gross Petersdorf/Dolní Vražné
Heinzendorf/Hynčice
Krakau/Krakow
Lemberg/Liu
Leipnik/Lipník nad Bečvou
Neutitschein/Nový Jičín
Neuhwiesdlitz/Nové Hvězdlice
Olmütz/Olomouc
Troppau/Opava
Wien/Wien
Znaim/Znojmo

Spelling of some names of individuals may not be identical due to historical sources being grammatically dependent on the German-Czech intercourse. Czech names were sometimes transcribed phonetically in German and Latin was still used in the nineteenth century. For example, Mendel occurred as Gregor in German, Řehoř in Czech, Gregorius in Latin.

Note: The German ß = ss.

Portrait of Gregor Mendel (1822–1884)

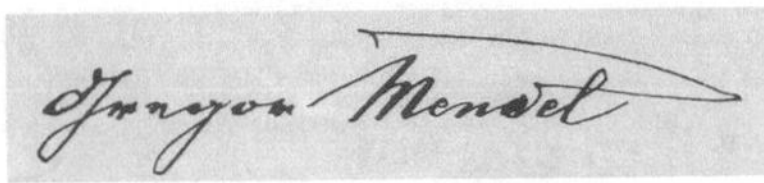

Signature Gregor Mendel

Brno, Czech Republic

Anna Matalová
Eva Matalová

Acknowledgements

Our special thanks go to Professor Daniel J. Fairbanks for valuable collaboration, especially in the preparation of the manuscript, and also to Professors Donna Fairbanks and Hilary Demske, who keep the Mendel-Janáček relationship alive. We are indebted to Professor Scott Abbott for his help in translation of Mendel's original documents.

Our thanks are also due to the past and present staff of the Mendelianum in the Moravian Museum: Dr. Jiří Sekerák, Head of the Department for the History of Biological Sciences in the Moravian Museum for granting permission to publish reproductions of the originals from the Mendelianum collection, Mgr. Pavlína Pončíková and Jana Daňková for providing digital copies of the originals and also to Lenka Mecerodová and Mgr. Marcela Kusáková for their earlier cooperation. Many thanks belong to Professors Christina Laukaitis and Robert Karn who documented Mendel's artefacts archived in the USA.

We are indebted to the world scientists, Professors Jan Klein, James D. Watson, Marshall W. Nirenberg, to the members of the International Mendel Team and many others for their support of Mendel's research. We appreciate all who were courageous enough to establish the Mendelianum, especially in times of anti-Mendelism in Czechoslovakia (1950–1966). Originally the Mendelianum was located in the Augustinian Monastery (1965–2000). After restitution of the buildings to the church, the Mendelianum was relocated to the historic buildings of the Moravian Museum, acquired in 1817 and used by the Agricultural Society as its central residence.

Thus, in the twenty-first century, Mendel has returned to the historic place of the Scientific Society, where he worked for 33 years. His scientific accomplishments made Brno famous also thanks to all Mendelianum visitors and Mendelian scholars disseminating Mendel's scientific ideas and elaborating his cultural legacy.

Contents

About the Authors

Anna Matalová is an Emeritus Head of the Mendelianum of the Moravian Museum that continues the historical research started by the Franz (now Moravian) Museum founded by the learned Agricultural Society in 1817. She was the first (and the last) assistant to Jaroslav Kříženecký after his release from the communist prison. Since then, she has devoted her life to research into Mendelian history and its presentation. She was an editor of the academic journal *Folia Mendeliana*. After the Velvet Revolution, and until her retirement, she was the Head of the Mendelianum in Brno and Chairperson of the Czech Committee for the History of Science and Technology in Prague. Shortly after the fall of the so-called Iron Curtain, Anna began organizing the international Mendel Forum conferences to facilitate intensive and open communication among researchers and experts from the West and the East. Thanks to her endeavour, the Mendelianum has now been located in the authentic rooms of Mendel's Agricultural Society in the historical heart of Brno. She also has highlighted other places connected with Mendel's life and authored Mendel's Path in Brno. She was instrumental in the establishment of the Fund for Mendel's native house in Vražné-Hynčice, allowing for its reconstruction. Now the reconstructed farmstead of Mendel's parents in Hynčice is a cultural meeting place. In her exhibitions and scientific publications, she has presented Mendel's accomplishments in context with actual cultural, social and scientific topics. *Her husband, son and daughter consider Mendel as a member of the family.*

Eva Matalová is a leading Research Scientist at the Institute of Animal Physiology and Genetics of the Czech Academy of Sciences and a Professor at the University of Veterinary Sciences in Brno. She has "inherited" her interest in Mendel from her mother, Anna. Eva has been active in the Mendelianum since her Ph.D. studies, organizing annual Mendel Forum conferences in the twenty-first century and other activities focused on students and youth (Junior Mendel Forums, Wednesdays with Mendel, DNA afternoons, Science in Action, etc.). Eva proposed the idea of the International Mendel Day on March 8, and established an International Mendel Network consisting of expert scientists and historians. She was the initiator, facilitator and guarantor of the project *Centrum Mendelianum*, based on Anna's concept of Mendel's accomplishments as a naturalist. Eva feels honoured to assist with the publication of this book on the occasion of the bicentennial of Mendel's birth. *With love to our genetic offspring, Tommy and Luky.*

Abbreviations

Ackerbaugesellschaft (Agriculture Society)	Kaiserlich-Königlich Mährisch-Schlesische Gesellschaft zur Beförderung des Ackerbaues, der Natur- und Landeskunde in Brünn. It supplied some functions of the Academy.
CV	Curriculum vitae
D	Registration numbers of the document in the Mendelianum collection in the Moravian Museum in Brno.
fl.	Florin
FM	Folia Mendeliana; Folia Mendeliana Musei Moraviae; Folia Mendeliana-Supplementum ad Acta Musei Moraviae.
Gymnasium	Secondary grammar school, a prep school for university
IM	Iconographia Mendeliana. Pictorial publication issued for the Centenary of the publishing of the discovery of the principles of heredity by Gregor Johann Mendel in Brno.
i.-r.	imperial-royal
k. k.	kaiserlich-königlich
Mittheilungen	Mittheilungen der kaiserlich-königlichen Mährisch-Schlesischen Gesellschaft zur Beförderung des Ackerbaues, der Natur- und Landeskunde in Brünn.
MM	Moravian Museum Brno
MMM	Mendelianum of the Moravian Museum
O	Registration numbers of photos and pictures in the Mendelianum collection of the Moravian Museum
Realschule	A novel type of secondary-level industrial high school at that time.

Chapter 1
Assistant in the Old Brno Parish 1848

Johann Mendel wanted to be a teacher, following the example of his great-uncle Anton Schwirtlich (1758–1808) who was the first private teacher in Heinzendorf/Hynčice and deserved for building the elementary school in Mendel's native village. Anton was known for high intelligence, thirst for knowledge, autodidact skills, pedagogical talent, industry, and perseverance. Johann might have inherited the main features of his character from his great-uncle through his mother Rosina, born Schwirtlich.

Already during his *Gymnasium* studies in Opava Mendel attended the course for candidates of teaching and private teachers. At the age of 16, Johann got the certificate for private teachers and candidates of teaching with best results and recommendations. He started to make his living as a private teacher giving additional instruction to interested students in Opava and later in Olomouc.

When his father was invalidated, the economic situation became untenable. For poverty and health reasons Mendel gave up his philosophical studies in Olomouc at the end of the first semester and recovered in the circle of his family. His father transferred the farmstead to his older daughter Veronika and her husband that enabled Mendel to make a new start as a repetent of the 2-year philosophy. He accomplished his studies at the Philosophical Institute in 1843 with excellent results. His teacher of physics in Olomouc F. Franz recommended Johann to enter the Old Brno Augustinian monastery to get rid of his material troubles. In the brief sketch of his life of 1850 Mendel states that *his circumstances decided his vocational choice*. He was accepted to the Old Brno Augustinian community in 1843 at the age of 21.

After 1 year of noviciate and 4-year study of theology, Anton Ernst Schaaffgotsche (1804–1870), Bishop of the Brno Diocese, appointed the Augustinian priest Gregor Mendel as chaplain (*Kooperator*). The Old Brno Roman catholic parish administration was situated on the ground floor of the Augustinian monastery in the former Probst House of the Cistercian nuns who lived in the monastery until 1782. The appointment was made on the proposal of Mendel's superior, Franz Cyrill Napp (1792–1867), Abbot and Prelate of the Old Brno monastery. Napp filled the gap caused by the early deaths of two young *Kooperators*, Ferdinand Schaumann in

A. Matalová, E. Matalová, *Gregor Mendel - The Scientist*, Springer Biographies,
https://doi.org/10.1007/978-3-030-98923-1_1

1846 and Fulgenz Süsser in 1847. The parish priest and his associate *Kooperators*, usually four, oversaw pastoral care of the Old Brno parishioners, visited ill and dying people in the Old Brno St. Anna Hospital, and administered catholic rites (such as births, weddings, funerals, and mass services). Mendel accepted this position and exerted his best efforts to fulfil these duties. However, his attacks of psychosomatic nature, afflicted him while serving as a parish priest. He had suffered similar recurring attacks during his secondary schooling when working under stress. Mendel became overly exhausted in his new parish function and collapsed. His indisposition opened the way to Mendel as a teacher of natural science.

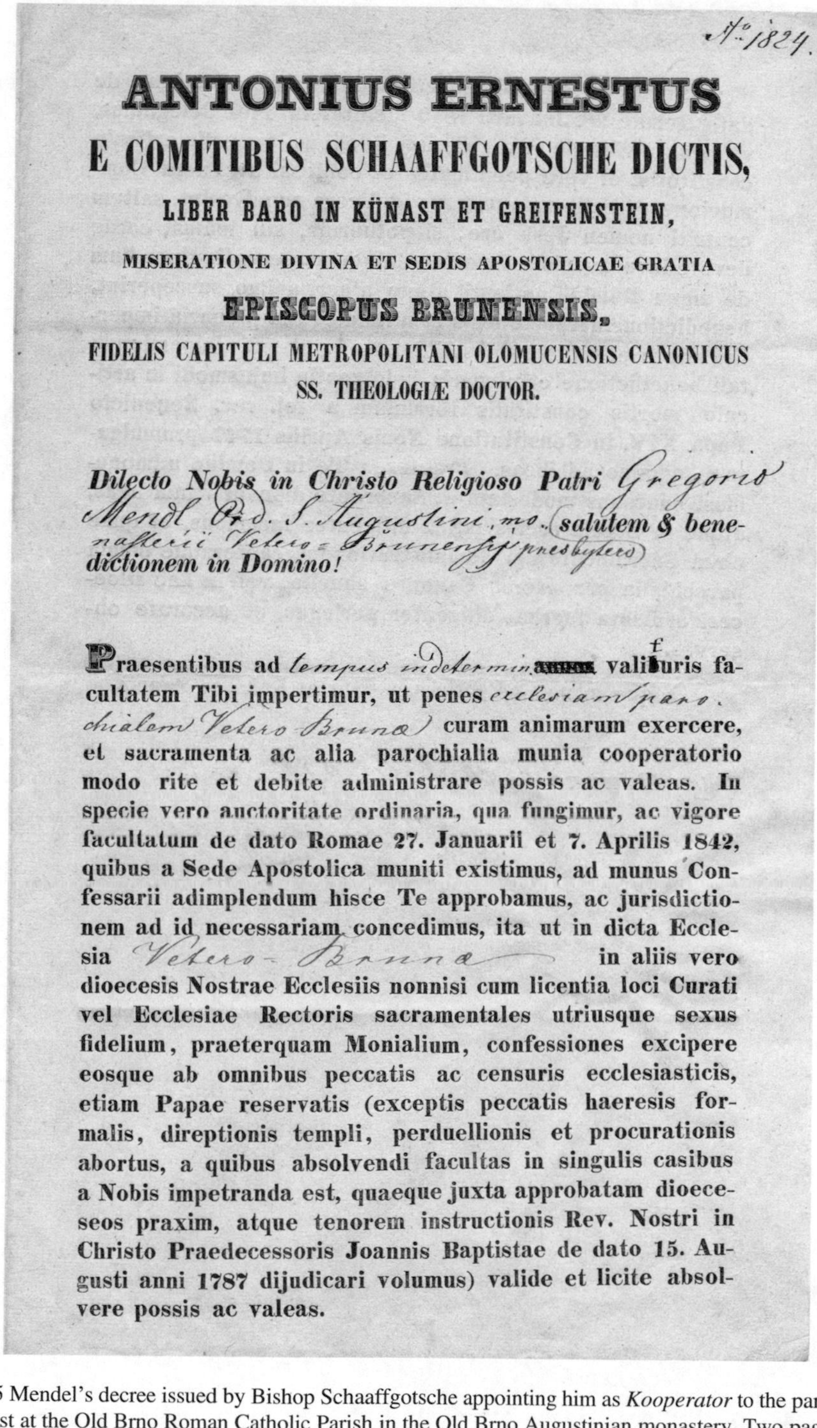

N° 1824.

ANTONIUS ERNESTUS

E COMITIBUS SCHAAFFGOTSCHE DICTIS,

LIBER BARO IN KÜNAST ET GREIFENSTEIN,

MISERATIONE DIVINA ET SEDIS APOSTOLICAE GRATIA

EPISCOPUS BRUNENSIS,

FIDELIS CAPITULI METROPOLITANI OLOMUCENSIS CANONICUS

SS. THEOLOGIÆ DOCTOR.

Dilecto Nobis in Christo Religioso Patri Gregorio Mendl Ord. S. Augustini, monasterii Vetero-Brunensis presbytero *salutem & benedictionem in Domino!*

Praesentibus ad tempus indetermin. valituris facultatem Tibi impertimur, ut penes ecclesiam parochialem Vetero-Brunae curam animarum exercere, et sacramenta ac alia parochialia munia cooperatorio modo rite et debite administrare possis ac valeas. In specie vero auctoritate ordinaria, qua fungimur, ac vigore facultatum de dato Romae 27. Januarii et 7. Aprilis 1842, quibus a Sede Apostolica muniti existimus, ad munus Confessarii adimplendum hisce Te approbamus, ac jurisdictionem ad id necessariam concedimus, ita ut in dicta Ecclesia Vetero-Brunae in aliis vero dioecesis Nostrae Ecclesiis nonnisi cum licentia loci Curati vel Ecclesiae Rectoris sacramentales utriusque sexus fidelium, praeterquam Monialium, confessiones excipere eosque ab omnibus peccatis ac censuris ecclesiasticis, etiam Papae reservatis (exceptis peccatis haeresis formalis, direptionis templi, perduellionis et procurationis abortus, a quibus absolvendi facultas in singulis casibus a Nobis impetranda est, quaeque juxta approbatam dioeceseos praxim, atque tenorem instructionis Rev. Nostri in Christo Praedecessoris Joannis Baptistae de dato 15. Augusti anni 1787 dijudicari volumus) valide et licite absolvere possis ac valeas.

D15 Mendel's decree issued by Bishop Schaaffgotsche appointing him as *Kooperator* to the parish priest at the Old Brno Roman Catholic Parish in the Old Brno Augustinian monastery. Two pages. In the decree, Mendel's name is wrongly spelled Mendl. The decree is dated in Brno, 20 July 1848, signed by Antonius Ernestus

Simul ex speciali Sanctissimi Patris concessione de dato Romae 28. Januarii 1842 facultatem Tibi delegamus, vi cujus utriusque sexus Christi fidelibus in mortis articulo constitutis, si vere poenitentes et confessi ac Sacra Communione refecti, vel quatenus id facere nequiverint, saltem contriti nomen Jesu ore, si potuerint, sin minus, corde devote invocaverint et mortem tamquam peccati stipendium de manu Domini patienti atque alacri animo susceperint, benedictionem Apostolicam cum indulgentia plenaria impertiri libere et licite valeas, ita tamen, ut pro impertienda tali benedictione et plenaria indulgentia hujusmodi in articulo mortis constitutis formulam a fel. rec. Benedicto Papa XIV. in Constitutione Nonis Aprilis 1747 promulgatam, omnino adhibeas. Praeterea Te in Domino admonemus, quatenus modestam et Sacerdote dignam vitam agas, atque in Studia ecclesiastica diligenter incumbas, et quae circa Sacramentorum administrationem, caeteraque munia parochialia per sacros Canones sancita, vel in hac Dioecesi ordinata fuerint, diligenter perlegas, et accurate observes.

Idque in nomine Patris et Filii et Spiritus Sancti.

Datum **Brunae** *die* 20 Julii 848

Antonius Ernestus

(continued)

IM17 Mendel's superior Cyrill Franz Napp, Abbot, and Prelate of the Augustinian monastery in Old Brno

IM17/FM15 The Old Brno Augustinian friars in the first half of the 19th century

FM15 Gregor Mendel in his younger years after his admission into the Augustinian Order (Dubec and Orel 1980)

References and Historical Printed Sources

Dubec K, Orel V (1980) Gregor Mendel's scientific activity in meteorology. FM 15:215–237

Chapter 2
Petitioner for Freedom 1848

On 07 August, in the revolutionary year 1848, a few days after his appointment as parish cooperator, Mendel revived hope for a change in his unhappy position. He added his signature to those of five other Augustinian friars on a petition addressed to the Imperial-Royal Constitutional Assembly, pleading for more freedom for friars serving as teachers. The author of the petition was Mattheus Klácel (1808–1882), a former professor, dismissed from the Philosophical Institute in Brno in 1844 for spreading pantheistic views. He was Mendel's closest friend and mentor in the monastery. Mendel was the scribe of the petition.

The facsimile of the petition for freedom signed by six Augustinian teachers and candidates for teaching submitted to the National Assembly of Austria is dated in Brno, 08 August 1848, signed by the Augustinians Fr. Mattheus Klácel, Dr. Philipp Gabriel, Josef Lindenthal, Benedict Fogler, Gregor Mendel and J. Chrysostomus Cygánek.

Orel and Verbík (1984, pp. 223–233) published the reproduction and transcription of the handwritten original text deposited in the State Archives in Vienna. An English translation was first published by Klein and Klein (2013, pp. 280–281). This important document reveals the special social milieu in the Old Brno Augustinian monastery under Abbot Napp. An excerpt from the English translation of the petition in Klein and Klein (2013, p. 281) reads as follows:

> Consequently, the undersigned professors and pastoral workers of the order of Saint Augustine in Old Brno take the liberty of appealing to the imperial parliament to <u>grant them constitutional civil rights</u>, and request to be allowed to devote their entire efforts, according to their abilities and their past services to public teaching institutions and to free, united, and <u>indivisible citizenship</u>. The undersigned make it respectfully their mission to promote science and humanity in accordance with the spirit of constitutional progress.—Brno, August 8, 1848—Fr. Mattheus Klácel, former professor of philosophy—Dr. Philipp Gabriel, professor of mathematics in Brno. Head of the Countess Thurn Institution—Josef Lindenthal, Kooperator at the parish church in Old Brno—Benedict Fogler, professor of French language & literature & accredited teacher of Italian language—Gregor Mendel, Kooperator & teaching candidate—Chrysostomus Cygánek, teaching candidate

A. Matalová, E. Matalová, *Gregor Mendel - The Scientist*, Springer Biographies,
https://doi.org/10.1007/978-3-030-98923-1_2

Shortly after Mendel became abbot, Klácel joined Czech emigrants in the USA. He left Brno secretly in 1869, arrived in Belle Plaine, Iowa, abandoned his monastic vows and assumed the first name Ladimír (Lada—harmony, mír—peace). In the 1868 abbatial election, Mendel gave his vote to Klácel in all rounds. Mattheus called Mendel his freethinking friend ("svobodomyslný přítel můj"). Mendel must have been disappointed at losing his best friend in the monastery. Is it possible that he knew about Klácel's intention to emigrate?

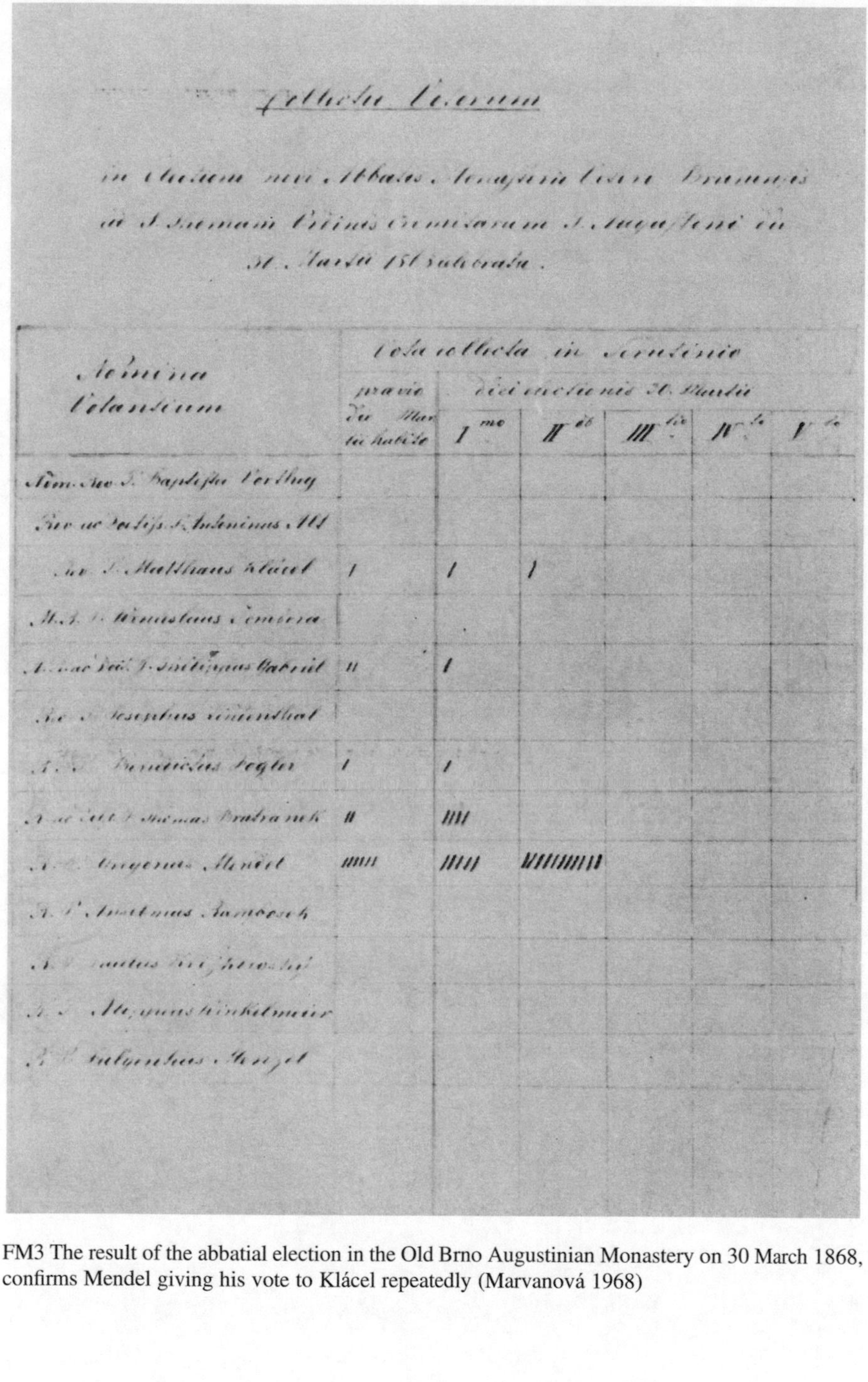

FM3 The result of the abbatial election in the Old Brno Augustinian Monastery on 30 March 1868, confirms Mendel giving his vote to Klácel repeatedly (Marvanová 1968)

IM18 Mendel's photograph enlarged from a group of unknown men

References and Historical Printed Sources

Klein J, Klein N (2013) Solitude of a Humble Genius – Gregor Johann Mendel: Volume 1 Formative Years. In: Klein P (ed) Springer, Berlin. ISBN 978-3-642-35254-6 (eBook). https://doi.org/10.1007/978-3-642-35254-6. Library of Congress Control Number 2013948128

Marvanová L (1968) Le centenaire de l'election abbatiale de Mendel. Folia Mendeliana 3:13–20

Orel V Verbík A (1984) Mendel's involvedment in the plea for freedom of teaching in the revolutionary year of 1848. FM 19:223–233

Chapter 3
Gymnasium Substitute Teacher 1849/1850

One of the consequences of the 1848 revolutionary year was the widespread school reform. The subjects taught originally at the philosophical institutes were transferred partially to the 6-year Gymnasia as the 7th and 8th grades, and partially to the technical secondary-level schools. Additional instructors were needed to teach in the extended grades. The 6-year *Gymnasium* in Znojmo planned to open a seventh grade. The Governor Lažansky invited Mendel to serve as a substitute (supply) teacher. Mendel's superior Napp welcomed the invitation for Mendel to teach in Znojmo. Originally, he wanted Mendel to pass a philosophical *rigorosum* and become a candidate of teaching at the Philosophical Institute in Brno that was the domain of the Old Brno Augustinians.

The first transcription of Lažansky's letter to Mendel was made by Iltis (1924, p. 30). Transcription Sajner (1967, p. 683). Translation by Eden and Cedar Paul from Iltis (1966, p. 57):

At Znaim High School [Znojmo Gymnasium] a seventh class is being established, and the local municipality is providing the extra funds required. In connexion with these changes there will be needed a supply teacher to give instruction in the fifth class in Latin, Greek, and German literature, and in the fifth and sixth classes instruction in mathematics. In view of the zeal you have displayed I think it well to appoint you as supply in these subjects at the Znaim High School, and I hereby instruct you to go to Znaim forthwith, to report yourself to the teaching staff there, and to take over your duties. Your travelling expenses will be refunded, and you will receive 60% of the salary of a teacher of the humanities (in addition to the ordinary salary of a supply) that makes 600 florins. The text in the brackets is not contained in the German original letter by Lažansky. According to the "Communkassa Hauptbuch of Znojmo" Mendel's salary was 360 florins. (Sajner 1967, p. 683)

The first transcription of Napp's writing to Schaaffgotsche was made by Iltis (1924, p. 31); IM, p. 21; Sajner (1967, p. 683). Translation Eden and Cedar Paul from Iltis (1966, p. 58). The excerpt of Napp's letter gives the reason for Mendel's dismissal from pastoral service:

A. Matalová, E. Matalová, *Gregor Mendel - The Scientist*, Springer Biographies,
https://doi.org/10.1007/978-3-030-98923-1_3

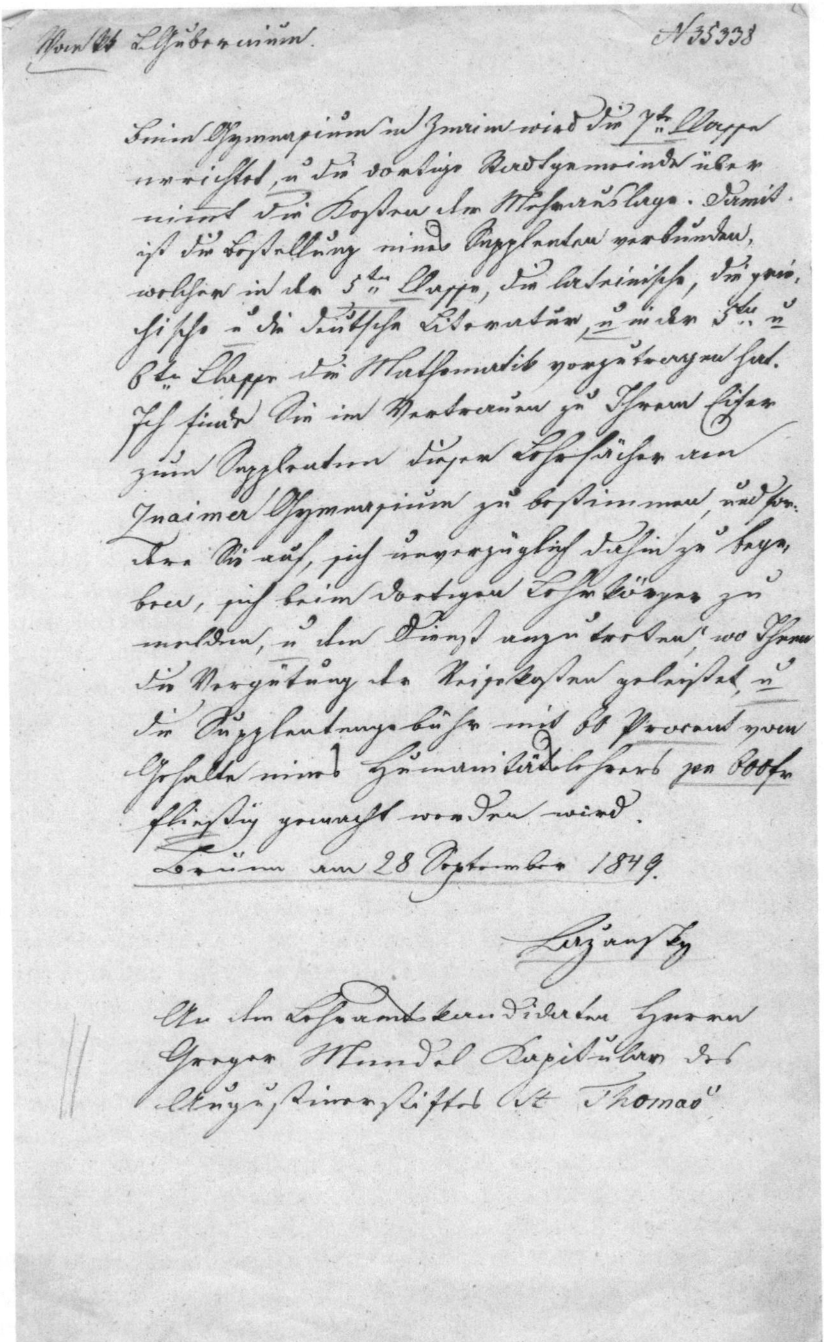

Mendel Lehrerernennung. № 35338

Beim Gymnasium in Znaim wird die 7te Classe errichtet, und die dortige Stadtgemeinde übernimmt die Kosten der Mehrauslage. Damit ist die Bestellung eines Supplenten verbunden, welcher in der 5ten Classe, die lateinische, die griechische und die deutsche Literatur, und in der 5ten und 6ten Classe die Mathematik vorzutragen hat. Ich finde Sie im Vertrauen zu Ihrem Eifer zum Supplenten dieser Lehrfächer am Znaimer Gymnasium zu bestimmen, und fordere Sie auf, sich unverzüglich dahin zu begeben, sich beim dortigen Lehrkörper zu melden, und den Dienst anzutreten, wo Ihnen die Vergütung der Reisekosten geleistet, und die Supplentengebühr mit 60 Procent vom Gehalte eines Humanitätslehrers per 600 fl. flüssig gemacht werden wird.

Brünn am 28 September 1849.

Lažansky

An den Lehramtskandidaten Herrn Gregor Mendel Kapitular des Augustinerstiftes St Thomas

D75 Letter by Lažansky asking Mendel to serve as a substitute teacher at the *Gymnasium* in Znojmo. The Znojmo Administration offered him 60% salary of a humanity teacher. Dated in Brünn/Brno, 28 September 1849, signed by Lažansky

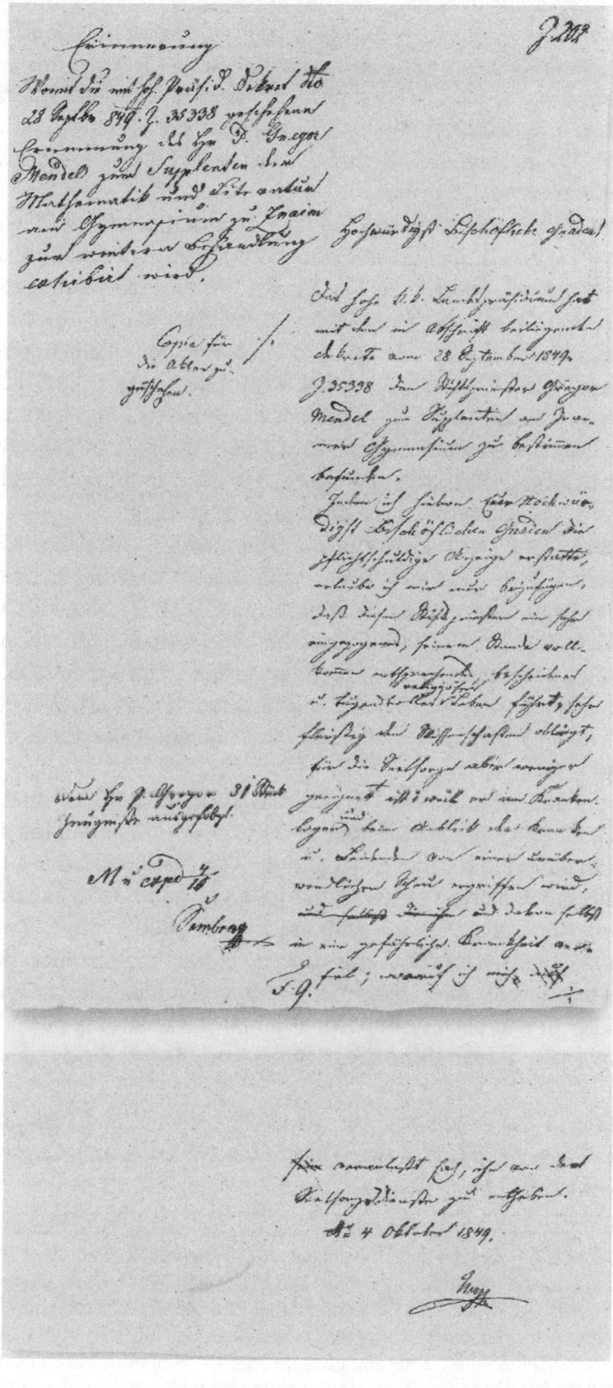

D102 Draft of a letter by Napp to Schaaffgotsche informing him about Mendel's invitation as a substitute teacher by the Znojmo Municipality. Dated in Brno, 04 October 1849, signed by Napp

I will content myself with adding that this collegiate priest lives a very retired life, modest, virtuous, and religious, thoroughly appropriate to his condition; also that he is very diligent in the study of the sciences; but that he is much less fitted for work as a parish priest, the reason being that he is seized by an unconquerable timidity when he has to visit a sick-bed or to see any one ill and in pain. Indeed, this infirmity of his has made him dangerously ill; and that was why I found it necessary to relieve him from service as a parish priest.

Minutes from the meeting of the Znojmo *Gymnasium* staff on 08 October 1849, in the presence of *Gymnasium* Director Ambros Spallek, Professors Johann Schäfert, Franz Sedleczko, Josef Jetschmen, Dr. Ignatz Winter. Substitute professors Carl Willmann—teacher of religion, Franz Pekárek, Wenzel Marek—teachers of history in the seventh grade, the chief teacher of religion Friedrich Heinemann and Gregor Mendel, capitular of St. Thomas Abbey in Old Brno, were first published by Richter (1943, pp. 75–6). Sajner (1967) reported that the document had been lost.

In Znojmo, Mendel was tutoring mathematics and Greek in the lower grades of the 6-year *Gymnasium*, 20 lessons per week. He lived in a private house in Upper Česká Street No. 42, not far from the *Gymnasium* building.

The Imperial Law and Government Regulations Annual, volume 1849, No. 36, prohibited uncertified substitute teachers from teaching at gymnasia, lycea and universities beginning with the school year 1850/1851. Mendel's studies at the Philosophical Institute and the Theological Institute were henceforth inadequate for the position of a substitute teacher at the *Gymnasium*. The Ministry of Culture and Education established a special Scientific Examination Commission at the University of Vienna in 1849 for teachers to undergo certification examinations. Transcription by Czihak (1984, p. 15).

Mendel applied to the Scientific *Gymnasium* Examination Commission in Vienna to qualify for natural history for all grades of the *Gymnasium* and for physics in the lower grades. In applying for that examination, Mendel appended to the application an outline of his life, which he called "a short sketch" (*eine kurze Skizze*), now called his *curriculum vitae* or, by some authors, his "autobiography", even though it consisted of only four handwritten pages. The German transcription of the outline, along with its translations into Latin and Czech, was published by Sajner (1965) to mark the hundredth anniversary of Mendel's oral presentations of his discovery to the *Naturforschender Verein* titled "Experiments on Plant Hybrids" and republished in 1972.

Transcription of the testimonial on Mendel's personal and pedagogic qualities was made by Sajner (1967, p. 682). Translation by Eden and Cedar Paul in Iltis (1966, pp. 60–1):

> Since the day of his entry into the teaching office assigned to him, the same has day by day developed better and better the most advantageous qualities of an exemplary and thorough instructor of youth, and moreover, by a vivid and lucid method of teaching, by an unresting application thereof, and by such results as might therefrom be expected, the same has, day by day, demonstrated, that he was not only well acquainted with the subject of instruction, but also that he was striving with all his energies, no less, to distinguish himself pre-eminently by a sustained zeal and tenacity in the presentation and inculcation of the object of study, and, no less, by the most effective influence upon the pure morality and the religion of his pupils. With regard to his conduct in matters of morality and religion and with regard to the

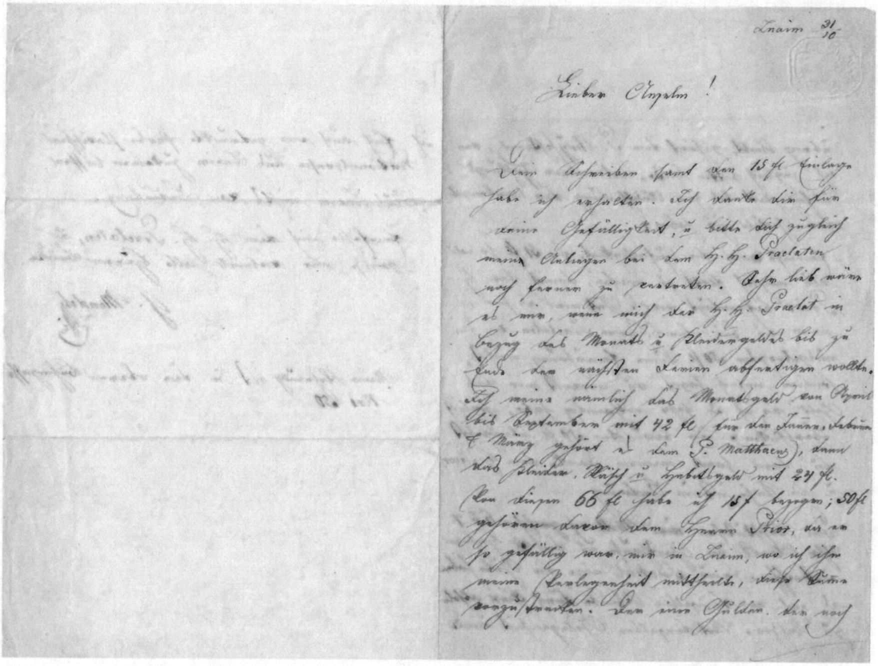

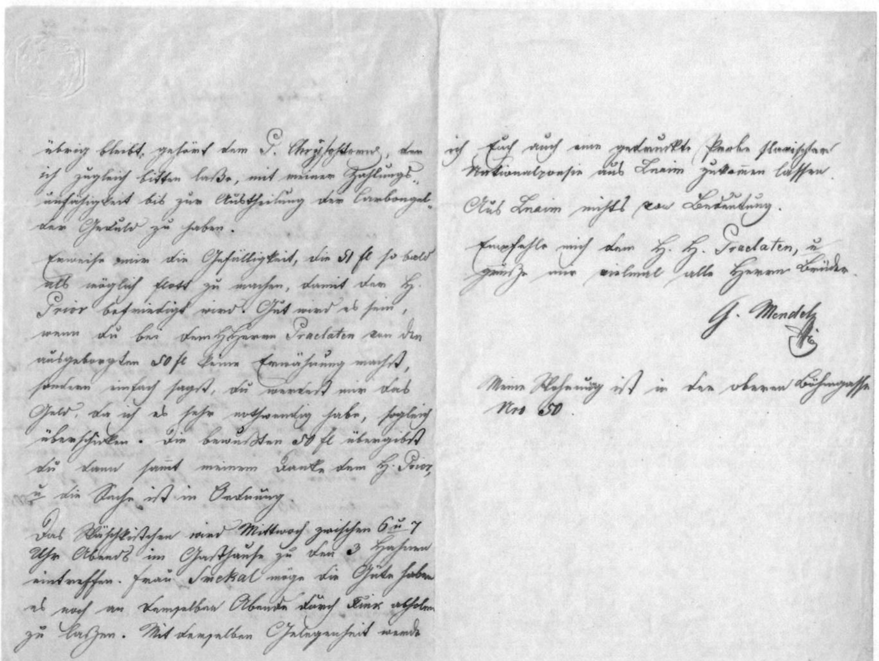

D104 Letter by Mendel to his fellow friar, Anselm Rambousek, in the Old Brno Augustinian monastery regarding his precarious financial situation. Three pages. Dated in Znojmo, 31 October 1849. Transcription by Sajner (1967, pp. 684–5)

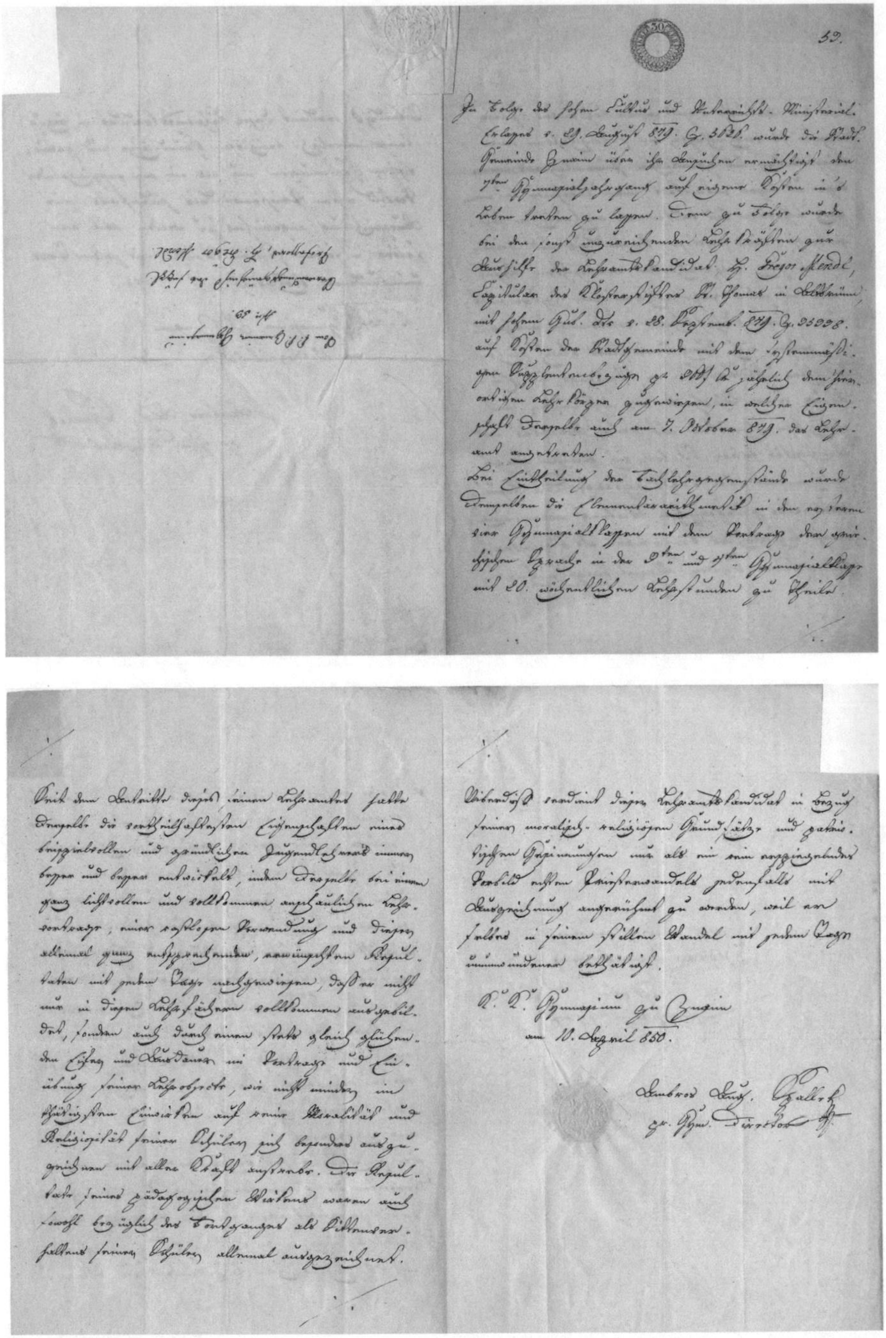

D83 Letter by the Director of the *Gymnasium* of Znojmo presenting Gregor Mendel to the Scientific *Gymnasium* Examination Committee of the University of Vienna as a candidate for the examination of proficiency in teaching. Four pages. Dated in Znojmo, 10 April 1850, signed by Ambros Aug. Spallek, sealed. Transcription Sajner (1967, pp. 681–2)

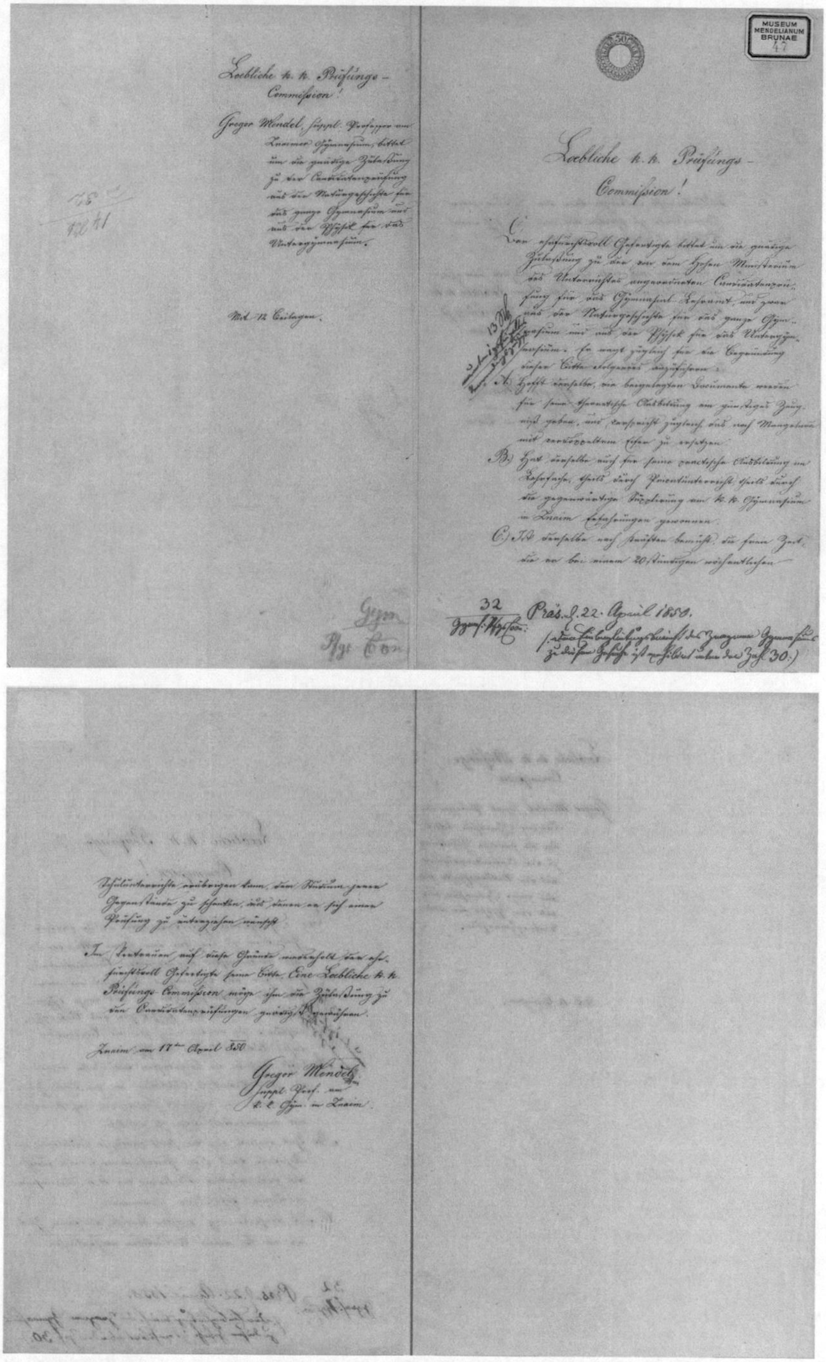
Löbliche k. k. Prüfungs-Commission!

Gregor Mendel

Löbliche k. k. Prüfungs-Commission!

Pres. d. 22. April 1850.

Gregor Mendel

D36 Mendel's application to the Examination Commission of the University of Vienna to qualify for natural history for the entire *Gymnasium* grades, and for physics in the lower grades only. Three pages. Znojmo, 17 April 1850, sign. Gregor Mendel. Transcription Sajner (1967, p. 685)

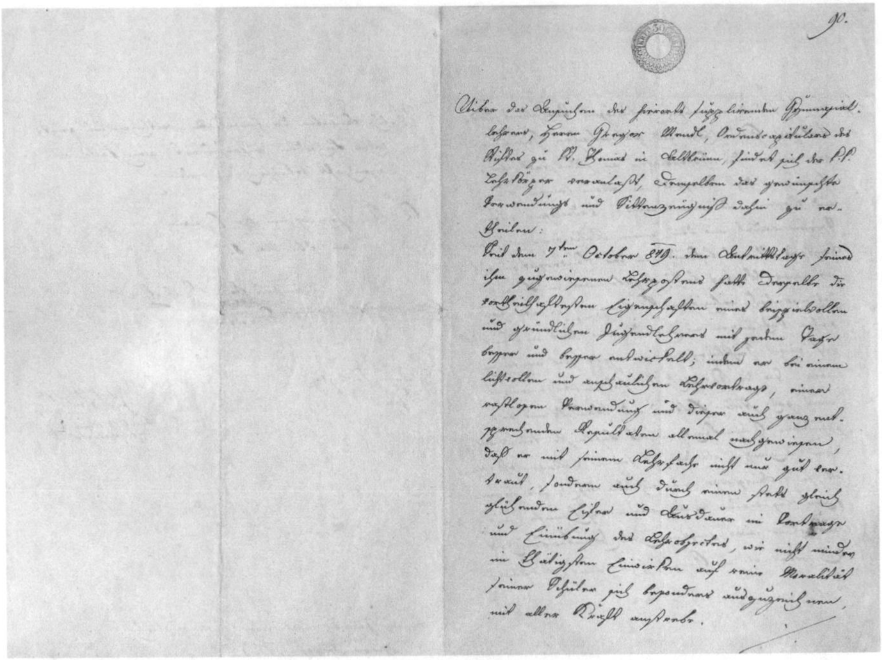

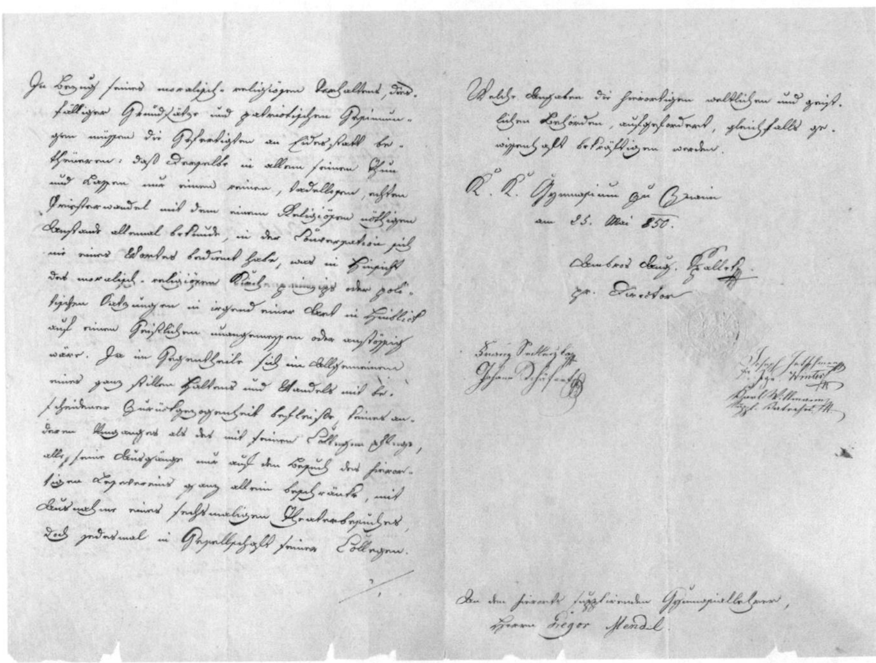

D85 Testimonial by Mendel's fellow teachers at the Znojmo *Gymnasium* written in support of his examination application. It is addressed to Mendel's hand. Three pages. Dated in Znojmo, 25 May 1850, signed by Ambros Aug. Spallek, Joseph Jetschmen, Dr. Ing. Winter, Karl Willmann, Franz Sedleczko, Johann Schäfert. Sealed

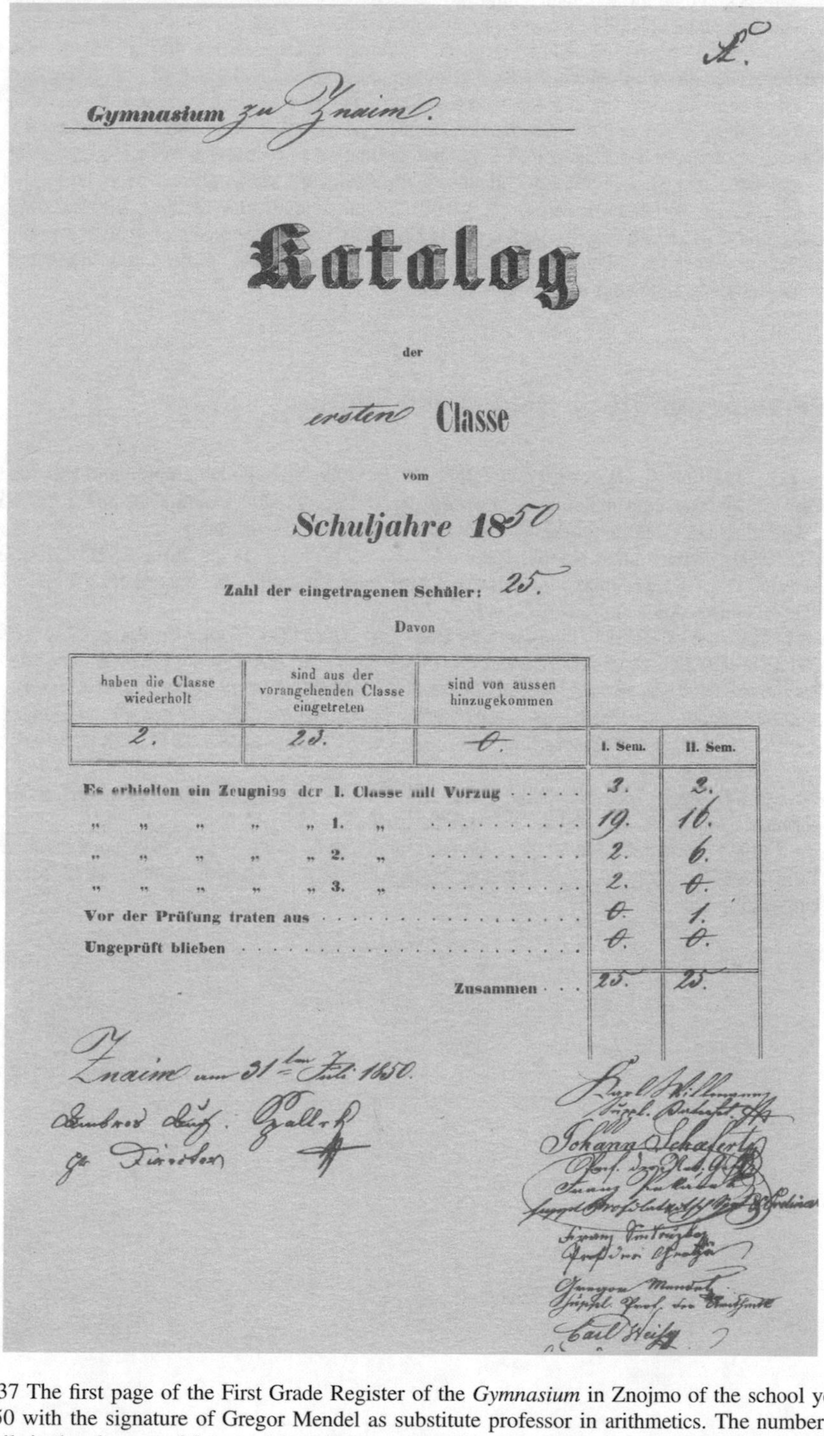

Gymnasium zu Znaim.

Katalog

der

ersten Classe

vom

Schuljahre 1850

Zahl der eingetragenen Schüler: 25.

Davon

haben die Classe wiederholt	sind aus der vorangehenden Classe eingetreten	sind von aussen hinzugekommen
2.	23.	0.

	I. Sem.	II. Sem.
Es erhielten ein Zeugniss der I. Classe mit Vorzug	3.	2.
" " " " " 1. "	19.	16.
" " " " " 2. "	2.	6.
" " " " " 3. "	2.	0.
Vor der Prüfung traten aus	0.	1.
Ungeprüft blieben	0.	0.
Zusammen	25.	25.

Znaim am 31ten Juli 1850.

Gregor Mendel

IM37 The first page of the First Grade Register of the *Gymnasium* in Znojmo of the school year 1850 with the signature of Gregor Mendel as substitute professor in arithmetics. The number of pupils in the class was 25, two of them were repeats. The classification scale is represented as four grades: first class with distinction, first class, second class and third class. The document is dated in Znaim/Znojmo, 31 July 1850, signed by Ambros Spallek

principles and patriotic sentiments appropriate thereto, the undersigned must asseverate that the same in all he did and left undone invariably manifested a pure and blameless genuinely priestly behaviour with the demeanour proper to a man of religious life, that in conversation he never used any word which in respect of moral or religious ecclesiastical principles or of political considerations could have been considered in any way inappropriate for a cleric to use or improper in a cleric's mouth, nay, on the contrary, that he invariably manifested a thoroughly tranquil demeanour and a modest inclination to retirement, avoiding any other associations than those with his colleagues, restricting his walks abroad to visits to the reading circle of this town in company with the other reputables of the place, the only exception to this rule being that he has on six occasions been to the theatre, but always in the society of one of his colleagues, which data the local secular and spiritual authorities will conscientiously confirm if requested to do so.

References and Historical Printed Sources

Czihak G (1984) Johann Gregor Mendel (1822-1884). Dokumentierte Biographie und Katalog zur Gedächtnisausstellung anläßlich des hundertsten Todestages mit Facsimile seines Hauptwerkes "Versuche über Pflanzenhybriden" 1866. Druckhaus Nonntal, Salzburg

Iltis H (1924) Gregor Johann Mendel. Leben, Werk und Wirkung. Julius Springer, Berlin. Life of Mendel. English translation by Eden and Cedar Paul. First edition. Norton, New York 1932; Second edition Allen and Unwin London, 1966

Iltis H (1966) Life of Mendel. Translated by Eden and Cedar Paul. Hofner Publ. Co, New York

Richter O (1943) Johann Gregor Mendel wie er wirklich war. Neue Beiträge zur Biographie des berühmten Biologen aus Brünns Archiven mit 31 Abbildungen im Texte. Verhandlungen des Naturforschenden Vereines in Brünn, Abteilung für Naturforschung der Deutschen Gesellschaft für Wissenschaft und Volkstumforschung in Mähren 74 (für das Jahr 1942):1-262. Druck von Josef Klär, Brno

Sajner J (1967) Gregor Johann Mendel und Znaim. Forschung, Praxis, Fortbildung. Organ für die gesammte praktische und theoretische Medizin 18:677–685

Sajner J (1965) Gregorii Mendel Autobiographia iuvenilis. Universitas Purkyniana Brunensis. 1965. Second edition Ad centesimum quinquagesimum J. G. Mendel natalem. 1972, Masaryk University, Brno

Chapter 4
Author of a Brief Sketch of His Life 1850

Mendel appended his short sketch (so-called curriculum vitae) to his application addressed to the Imperial-Royal Scientific Examination Commission at the University of Vienna. He presented himself in the third person singular, as was customary at the time. The CV is handwritten by Mendel.

The first page of the manuscript was first published by Iltis (1924, pp. 120–121). The last page of Mendel's CV manuscript appeared in Iltis (1924, pp. 34–35). From March 1938 to November 1941 authentic documents pertaining to Mendel's life appeared anonymously on serialization in English translation in the newsletter Messenger. Their translator was most probably P. E. Ryan from the diocese Covington in Kentucky where the Messenger was published. Anne Iltis was the translator of the English version of CV (1954). J. Sajner (1965) prepared a bibliophilic edition of Mendel's CV manuscript with German transcription and translations into Latin and Czech. The bibliophilic print was republished in 1972 at the sesquicentennial of Mendel's birth.

Transcription of Mendel's CV, which was appended to his application to the Scientific Examination Commission of the University of Vienna is given here, translated by Anne Iltis (1954, pp. 231–234).

> Praiseworthy Imperial and Royal Examination Commission!—In accordance with the high regulations of the Ministry of Public Worship and Education, the respectfully undersigned submits a short sketch of his life.—The same was (enclosure A) born in the year 1822 in Heinzendorf [now Hynčice] in Silesia, where his father was the owner of a small farm. After he had received elementary instruction at the local village school, and later in the Piarists' College in Leipnik [now Lipník], he was admitted in the year 1834 to the first grammatical class of the Imperial Royal Gymnasium in Troppau [now Opava].—Four years later, due to several successive disasters, his parents were completely unable to meet the expenses necessary to continue his studies, and it therefore happened that the respectfully undersigned, then only sixteen years old, was in the sad position of having to provide for himself entirely. For this reason, he attended the course for School Candidates and Private Teachers in the District Teachers' Seminary in Troppau. Since, following his examination, he was highly recommended in the qualification report (enclosure B) he succeeded by private tutoring during the time of his humanities studies in earning a scanty livelihood.—When he graduated from the Gymnasium in the year 1840, his first care was to secure for himself

A. Matalová, E. Matalová, *Gregor Mendel - The Scientist*, Springer Biographies,
https://doi.org/10.1007/978-3-030-98923-1_4

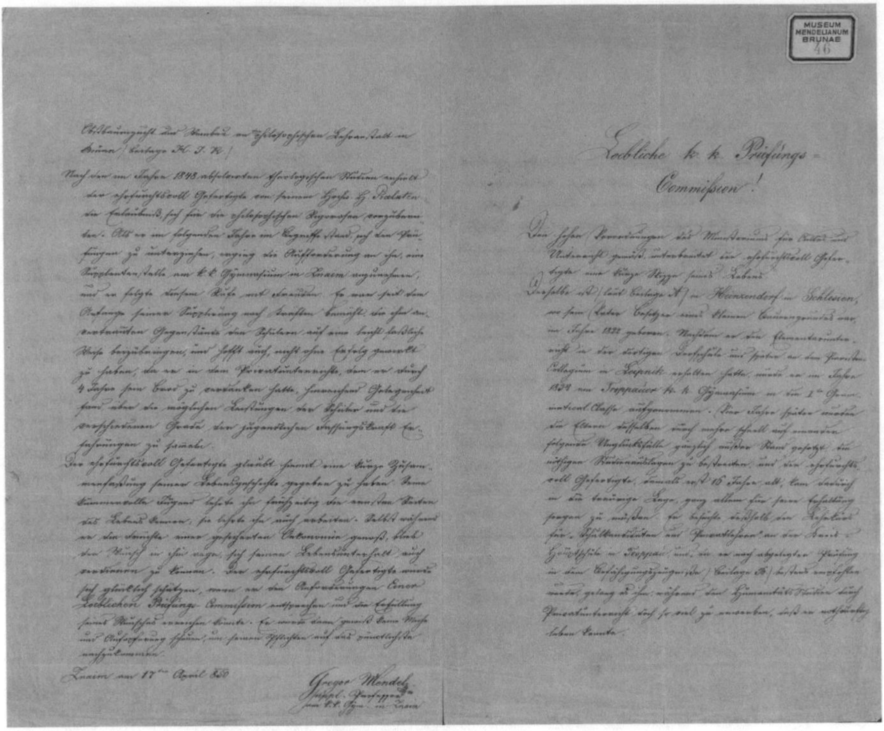

Mendel's brief sketch of his life is called Autobiography. Four pages. Dated in Znaim/Znojmo, 17 April 1850, signed by Gregor Mendel

the necessary means for the continuation of his studies. Because of this, he made repeated attempts in Olmütz [now Olomouc] to offer his services as a private teacher, but all his efforts remained unsuccessful because of lack of friends and recommendations. The sorrow over these disappointed hopes and the anxious, sad outlook which the future offered him, affected him so powerfully at that time, that he fell sick and was compelled to spend a year with his parents to recover.

In the following year the respectfully undersigned found himself finally placed in the desired position of being able to satisfy at least the most necessary wants by private teaching in Olmütz, and thus to continue his studies. By a mighty effort, he succeeded in completing the two years of philosophy (enclosure D, E, F, G). The respectfully undersigned realized that it was impossible for him to endure such exertions any further. Therefore, after having finished his philosophical studies, he felt himself compelled to step into a station of life, which would free him from the bitter struggle for existence. His circumstances decided his vocational choice. He requested and received in the year 1843 admission to the Augustinian Monastery St. Thomas in Altbrünn [now Old Brno].—Through this step, his material circumstances changed completely. With the comfortableness of his physical existence, so beneficial to any kind of study, the respectfully undersigned regained his courage and strength and he studied the classical subjects prescribed for the year of probation with much liking and devotion. In the spare hours, he occupied himself with the small botanical-mineralogical collection which was placed at his disposal in the monastery. His special liking for the field of natural science deepened the more he had the opportunity to become familiar with it. Despite his lack of any oral guidance in these studies, plus the fact that the auto-didactic method here, as perhaps in no other science, is extremely difficult and

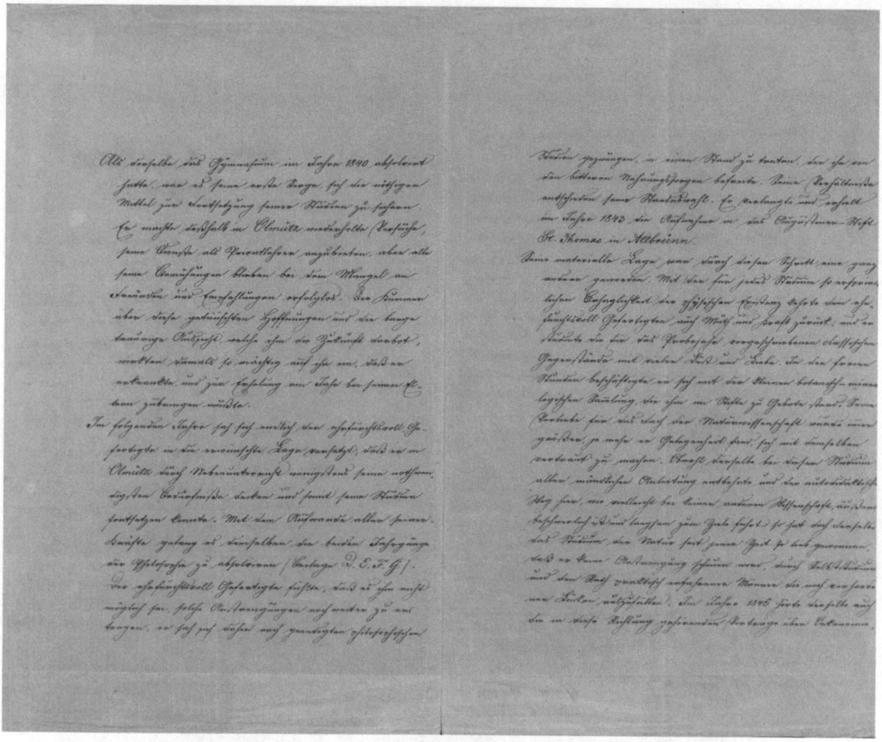

(continued)

leads to the goal only slowly, he became so attached to the study of nature from this time on that he will not spare any effort to fill the gaps that are still present through self instruction and the advice of practically experienced men. In the year 1846, he also attended courses in agriculture, pomiculture and vine-growing at the Philosophical Institute in Brünn (enclosure H, I, K).—After completing his theological studies in 1848, the respectfully undersigned received permission from his prelate to prepare himself for the philosophical rigorosum. In the following year in the time when he was about to undergo his examination, he was asked to accept the position of a substitute teacher at the Imperial Royal Gymnasium in Znaim [now Znojmo], and he followed this call with pleasure. From the beginning of his substitute teaching, he made all efforts to present his assigned subjects to the students in an easily comprehensible manner. He hopes his endeavour was not quite without success since, during the private tutoring to which he owed his bread for four years, he found sufficient opportunity to collect experience regarding the possible accomplishments of the students and the different grades of their mental capacity.—The respectfully undersigned believes to have rendered with this a short summary of his life's history. His sorrowful youth taught him early the serious aspects of life, and taught him also to work. Even while he enjoyed the fruits of a secure economic position, the wish remained alive within him to be permitted to earn his living. The respectfully undersigned would consider himself happy if he could conform with the expectations of the praiseworthy Board of Examiners and gain the fulfilment of his wish. He would certainly then shun no effort and sacrifice to comply with his duties most punctually—Znaim on the 17th April 1850, Gregor Mendel—, Subst. Professor—at the Imp.—Roy. Gym. in Znaim.

D43 The cover of the *Matrik* (register of births) in the parish of Gross Petersdorf/Dolní Vražné. Mendel's certificate of baptism was written according to the entry in the register of births

Transcription of the heading of Folio No. 86 in the parish register of birth in Gross Petersdorf/Dolní Vražné: 1822—ihr Geburt—Monat, Tag der Taufe—hat getauft—Haus No—Name des Täuflings—Religion: katholisch, protest. Geschlecht: männlich, weiblich—ehelich, unchclich—Eltern: Vater, Mutter—Pathen: Namen, Stand.

Translation of Mendel's entry in the parish Register of Births: born July 20, baptized July 20 by Johann A. E. Schreiber, parish priest, house No. 58, religion catholic, gender male, legitimate son—a note across the columns reads catholic baptist, No. 35—parents father Anton Mendel, farmstead owner, mother Rosina Schwirtlich, born to Martin Schwirtlich gardener—godparents Karl Kuntscher—farmstead owner, Juliana Walzel. The three crosses added to the names of godparents symbolize the signatures. There is a note across the columns in all entries in the folio listing the births from 12 May to 4 November 1822, catholic first name, No. 35 (Kath. Taufname etc. No. 35).

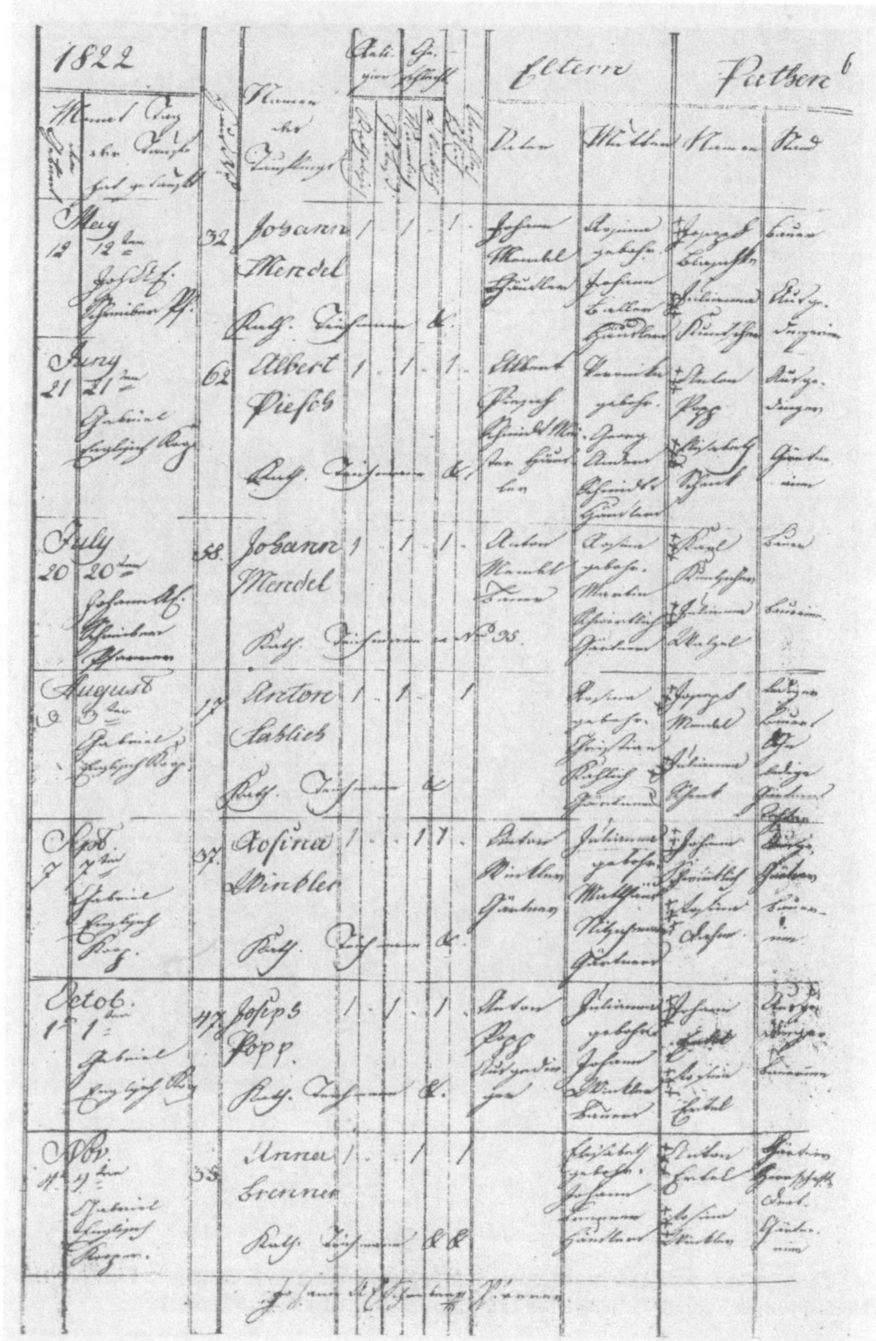

IM2 The entry of Mendel's birth on 20 July 1822 in the parish register of births in Gross Petersdorf/ Dolní Vražné

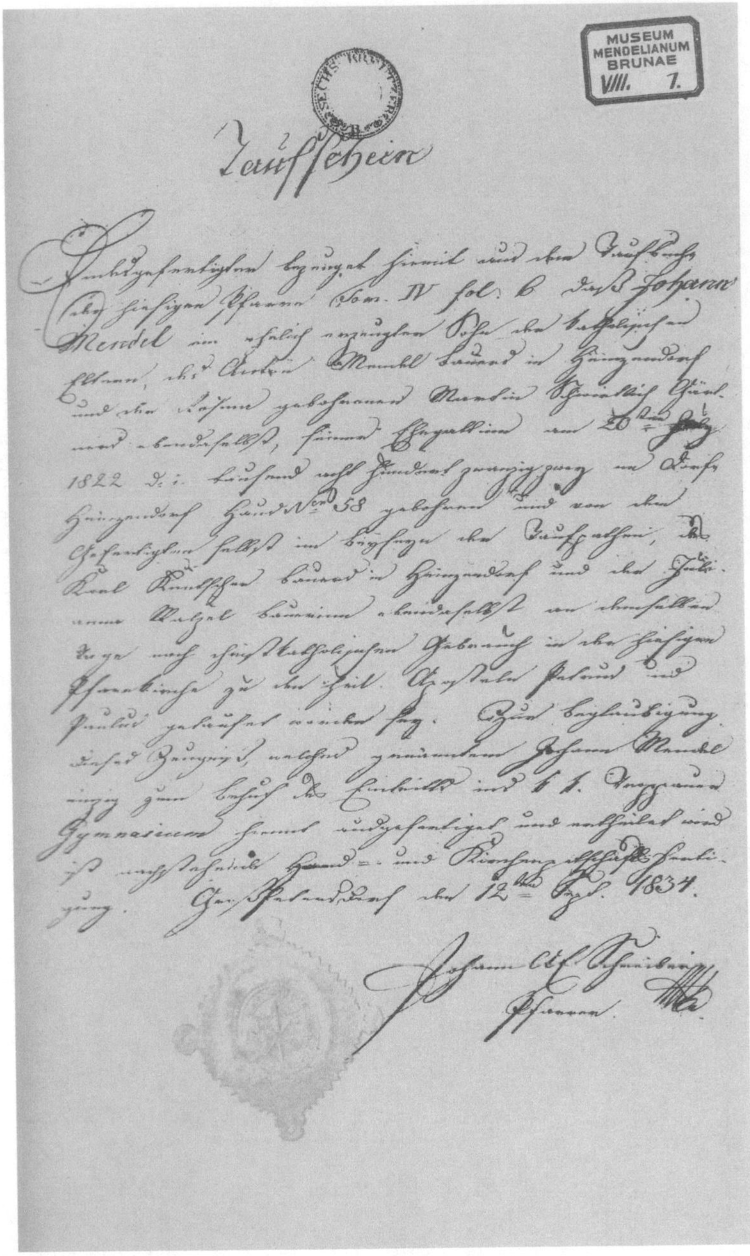

D43 The baptismal certificate issued by the parish priest Johann A. E. Schreiber. Dated in Gross Petersdorf/Dolní Vražné, 12 September 1834, signed by Johann A. E. Schreiber

Taufschein—Endesgefertigter bezeuget hiemit aus dem Taufbuche der hiesigen Pfarre Tom. IV Fol. 6, dass Johann Mendel ein ehelich erzeugter Sohn der katholischen Eltern, des Anton Mendel Bauers in Heinzendorf und der Rosina gebohrenen Martin Schwirtlich Gärtners ebendaselbst, seiner Ehegattin, am 20ten Juli 1822 d. i. Tausend acht hundert zwanzig zwey im Dorfe Heinzendorf Haus No 58 gebohren und von dem Gefertigten selbst im Beyseyn der Taufpathen, des Karl Kuntscher Bauers in Heizendorf und der Julianna Walzel Bauerin ebendaselbst an demselben Tage nach christkatholischem Gebrauch in der hiesigen Pfarrkirche zu den heil. Aposteln Petrus und Paulus getauft worden sey. Zur Beglaubigung dieses Zeugniss, welches genannten Johann Mendel einzig zu Behuf des Eintritts ins k. k. Troppauer Gymnasium hiemit ausgefertiget und ertheilt wird, ist nahstehende Hand—und Kirchenpetschafts Fertigung. Gross Petersdorf den 12ten Sept. 1834.—Johann A. E. Schreiber—Pfarrer.

The Birth Certificate confirms that Mendel was baptized on the day of his birth in the parish church of Sts. Peter and Paul. Formally, the certificate of baptism (*Taufschein*) required baptism on the day of birth. *Taufschein* and *Geburtsschein* (certificate of birth) were two different papers. Mendel's certificate of baptism, written by Schreiber on 12 September 1834, was issued on the occasion of the entrance of Johann Mendel in the Imperial-Royal *Gymnasium* in Opava.

Klein and Klein (2013, pp. 121–125) provided an analysis of the uncertainty of Mendel's day of birth. In Mendel's childhood house, two plaques have been installed. The German plaque of 1902 gives his birthday as July 22. The Czech plaque was unveiled in 1965 as of 20 July 1822.

The Directorium, which is the official printed document reporting on the state of the Augustinian monastery, published each year, reports Mendel's birth date exclusively as 22 July 1822. July 22 is the Roman Catholic feast day dedicated to Mary Magdalene, which was easy to remember at the time when the lives of the people were centred around the church.

Mendel received his elementary education at the local village school in Hynčice. In 1833/1834, he was a pupil of the regional main school in Lipník. Mendel attended only the last grade and left the school with highest accolades in 1834. Mendel studied at the *Gymnasium* in Opava from 1834 to 1840. Despite the fact that Mendel's poor health interfered with his *Gymnasium* studies, he finished the six-year-*Gymnasium* curriculum with excellent results.

In 1840, Mendel began his studies at the Philosophical Institute in Olomouc. He interrupted his studies at the end of the first semester in 1841 for health reasons coming from the invalidation of his father (Křížənecký 1965a, b, p. 179). After his recovery Mendel made a fundamental decision in the summer of 1841, choosing to forego taking over his father's farmstead. On 07 August 1841, Alois Sturm, the husband of Mendel's older sister Veronika, purchased the Mendel estate in Hynčice, number 58 on the village map. Johann Mendel was the only son (he had two surviving sisters) and traditionally a son continued to maintain the house and the fields attached to it, perpetuating the legacy of his father. Johann's talent for study and his poor health was decisive in his determination to not continue in his father's footsteps.

An excerpt from the Mendel-Sturm transaction states: *The purchaser shall pay to the son of the seller, Johann by name, if the latter as he now designs should enter the*

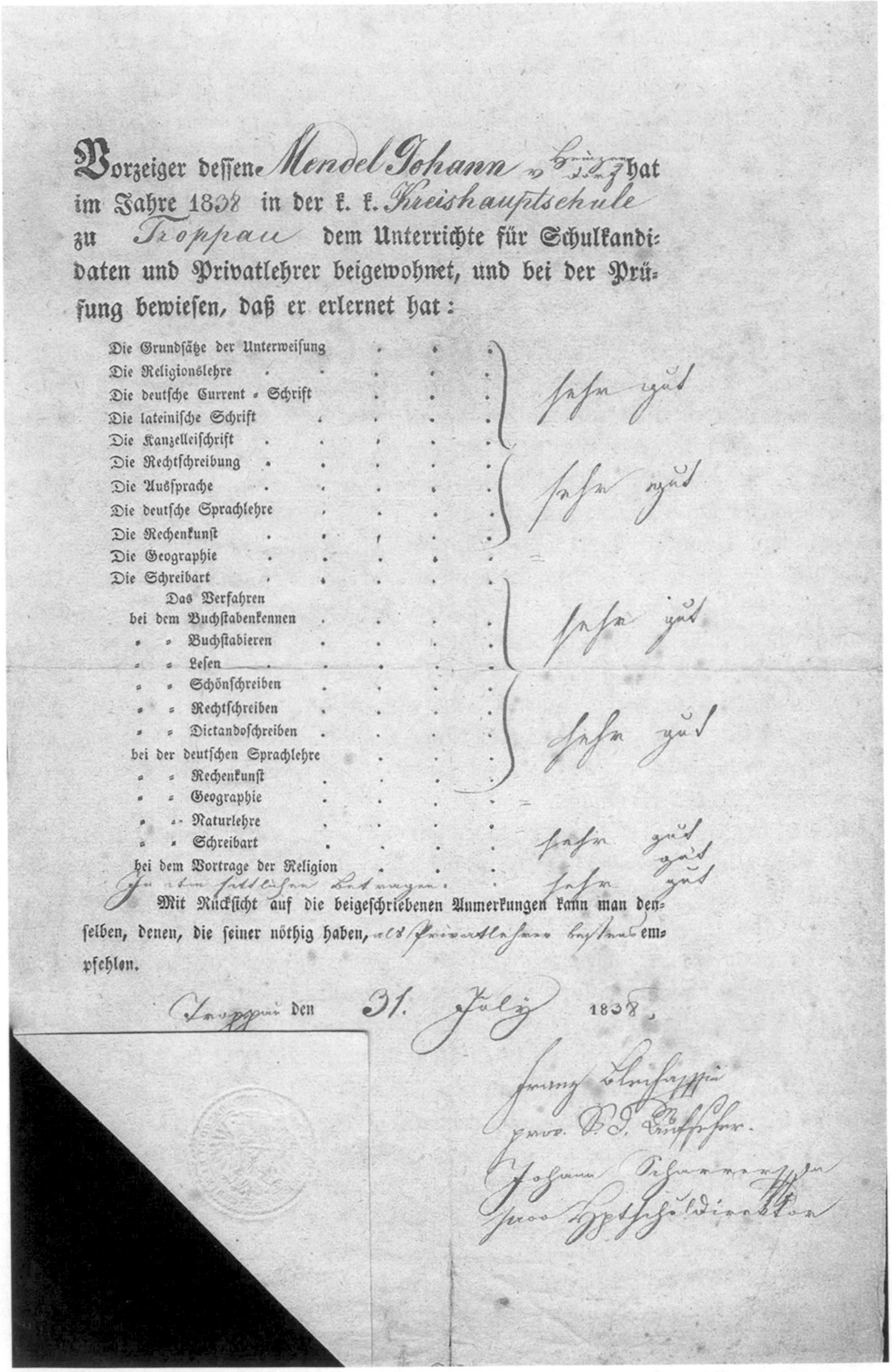

Vorzeiger dessen Mendel Johann hat im Jahre 1838 in der k. k. Kreishauptschule zu Troppau dem Unterrichte für Schulkandidaten und Privatlehrer beigewohnet, und bei der Prüfung bewiesen, daß er erlernet hat:

Die Grundsätze der Unterweisung Die Religionslehre Die deutsche Current - Schrift Die lateinische Schrift Die Kanzelleischrift	sehr gut
Die Rechtschreibung Die Aussprache Die deutsche Sprachlehre Die Rechenkunst	sehr gut
Die Geographie	–
Die Schreibart Das Verfahren bei dem Buchstabenkennen " " Buchstabieren " " Lesen	sehr gut
" " Schönschreiben " " Rechtschreiben " " Dictandoschreiben bei der deutschen Sprachlehre " " Rechenkunst	sehr gut
" " Geographie	–
" " Naturlehre	
" " Schreibart	sehr gut
bei dem Vortrage der Religion	sehr gut
In den sittlichen Betragen	sehr gut

Mit Rücksicht auf die beigeschriebenen Anmerkungen kann man denselben, denen, die seiner nöthig haben, als Privatlehrer bestens empfehlen.

Troppau den 31. Julii 1838.

D57 Mendel's certificate for candidates and private teachers (*Schulkandidaten und Privatlehrer*) qualifying him to tutor other students. Dated in Troppau/Opava, 31 July 1838. Sealed

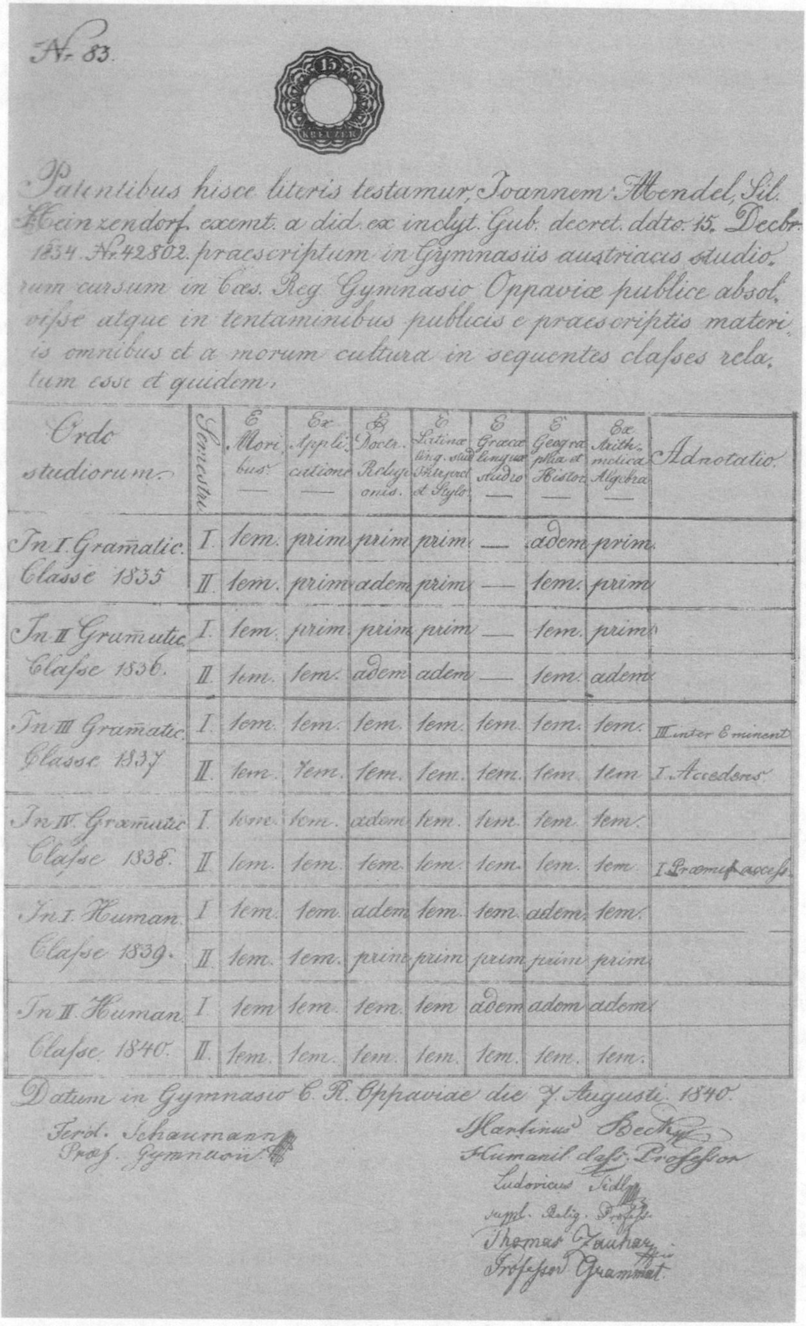

Nr. 83.

Patentibus hisce literis testamur, Joannem Mendel, Sil. Heinzendorf. exemt. a did. ex inclyt. Gub. decret. ddto. 15. Decbr. 1834. Nr. 42802. praescriptum in Gymnasiis austriacis studiorum cursum in Caes. Reg. Gymnasio Oppaviae publice absolvisse atque in tentaminibus publicis e praescriptis materiis omnibus et a morum cultura in sequentes classes relatum esse et quidem,

Ordo studiorum	Semestre	E Moribus	Ex Applicatione	E Doctr. Religionis	E Latinae ling. stud. Interpret. et Stylo	E Graecae linguae studio	E Geographia et Historia	Ex Arithmetica Algebra	Adnotatio
In I. Gramatic. Classe 1835	I.	1em.	prim.	prim.	prim.	—	adem	prim.	
	II.	1em.	prim.	adem	prim.	—	1em.	prim.	
In II. Gramatic. Classe 1836.	I.	1em.	prim.	prim.	prim.	—	1em.	prim.	
	II.	1em.	1em.	adem	adem	—	1em.	adem	
In III. Gramatic. Classe 1837	I.	1em.	1em.	1em.	1em.	1em.	1em.	1em.	III. inter Eminent.
	II.	1em.	1em.	1em.	1em.	1em.	1em.	1em.	I. Accedens
In IV. Gramatic. Classe 1838.	I.	1em.	1em.	adem	1em.	1em.	1em.	1em.	
	II.	1em.	1em.	1em.	1em.	1em.	1em.	1em.	I. Praemii accessit
In I. Human. Classe 1839.	I	1em.	1em.	adem	1em.	1em.	adem.	1em.	
	II.	1em.	1em.	prim.	prim.	prim.	prim.	prim.	
In II. Human. Classe 1840.	I.	1em.	1em.	1em.	1em.	adem	adem	adem	
	II.	1em.	1em.	1em.	1em.	1em.	1em.	1em.	

Datum in Gymnasio C. R. Oppaviae die 7. Augusti 1840.

Ferd. Schaumann Prof. Gymnasii

Martinus Beck Humanit. class. Professor

Ludovicus Tindl suppl. Relig. Profess.

Thomas Zauhar Professor Grammat.

D16 Mendel's completion certificate from the *Gymnasium* in Opava from four grammatical courses and from two humanities courses from 1834 to 1840. Dated in Troppau (now Opava, Oppavia in Latin), 07 August 1840, signed by Ferd. Schaumann, Martinus Beck, Ludovicus Tindl. Thomas Zauhar. Transcription of the certificate by Křížanecký (1965a, b, p. 179)

priesthood, or should he in any other way begin to earn an independent livelihood, the sum of 100 fl., say one hundred gulden convention-coins, and also annually, so long as Johann is still engaged at his studies, shall pay the father the sum of 10 fl. convention-coins as an aid to the cost of study, and shall also defray all the expenses connected with the first mass.

Translation Eden and Cedar Paul, from Iltis (1966, p. 39).

Mendel received the best marks possible (Kříženecký 1965a, b, p. 179). He was certified in morals, religion, theoretical and practical philosophy, mathematics, physics, philology and pedagogy.

Transcription Iltis (1924. pp. 18–19), English translation Eden and Cedar Paul from Iltis (1966, pp. 42–43): *As a result of your letter of June 12, I have made known to my pupils the Right Reverend Prelate's decision to accept satisfactory candidates at your monastery. Up to now, two candidates have given me their names, but I can only recommend one of them. This is Johann Mendel, born in Heinzendorf [Hynčice] in Silesia. During the two-year course in philosophy he has had, almost invariably, the most unexceptionable reports, and is a young man of very solid character. In my own branch he is almost the best. He has some knowledge of Czech, but not sufficient, so he is willing to devote himself to the mastery of the language during the years of his theological study. Please convey this information to the Right Reverend Prelate, with my respects, and ask him what he would like me to do about the matter.*

Transcription: *Einwilligung.—Wir Endesgefertigten erklären hiemit, dass wir mit der Standeswahl unseres Sohnes Johann vollkommen einverstanden sind, und uns nichts anderes zu wünschen übrig bleibt, als dass er seinen gewählten Beruf treu u. gewissenhaft zu erfüllen sich bestrebe.*

In 1843, Mendel was admitted to the Old Brno Augustinian monastery for a one-year probation. He assumed the monastic name Gregor, derived in German from the Latin Gregorius. (In the catholic ecclesiastical tradition, the monastic name Gregor was his first name, placed before Johann, his baptismal name.) Abbot Napp officially accepted Mendel into the order in a ceremony held in the monastic church of the Assumption of the Virgin Mary in Old Brno on 09 October 1843.

At the Brno Theological Institute, Mendel studied ecclesiastical history, ecclesiastical archaeology, Hebrew language, exegesis of the Old Testament, introduction to the books of the Old Testament, ecclesiastical law, biblical hermeneutics, Greek language, exegesis of the New Testament, introduction to the books of the New Testament, pedagogy, dogmatic theology, moral theology, pastoral theology, catechism and methodology of elementary school education, Chaldaic language, Syriac and Arabic. Mendel studied pedagogy for two semesters at the Philosophical Institute in Olomouc in 1843, and rural economy for two semesters at the Philosophical Institute in Brno in 1846 (i.e. agriculture, horticulture, pomiculture and viniculture).

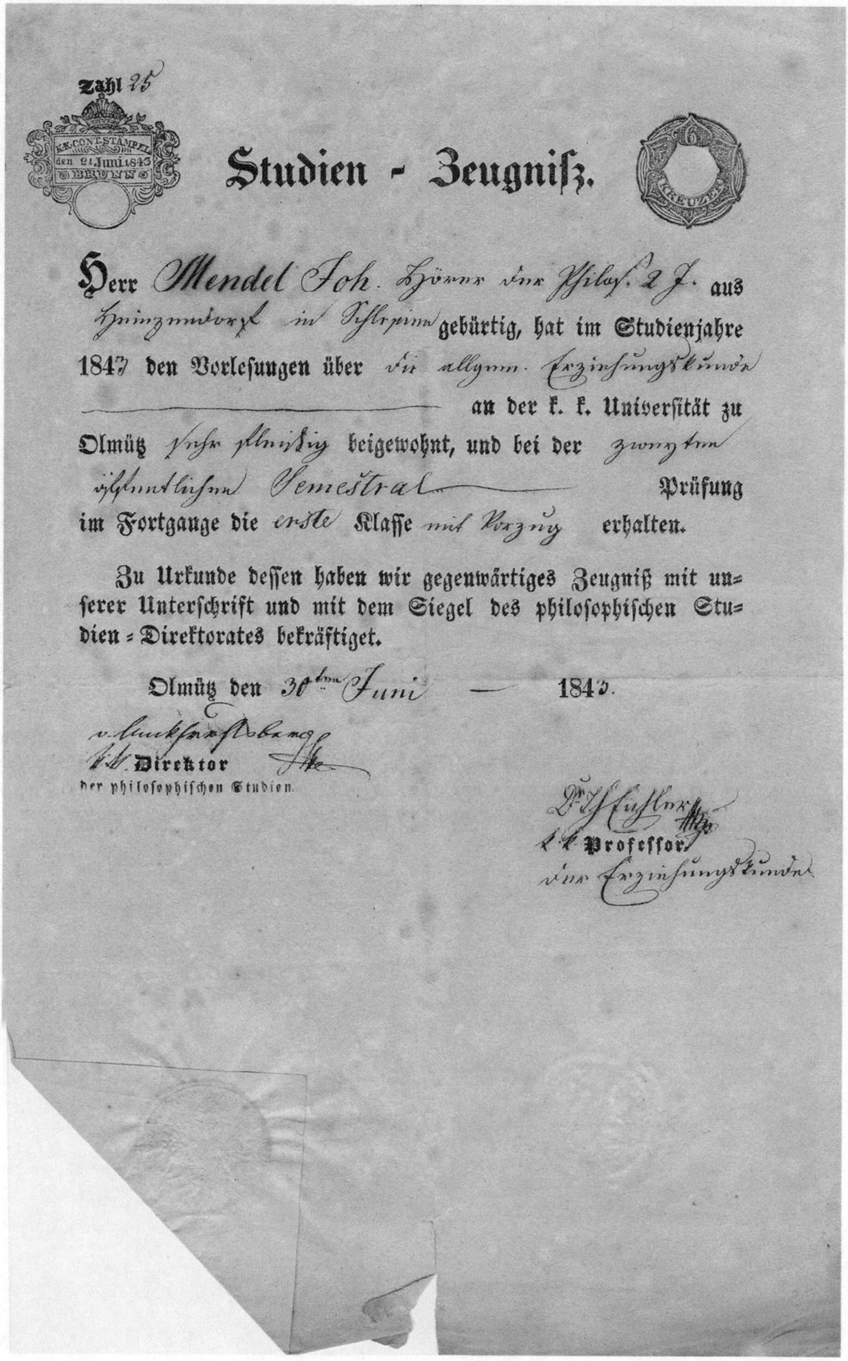

Zahl 25

Studien - Zeugniß.

Herr Mendel Joh. Hörer der Philos. 2 J. aus Heinzendorf in Schlesien gebürtig, hat im Studienjahre 1843 den Vorlesungen über die allgem. Erziehungskunde an der k. k. Universität zu Olmütz sehr fleißig beigewohnt, und bei der zweyten öffentlichen Semestral Prüfung im Fortgange die erste Klasse mit Vorzug erhalten.

Zu Urkunde dessen haben wir gegenwärtiges Zeugniß mit unserer Unterschrift und mit dem Siegel des philosophischen Studien-Direktorates bekräftiget.

Olmütz den 30ten Juni 1843.

k. k. Direktor der philosophischen Studien.

Dr Th Eichler
k. k. Professor der Erziehungskunde

D45 Mendel's graduation certificate from the final semester of the second year of study at the Philosophical Institute of the University of Olomouc. Dated in Olmütz/Olomouc, 30 June 1843, signed by Dr. Th. Eichler

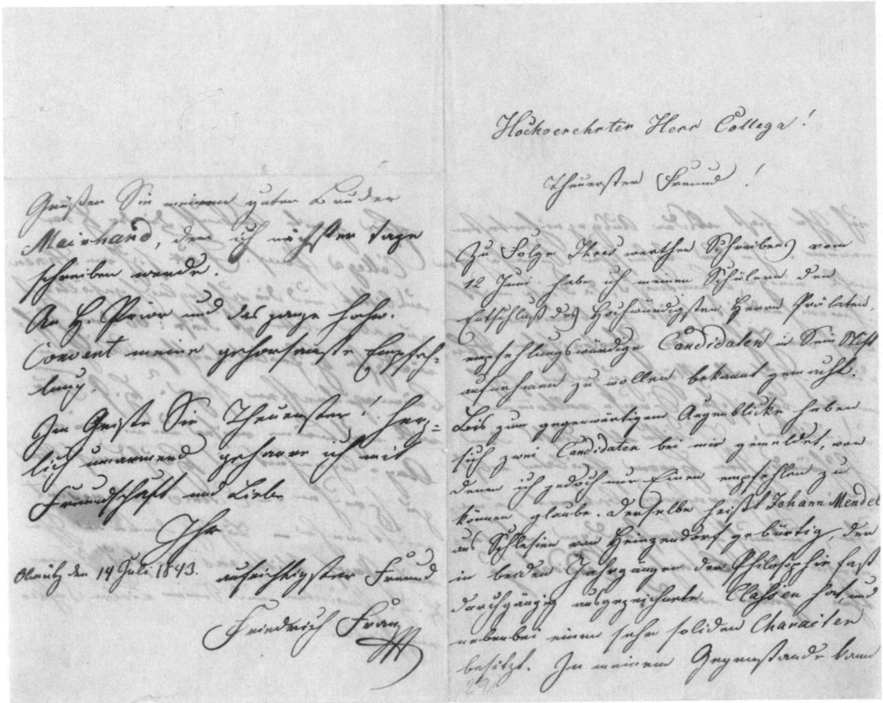
Grüßen Sie meinen guten Bruder
Meinhard, dem ich nächsten Tage
schreiben werde.
An H. Prior und das ganze hochw.
Convent meine gehorsamste Empfeh-
lung.
Im Geiste Sie Theuerster! herz-
lich umarmend verharre ich mit
Freundschaft und Liebe
Ihr
Olmütz den 14 Juli 1843. aufrichtigster Freund
Friedrich Franz

Hochverehrter Herr Collega!
Theuerster Freund!
Zu Folge Ihres werthen Schreibens vom
12 Juni habe ich meinen Schülern den
Entschluß des Hochwürdigsten Herrn Prälaten,
empfehlungswürdige Candidaten in den Stift
aufnehmen zu wollen bekannt gemacht.
Bis zum gegenwärtigen Augenblicke haben
sich zwei Candidaten bei mir gemeldet, von
denen ich jedoch nur Einen empfehlen zu
können glaube. Derselbe heißt Johann Mendel
aus Schlesien von Heinzendorf gebürtig, der
in beiden Jahrgängen der Philosophie fast
durchgängig ausgezeichnete Classen hat, und
nebenbei einen sehr soliden Character
besitzt. In meinem Gegenstande kann

D61 Friedrich Franz, Mendel's teacher of physics at the Philosophical Institute of the Olomouc University, recommended Mendel to become a candidate to the Old Brno Augustinian monastery. Four pages. The addressee of Franz' letter is not known (most probably Johann Baptist Vorthey or Václav Šembera). Dated in Olmütz/Olomouc, 14 July 1843, signed by Friedrich Franz

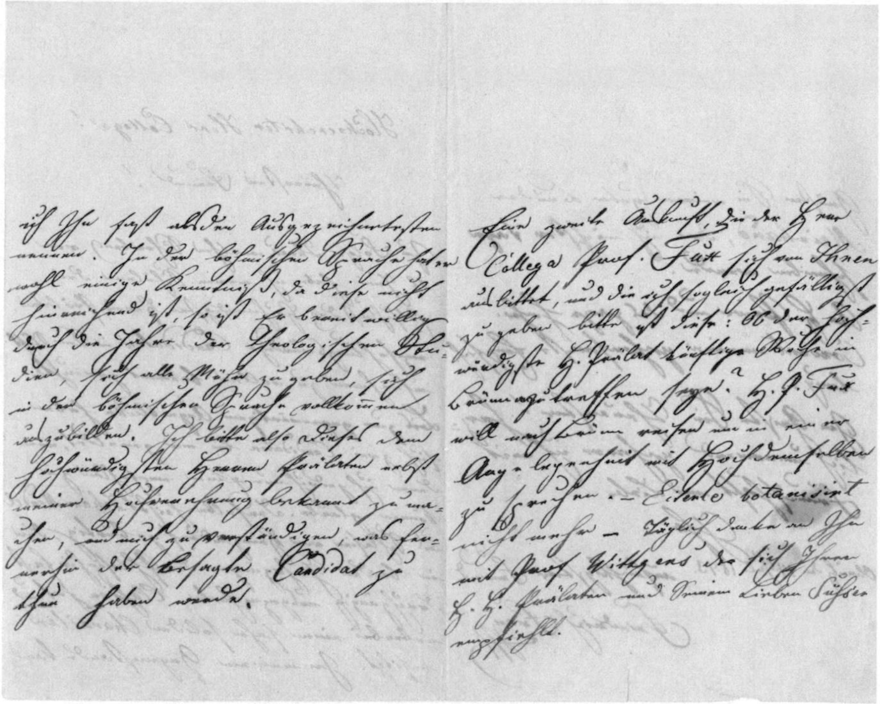

(continued)

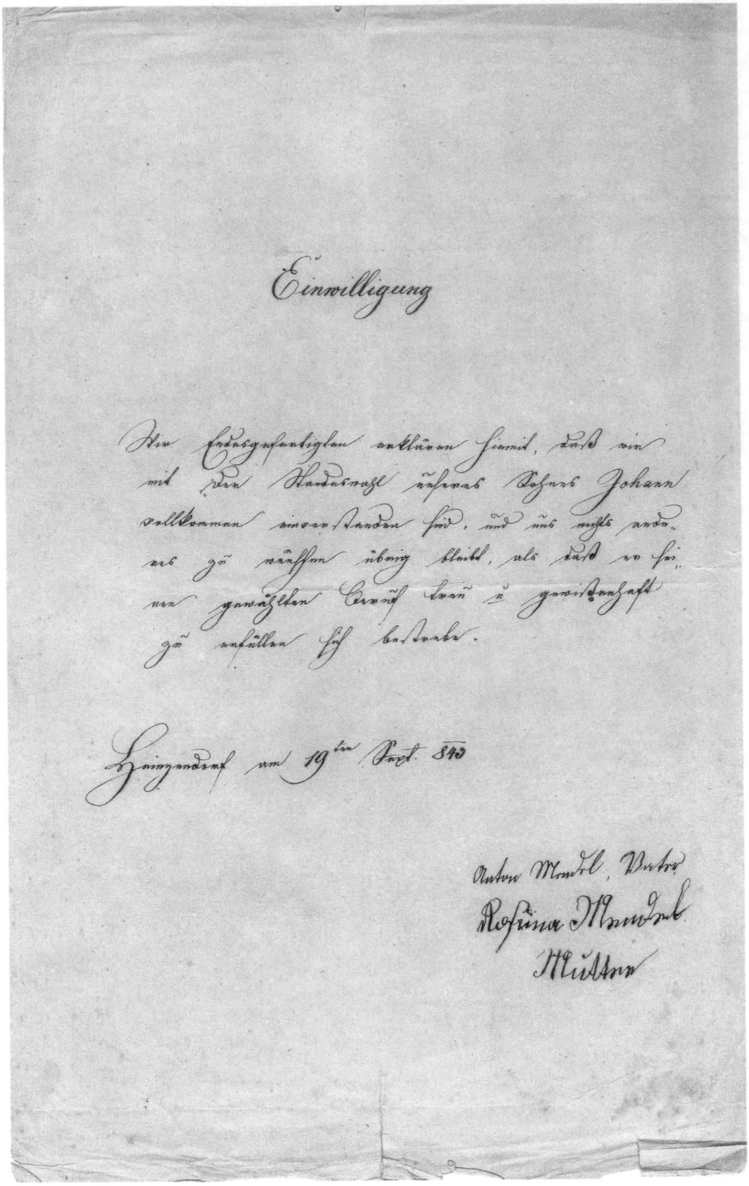

Einwilligung

Wir Endesgefertigten erklären hiemit, daß wir mit der Standeswahl unseres Sohnes Johann vollkommen einverstanden sind, und uns nichts anderes zu wünschen übrig bleibt, als daß er seinen gewählten Beruf treu u. gewissenhaft zu erfüllen sich bestrebe.

Heinzendorf am 19ten Sept. 843

Anton Mendel, Vater
Rosina Mendel
Mutter

D35 Mendel's parents' agree with their son's vocational choice. Dated in Heinzendorf/Hynčice, 19 September 1843, signed by Anton Mendel, father/*Vater*/, Rosina Mendel, mother/*Mutter*/

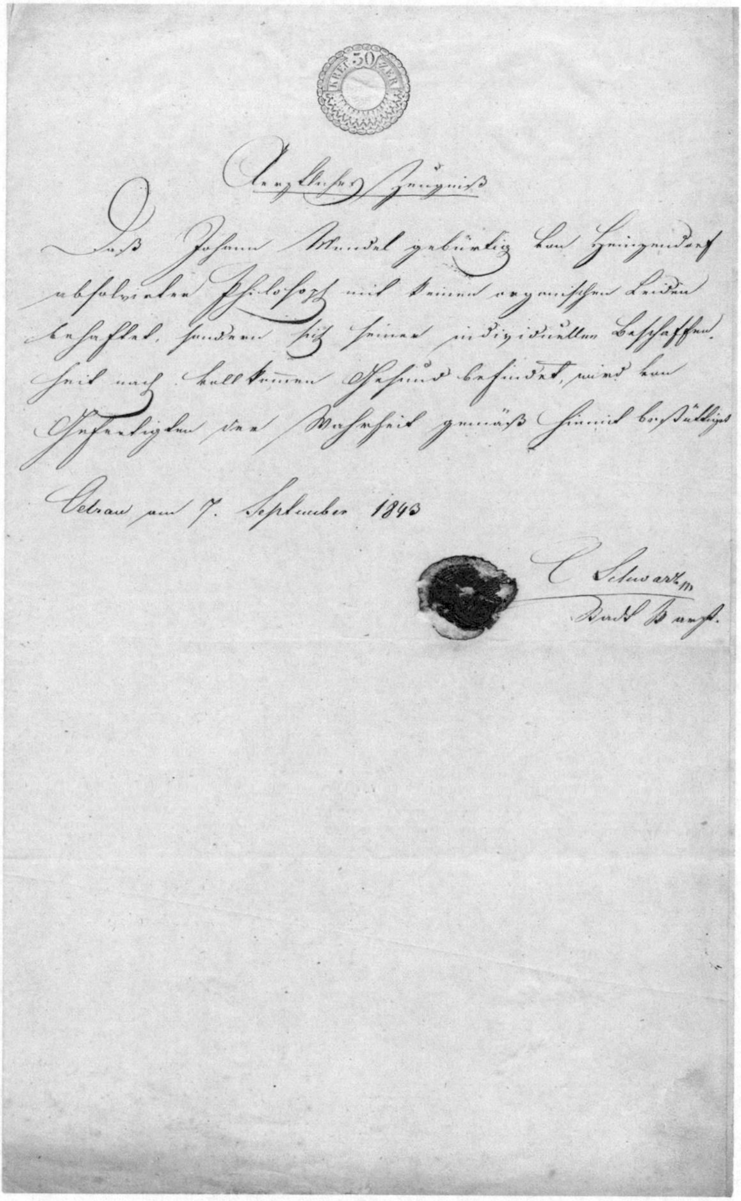
Ärztliches Zeugniß

Daß Johann Mendel gebürtig von Heinzendorf absolvirter Philosoph mit keinem organischen Leiden behaftet, sondern sich seiner individuellen Beschaffenheit nach vollkommen gesund befindet, wird von Unterfertigten der Wahrheit gemäß hiemit bestättiget

Odrau am 7. September 1843

C. Schwarz m.p.
Stadt Haupt.

D62 Medical certificate of Mendel required for his entry into the monastery. Dated in Odrau/Odry, 07 September 1843, signed by C. Schwarz, Stadt Haupt. Sealed

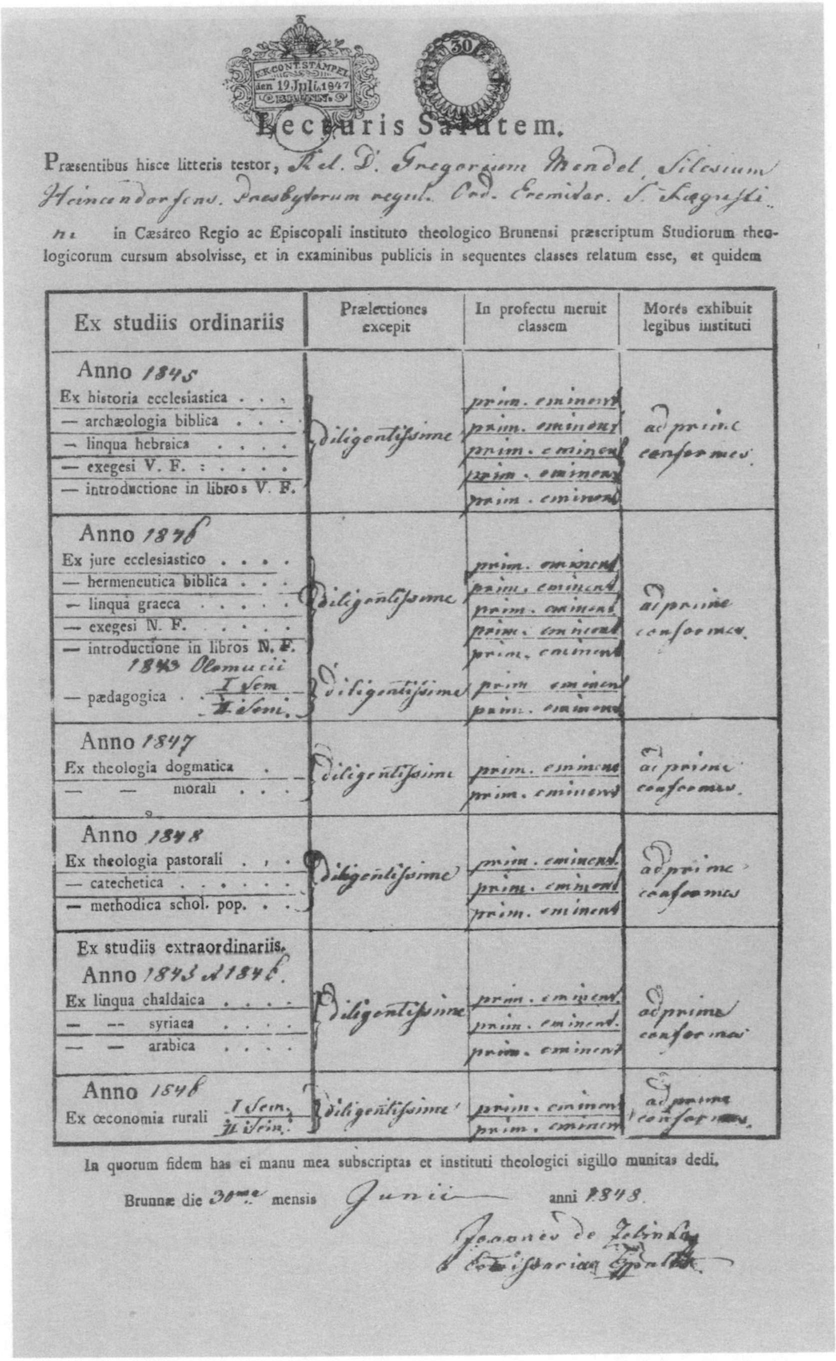

Lectoris Salutem.

Præsentibus hisce litteris testor, Rel. D. Gregorium Mendel, Silesium Heinzendorfensem, Presbyterum regul. Ord. Eremitar. S. Augustini in Cæsareo Regio ac Episcopali instituto theologico Brunensi præscriptum Studiorum theologicorum cursum absolvisse, et in examinibus publicis in sequentes classes relatum esse, et quidem

Ex studiis ordinariis	Prælectiones excepit	In profectu meruit classem	Morés exhibuit legibus instituti
Anno 1845			
Ex historia ecclesiastica	diligentissime	prim. eminent.	adprime conformes.
— archæologia biblica		prim. eminent.	
— linqua hebraica		prim. eminent.	
— exegesi V. F.		prim. eminent.	
— introductione in libros V. F.		prim. eminent.	
Anno 1846			
Ex jure ecclesiastico	diligentissime	prim. eminent.	adprime conformes.
— hermeneutica biblica		prim. eminent.	
— linqua graeca		prim. eminent.	
— exegesi N. F.		prim. eminent.	
— introductione in libros N. F.		prim. eminent.	
1843 Olomucii			
— pædagogica I Sem.	diligentissime	prim. eminent.	
II Sem.		prim. eminent.	
Anno 1847			
Ex theologia dogmatica	diligentissime	prim. eminent.	adprime conformes.
— — morali		prim. eminent.	
Anno 1848			
Ex theologia pastorali	diligentissime	prim. eminent.	adprime conformes
— catechetica		prim. eminent.	
— methodica schol. pop.		prim. eminent.	
Ex studiis extraordinariis. Anno 1843 et 1846.			
Ex linqua chaldaica	diligentissime	prim. eminent.	adprime conformes
— — syriaca		prim. eminent.	
— — arabica		prim. eminent.	
Anno 1846			
Ex œconomia rurali I Sem.	diligentissime	prim. eminent.	adprime conformes.
II Sem.		prim. eminent.	

In quorum fidem has ei manu mea subscriptas et instituti theologici sigillo munitas dedi.

Brunnæ die 30ma mensis Junii anni 1848.

Joannes de Zelinka
Episcopalis Consistorii ...

D17 Mendel's completion certificate from the Theological Institute in Brünn/Brno for the years 1845–1848. Dated in Brünn/Brno, 30 June 1848, signed by Joannes de Zelinka

The botanical exsicata collection in the Augustinian monastery, Mendel had at his disposal, was prepared by Aurelius Thaler, a botanist who died in 1843. In the same year, Mendel joined the Augustinian friars. Items from Thaler's botanical collection represented a substantial part of the twentieth-century exhibition in the monastic Mendel Museum. The oval picture of Prelate Mendel is from the Moravian Mortgage Bank. One of the two chairs was donated to Villanova University in Pennsylvania along with Mendel's signature on the cover of Strumpf's Chemistry

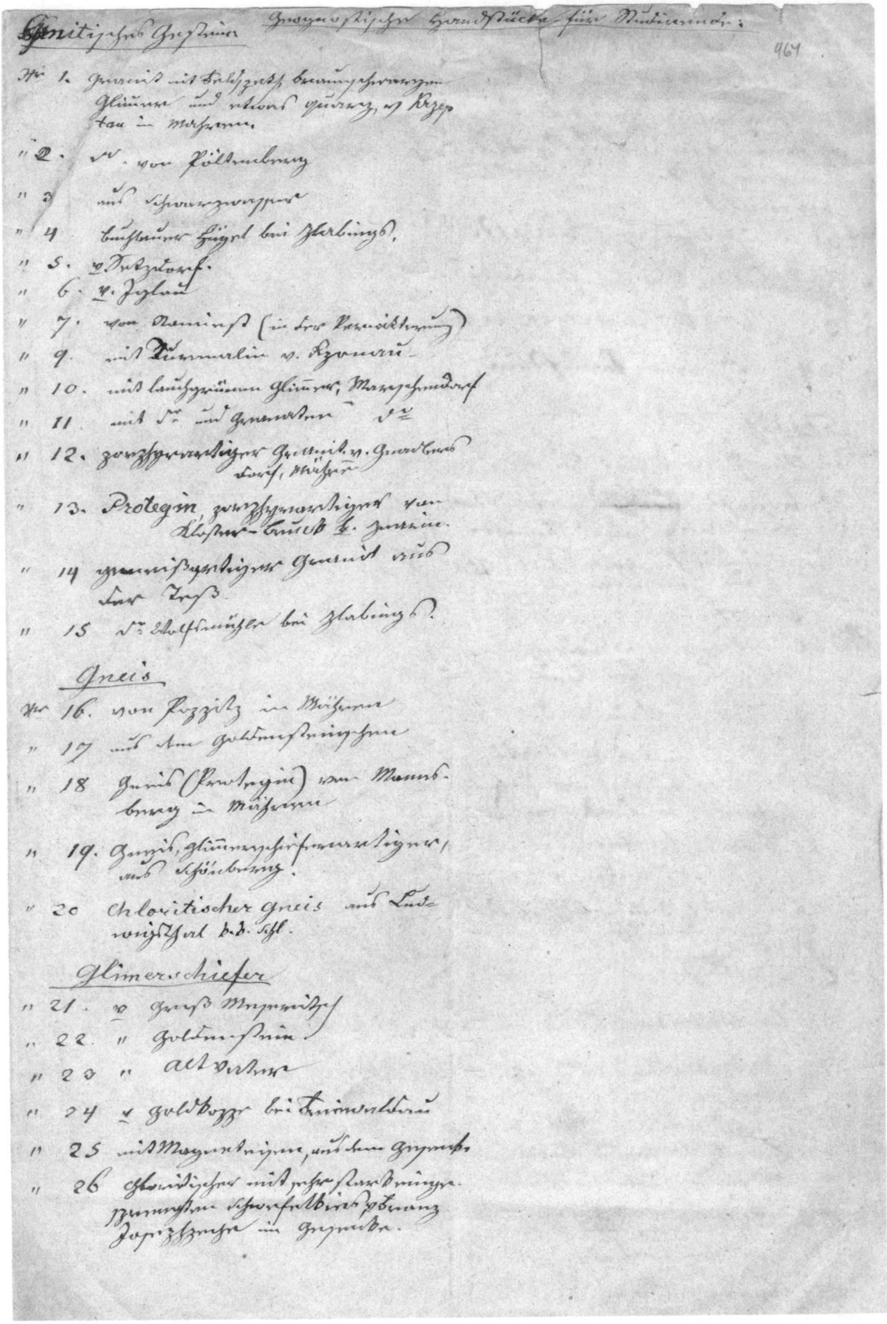

D964 The first page of the list of 162 items of the monastic mineralogical collection that Mendel had at his disposal. Eight pages. Dated in Brünn/Brno, 17 June 1836

References and Historical Printed Sources

Iltis A (1954) Gregor Mendel's autobiography. J Heredity 45:231–234

Iltis H (1924) Gregor Johann Mendel. Leben, Werk und Wirkung. Julius Springer, Berlin. Life of Mendel. English translation by Eden and Cedar Paul. First edition. Norton, New York 1932; Second edition Allen and Unwin London, 1966

Iltis H (1966) Life of Mendel. Translated by Eden and Cedar Paul. Hofner Publ. Co, New York

Klein J, Klein N (2013) Solitude of a Humble Genius – Gregor Johann Mendel: Volume 1 Formative Years. In: Klein P (ed) Springer, Berlin. ISBN 978-3-642-35254-6 (eBook). https://doi.org/10.1007/978-3-642-35254-6. Library of Congress Control Number 2013948128

Kříženecký J (1965a) Gregor Johann Mendel 1822-1884 Texte und Quellen zu seinem Wirken und Leben. In: Zaunick R (ed) Lebensdarstellungen deutscher Naturforscher. Deutsche Akademie der Naturforscher Leopoldina. Johann Ambrosius Barth Verlag, Leipzig

Kříženecký J (1965b) Fundamenta genetica. The revised edition of Mendel's printed paper with a collection of 27 reprinted original papers published during the rediscovery era. Academia, Prague

Sajner J (1965) Gregorii Mendel Autobiographia iuvenilis. Universitas Purkyniana Brunensis. 1965. Second edition Ad centesimum quinquagesimum J. G. Mendel natalem. 1972, Masaryk University, Brno

Chapter 5
Applicant for a University Accreditation 1850

Prof. Baumgartner chaired the Scientific Examination Commission in 1850 and was Mendel's examiner in physics in Vienna in 1850. Photographer unknown dated approximately 1855 (Weiling 1993, p. 45).

Mendel's university examination consisted of two homeworks (*eingesandte Arbeiten*), one in physics and one in natural history. The on-site written examination (*Klausurprüfung*), in physics and natural history, took place in Vienna, where Mendel wrote his answers in a closed room with no opportunity for him to consult books or notes. The final part of the examination was an oral examination (viva voce), in physics and natural history. Professor Baumgartner examined Mendel in physics and Professor Kner in natural history. A protocol was recorded from the oral examination in two copies. One copy has been preserved in the Mendelianum of the Moravian Museum and the other in the archives of the University of Illinois in Urbana-Champaign. The final certificate of the entire set of examinations is signed by the members of the Scientific Examination Commission of the University of Vienna and dated on 17 October 1850.

The written homeworks were evaluated from the viewpoint of the candidate's knowledge of facts, the latest scientific terminology, composition and style, explanatory clarity and literary language.

Mendel's application for the examination in physics for lower grades and natural history for all *Gymnasium* grades in the Registration Book of the Scientific Examination Commission of the University of Vienna was entered on 17 April 1850 under registration number 32 (Gicklhorn 1969, p. 146). Gicklhorn (1969, p. 145) transcribed the text of the entry: *Gregor Mendel suppl. Professor am Znaymer Gymnasium bittet um Zulassung zur Candidaten Prüfung aus der Naturgeschichte für das ganze Gymnasium und aus der Physik für das Untergymnasium.*

Mendel received the theme for his homework on physics on 12 May 1850. The theme was formulated by Baumgartner (Gicklhorn 1969, p. 147). On 12 May 1850, Mendel confirmed his receipt of the topic for his homework in natural history given by Kner (Gicklhorn 1969, p. 146). The originals are deposited in the archives of the University of Illinois in Urbana-Champaign.

A. Matalová, E. Matalová, *Gregor Mendel - The Scientist*, Springer Biographies,
https://doi.org/10.1007/978-3-030-98923-1_5

The members of the Academy of Sciences in Vienna, the class of mathematics and natural science. Seated from left (Mendel's professors in boldface): *Ettingshausen*, *Baumgartner*, Schrötter, Petzval. Standing from left: Rochleder, Burg, Littrow, *Redtenbacher*, *Fenzl*, Hyrtl, Stampfer, Prechtl, Koller, Rokitansky, *Unger*

Transcription of the title page and the first page of Mendel's work in physics reads: *Beantwortung der Prüfungsfrage aus der Physik von Gregor Mendel, Suppl. Prof. am Znaimer Gymnasium.* In the bottom it bears the date of Mendel's handing over the manuscript to the examination commission procedure. 26 July 1850.

The headline of the first page reads *Beantwortung der Prüfungsfrage aus der Physik: "Es sind die mechanischen und chemischen Eigenschaften der atmosphärishen Luft nachzuweisen und aus ersteren die Winde zu erklären"*. Eighteen pages.

Baumgartner's evaluation of Mendel's work in physics on the theme: *the mechanical and chemical properties of atmospheric air are to be demonstrated, and from the former the winds are to be explained.* English translation by Eden and Cedar Paul from Iltis (1966, p. 64). Dated in Vienna, 24 July 1850, signed by A. Baumgartner. Co-signed by Doppler with a note: *I am in full accord with the foregoing opinion*, H. Bonitz, W. Grauert, Kner and Lott.

Transcription of the document was published by Gicklhorn (1973, p. 78).

Baumgartner was of the opinion that Mendel gave a clear description of the mechanical and chemical relationships in the atmosphere and of the experiments documenting them, that he applied the data accurately, and that he fully explained the nature of the winds. From the following excerpt from Baumgartner's evaluation, Mendel seems to be a prospective teacher: English translation by Eden and Cedar Paul from Iltis (1966, p. 65): *In view of the fact that the candidate is only desirous in qualifying, in this field, as a teacher in the lower schools, the essay must be regarded*

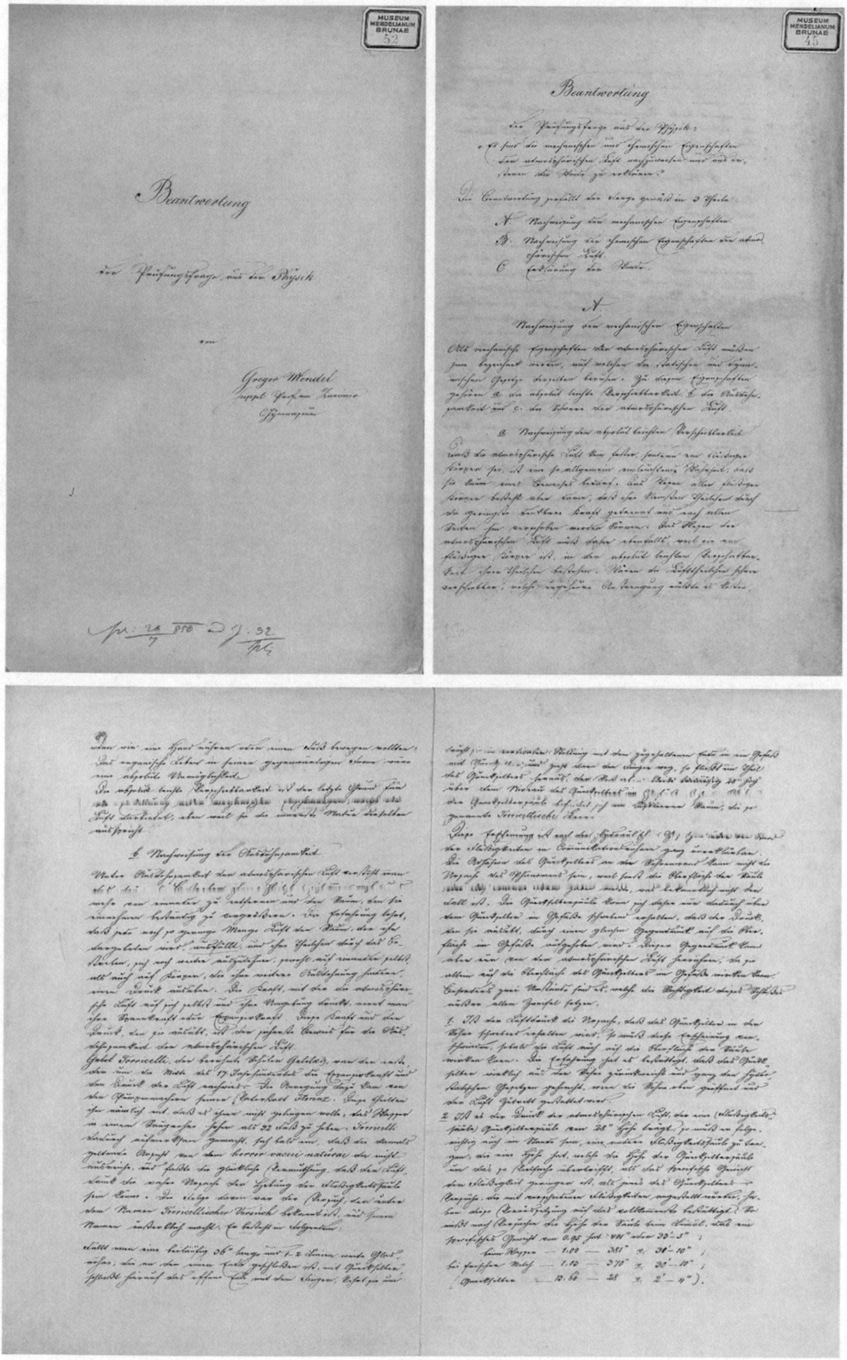

Beantwortung

der Prüfungsfrage aus der Physik

von

Gregor Mendel
suppl. Professor am Znaimer
Gymnasium

Beantwortung

D34 Mendel's homework in physics signed Gregor Mendel, 26 July 1850: published in full for the first time

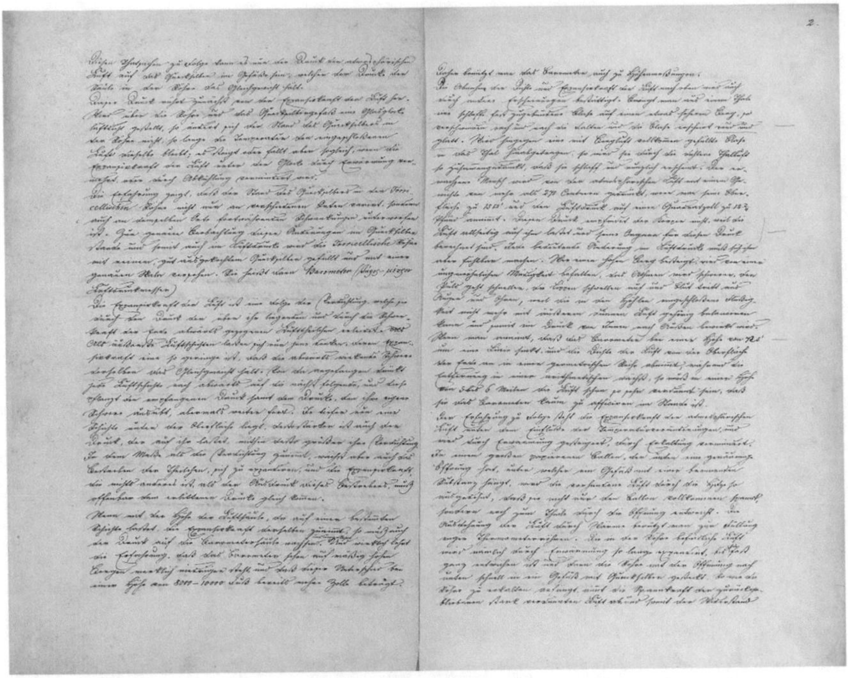

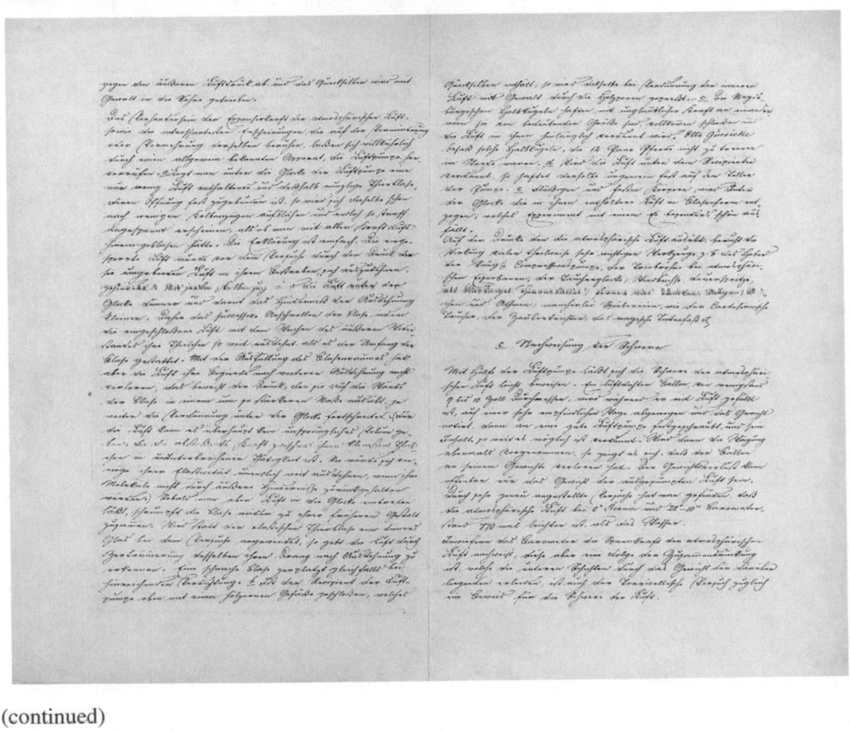

(continued)

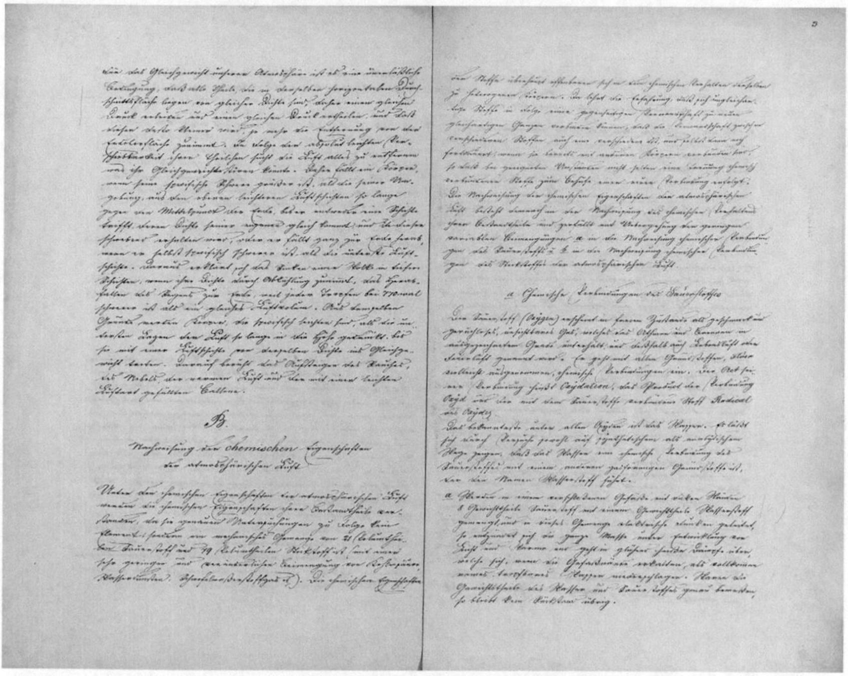

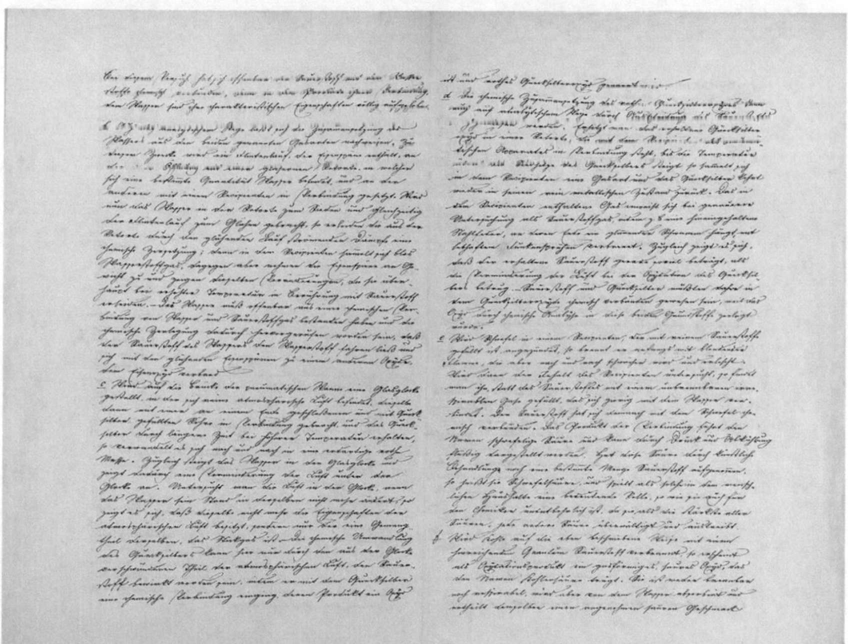

(continued)

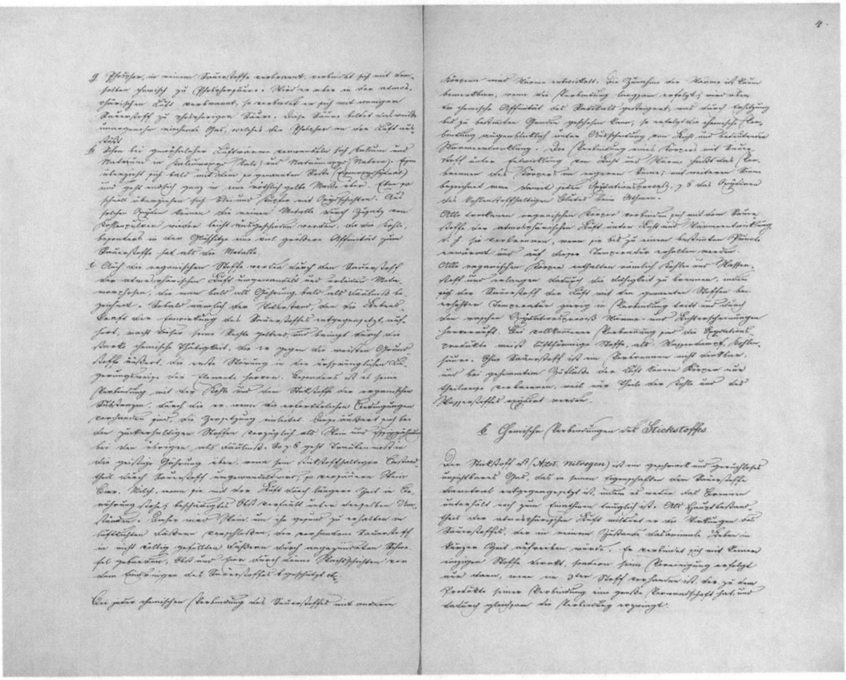
4.
6. Chemische Verbindungen des Stickstoffs

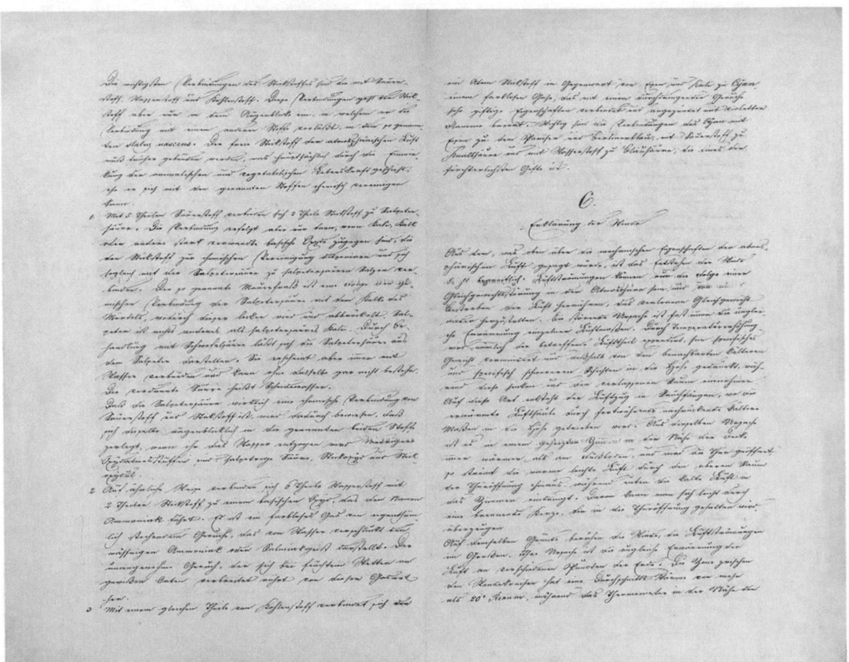
C.
Entstehung der Winde

(continued)

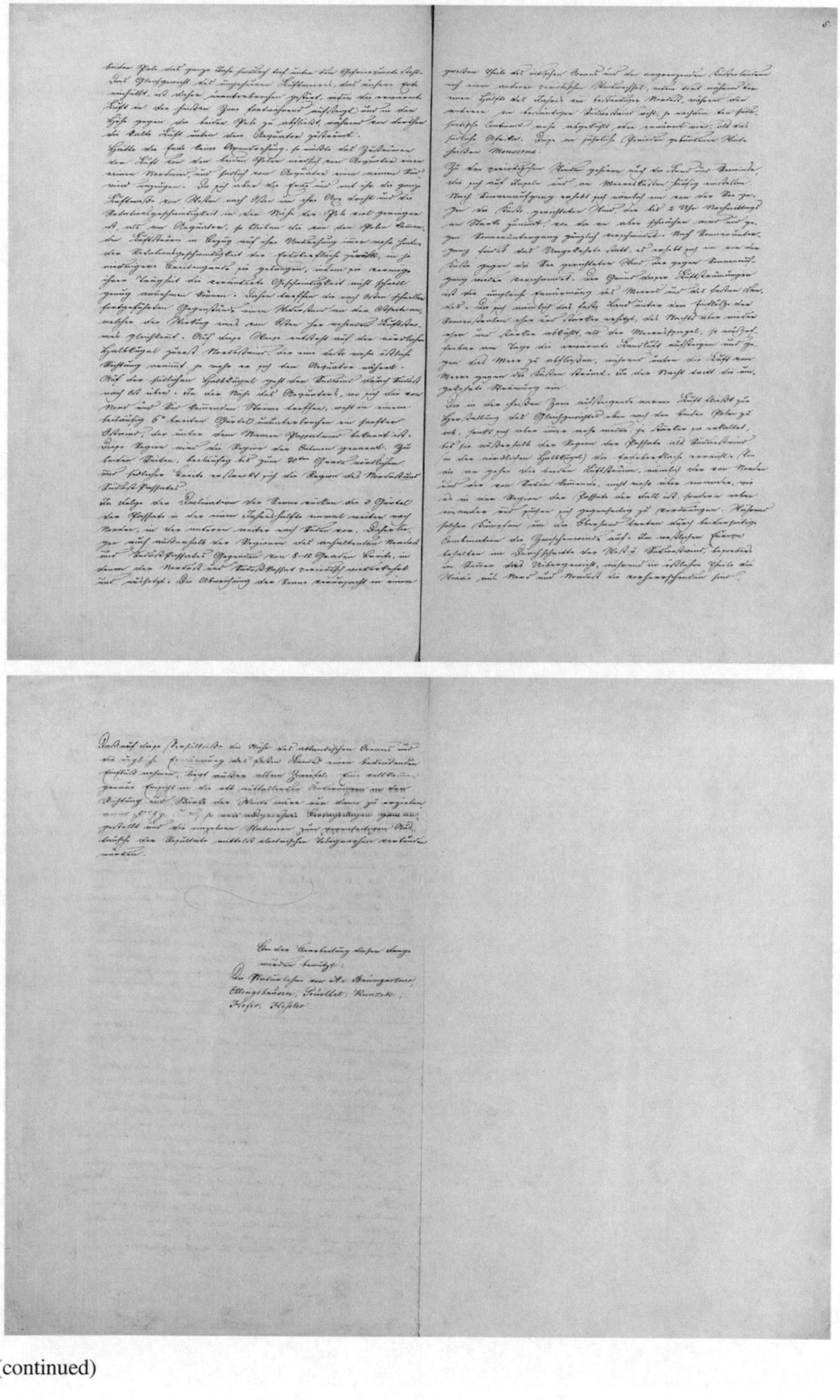

(continued)

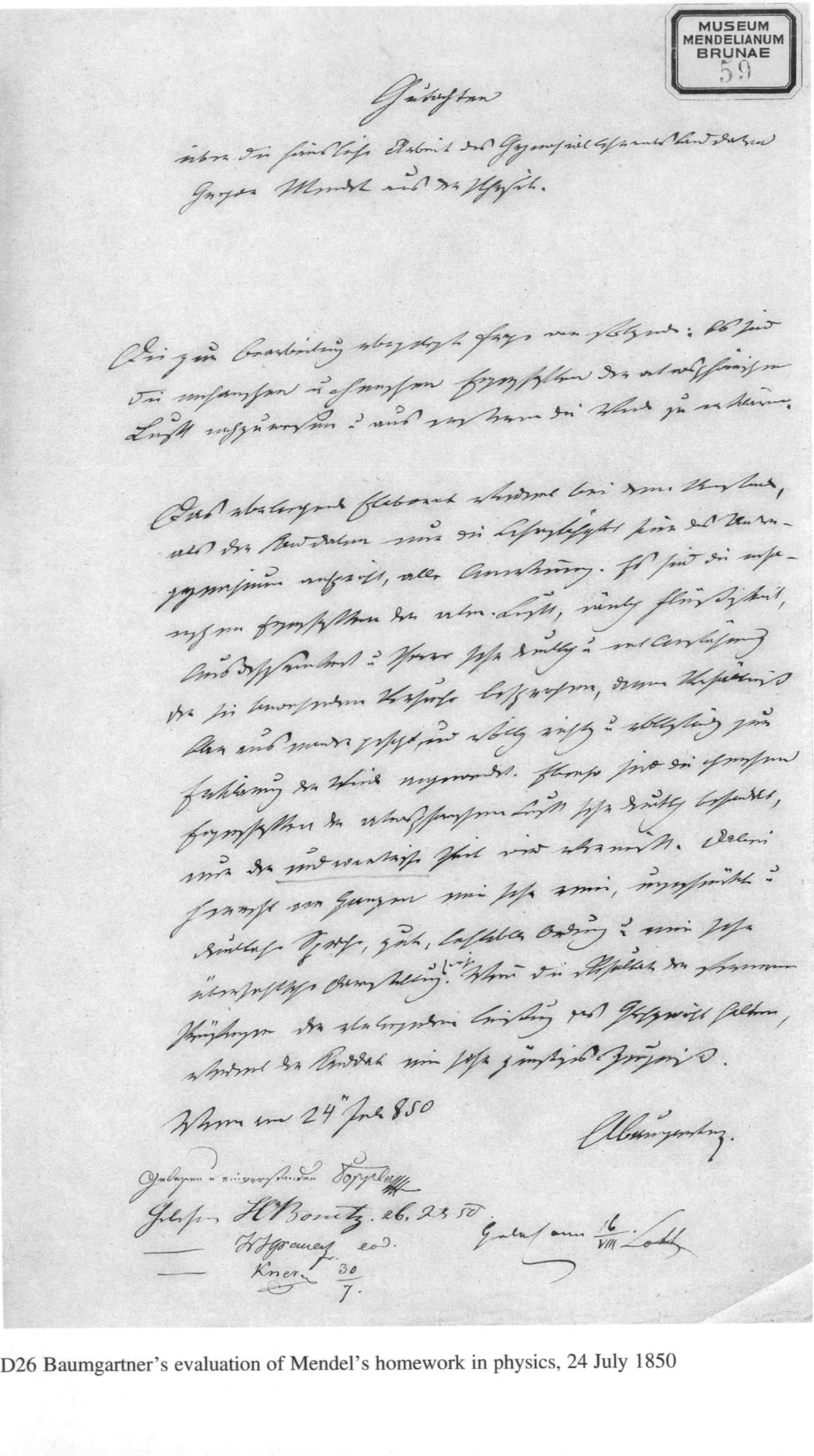

D26 Baumgartner's evaluation of Mendel's homework in physics, 24 July 1850

as satisfactory. He gives a clear description of the relationships of the atmosphere and of the experiments bearing upon these, applying the data separately, accurately, and fully to the explanation of the winds. Speaking generally, he writes simply, plainly, and clearly, his method of exposition being orderly and lucid. If the other examiners are as well satisfied as I, the candidate should have a very favourable testimonial.

Mendel's natural history homework was published in facsimile and transcription by Orel, Czihak, Wieseneder (FM18, pp. 227–72).

The title of Mendel's homework in natural history was prepared by R. Kner, English translation by Eden and Cedar Paul from Iltis (1966, p. 64): *Describe the chief differences between the rocks formed by water and those formed by fire, detailing the main varieties of the Neptunian strata in serial order according to their age and giving a short characterisation thereof, and, to conclude, giving in like manner a survey of the igneous rocks, both Plutonian and Vulcanian.*

A facsimile of Kner's evaluation of Mendel's essay in natural history was published by Gicklhorn (1969, pp. 148–9). It is dated in Wien/Vienna, 26 July 1850, signed by Rud. Kner; vidi H. Bonitz 27 July 1850, W. Grauert 29 July 1850, Doppler Chr. 29 July 1850, Karajan 01 August 1850, forwarded on 01 August 1850, vidi Lott 16 August. The original is deposited at the University of Illinois Archives in Urbana-Champaign.

The following were the specialties of Mendel's examiners: Hermann Bonitz was a philologist, Wilhelm Grauert a historian, Christian Doppler a physicist, Theodor Georg von Karajan a linguist and professor of German literature, and Franz Karl Lott a philosopher.

Transcription of Kner's evaluation of Mendel's homework essay in natural history was published by Gicklhorn (1969, pp. 156–7).

The following excerpt from Kner's opinion, in English translation by Eden and Cedar Paul from Iltis (1966, pp. 65–6) must have lessened Mendel's hopes for the successful exam: *The wide scope of the question that was set makes it plain at the outset that the main object was to obtain a precise but clear general exposition, presenting the essential characteristics, thus giving the candidate an opportunity of showing very plainly the extent and accuracy of his knowledge. Neither in this respect, however, nor yet in any other, is the answer satisfactory. The candidate has, indeed, written about many things, briefly in many cases, but neither concisely nor clearly. He has been apt to miss the punctum saliens. His characterisations are anything but sharply defined, and erroneous statements are not infrequent. In the account of the Neptunian formations, too much space is given to the mineralogical description of the rocks, whereas the candidate ought to have known that in this field the differences between the rocks are of trifling and in most cases of a purely local significance. For this reason, the description of the various strata is likewise arid, obscure, and hazy—all the more seeing that the most characteristic fossils are scarcely mentioned, or, if mentioned, are alluded to wrongly. The style is pretty good, but the language is somewhat hyperbolic, and inappropriate expressions are not infrequent.—The candidate informs us that he has had at his disposal only a few books, but good ones. However, the use he has made of these works shows that he*

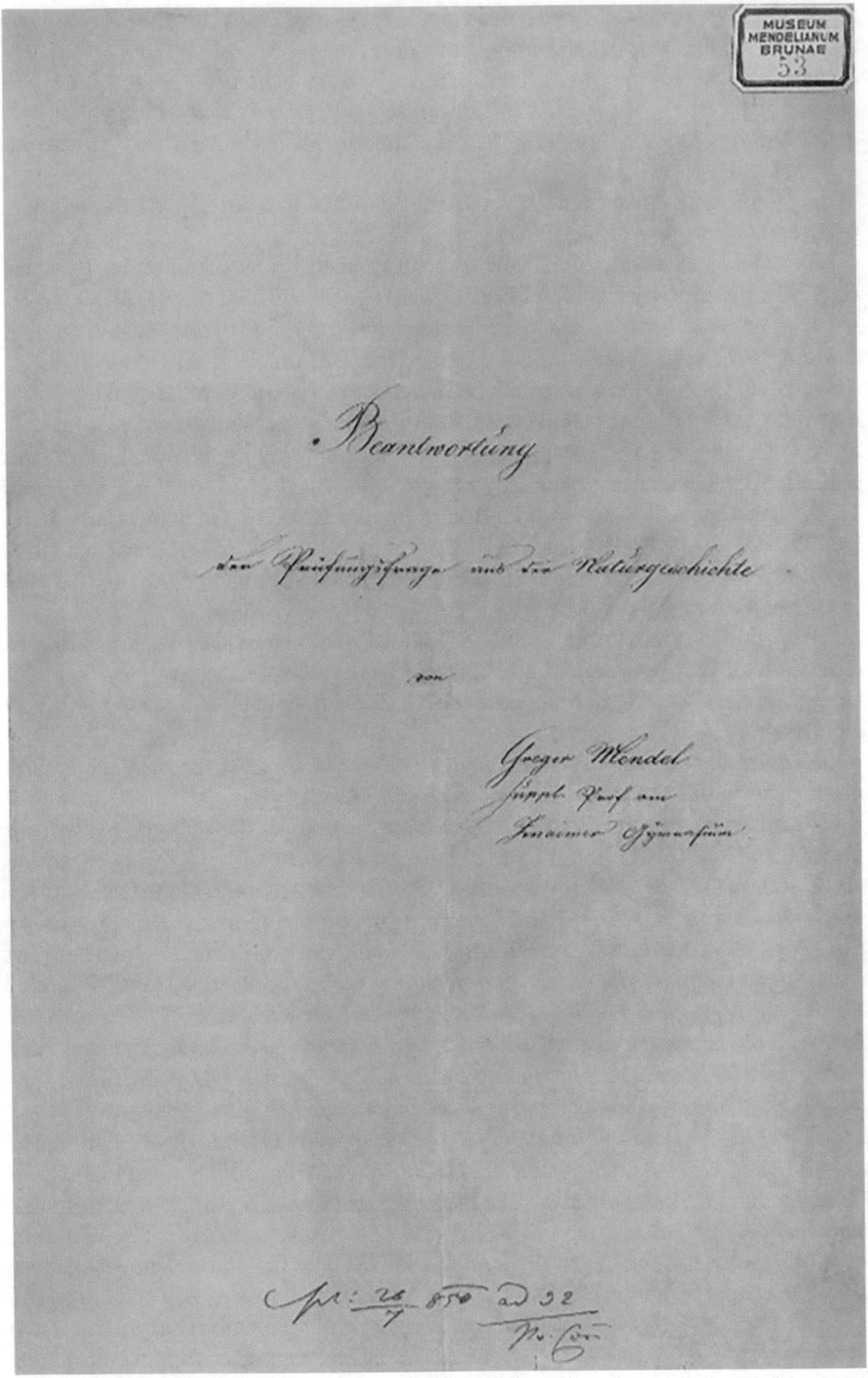
MUSEUM MENDELIANUM BRUNAE 53

Beantwortung

der Prüfungsfragen aus der Naturgeschichte

von

Gregor Mendel
suppl. Prof. am
Znaimer Gymnasium

D39 The title page of Mendel's homework in natural history

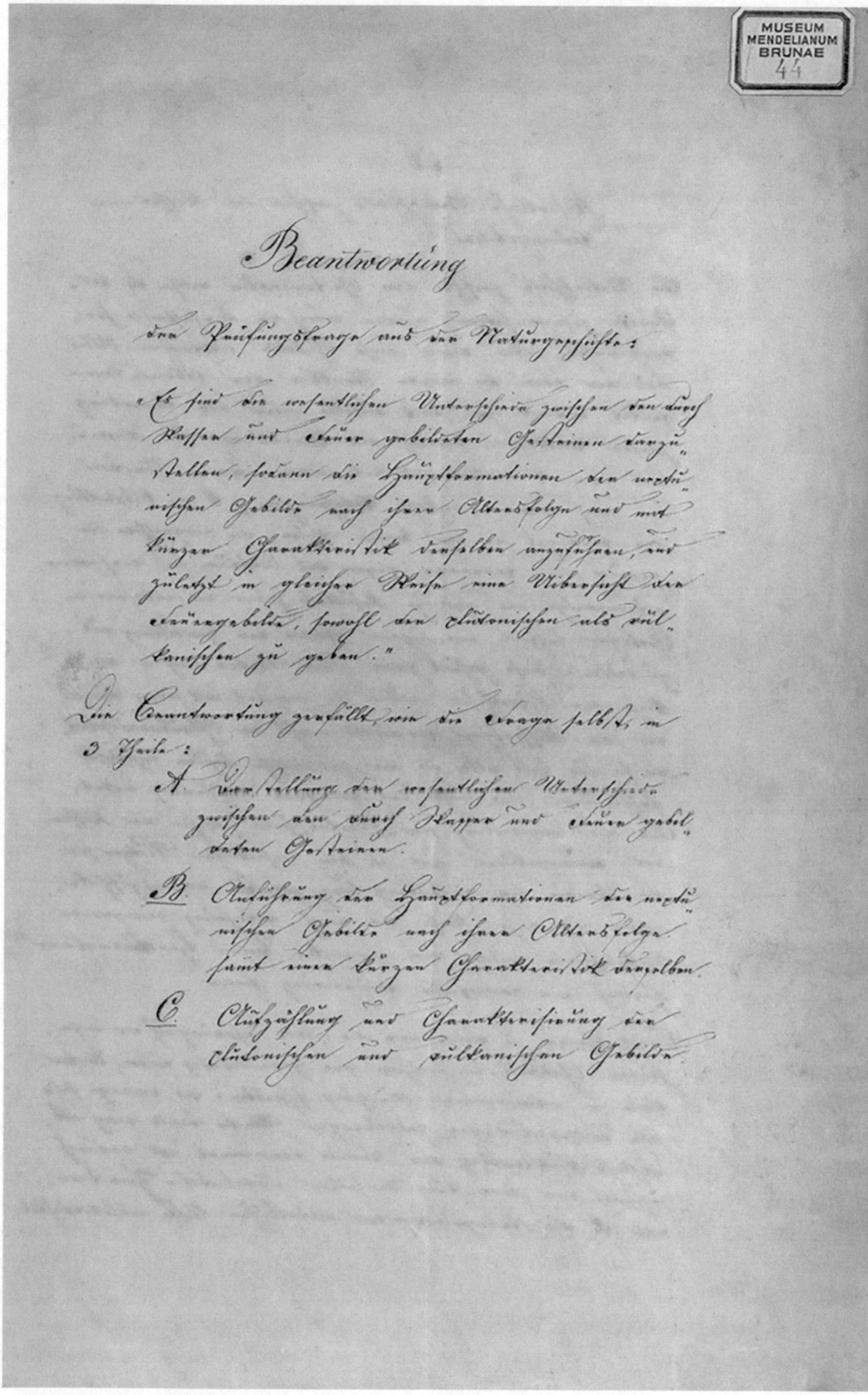

Beantwortung

der Prüfungsfrage aus der Naturgeschichte:

„Es sind die wesentlichen Unterschiede zwischen den durch Wasser und Feuer gebildeten Gesteinen darzustellen, sodann die Hauptformationen der neptunischen Gebilde nach ihrer Altersfolge und mit kurzer Charakteristik derselben anzuführen, und zuletzt in gleicher Weise eine Übersicht der Feuergebilde, sowohl der plutonischen als vulkanischen zu geben."

Die Beantwortung zerfällt, wie die Frage selbst, in 3 Theile:

A. Darstellung der wesentlichen Unterschiede zwischen den durch Wasser und Feuer gebildeten Gesteinen.

B. Anführung der Hauptformationen der neptunischen Gebilde nach ihrer Altersfolge sammt einer kurzen Charakteristik derselben.

C. Aufzählung und Charakterisirung der plutonischen und vulkanischen Gebilde.

D39 The first page of Mendel's homework essay in natural history

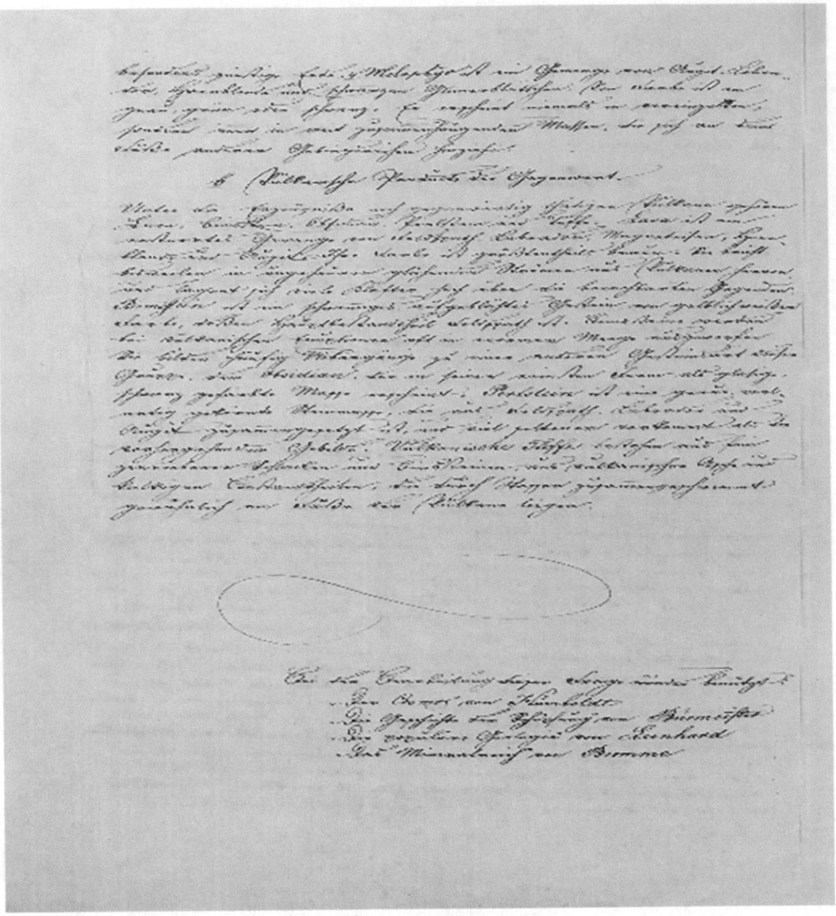

D39 The last page of Mendel's homework in natural history with the titles of books he used for information in writing the essay: *Der Kosmos von Humboldt, Die Geschichte der Schöpfung von Burmeister, Die populäre Geologie von Leonhard, Das Mineralreich von Bromme*

certainly has not mastered geognosy sufficiently to qualify him as a teacher in the higher schools. Still, I am prepared to approve the continuance of the examination, in the hope that the candidate may show, a fuller knowledge of other branches of natural history, and in view of the fact that geognosy belongs to the field of advanced instruction.

Mendel travelled to the University of Vienna to complete the on-site written (*Klausurprüfung*) and oral (*viva voce*) sessions in August of 1850. The question Mendel addressed in the on-site written examination in physics was, in English translation by Eden and Cedar Paul from Iltis (1966, p. 69): *By what means can steel be permanently magnetised, and in accordance with what laws is the magnetism distributed in a steel rod?* Mendel answered in five pages (D31) that are published in facsimile and transcribed by Gicklhorn (1973, pp. 75–7).

A transcription of Baumgartner's evaluation (D27) of Mendel's on-site written essay in physics was published by Gicklhorn (1973, p. 78). Baumgartner appreciated Mendel's treatment of the first part of the question. The second part of the question, however, he considered to be insufficient. Baumgartner was of the opinion that *still, as far as it goes, the answer is plain and definite, well arranged; the language is correct and simple, as it should be in the case of a beginner in the science. The answer shows that the candidate has a fair amount of formal knowledge, but has not yet got beyond the elements of physical science* (English translation by Eden and Cedar Paul from Iltis (1966, p. 69).

For his on-site written essay in natural history (D32), Mendel was asked to classify mammals and to give some examples of their utility for man. He answered in four pages. Dated in Wien/Vienna, 13 August 1850. Transcription by Gicklhorn (1973, pp. 80–2).

Kner was not willing to recommend that Mendel be granted certification for teaching. He referred to the fact that Mendel had not studied the very latest literature. Mendel learned from Gistel's book of 1848 *Natural History of Animal Kingdom for Higher Schools*. Kner was of the opinion that *the orders of the mammalian kingdom have been enumerated in accordance with a system now little in use, which does indeed appear simple at first sight, but is in reality extremely confusing, and seems to me by no means commendable. The characteristics selected for this classification are most unsuitable, not being what they pretend to be, and ill-adapted for descriptive purposes. The part of the question which relates to the utility of animals as yielding materials for industrial purposes or for use in medicine is answered in the most schoolboy fashion. The candidate seems to know nothing about technical terminology, naming all the animals to which he refers in colloquial German, and avoiding systematic nomenclature. Moreover, the style of the answer would be most unsuitable for a teacher in a high school. My judgment of this answer is, therefore, necessarily unfavourable, but the candidate may be allowed to enter for the viva voce examination, in the hope that a more favourable result will be achieved, although this written examination paper would hardly allow us to regard him competent to become an instructor even in a lower school.* [English translation by Eden and Cedar Paul in Iltis (1966, p. 71).]

The protocol documenting the viva voce examination from physics presided over by Baumgartner, and natural history by Kner in Vienna/Wien on 16 August 1850 was made in two copies. One has been preserved at the University of Illinois Archives in Urbana-Champaign. It is published in facsimile by Gicklhorn (1969, p. 150) and in transcription (Gicklhorn 1969, p. 157). The Protocol is without signature. The writer is neither Baumgartner nor Kner. Its conclusion is to not grant Mendel the decree of the teaching proficiency even for the lower grades of the *Gymnasium*.

The last paragraph of the Protocol recording the minutes of Mendel's oral examination from physics and natural history says that the protocol is made in two copies (*in duplo*), one for the commission and one for the candidate. Three pages.

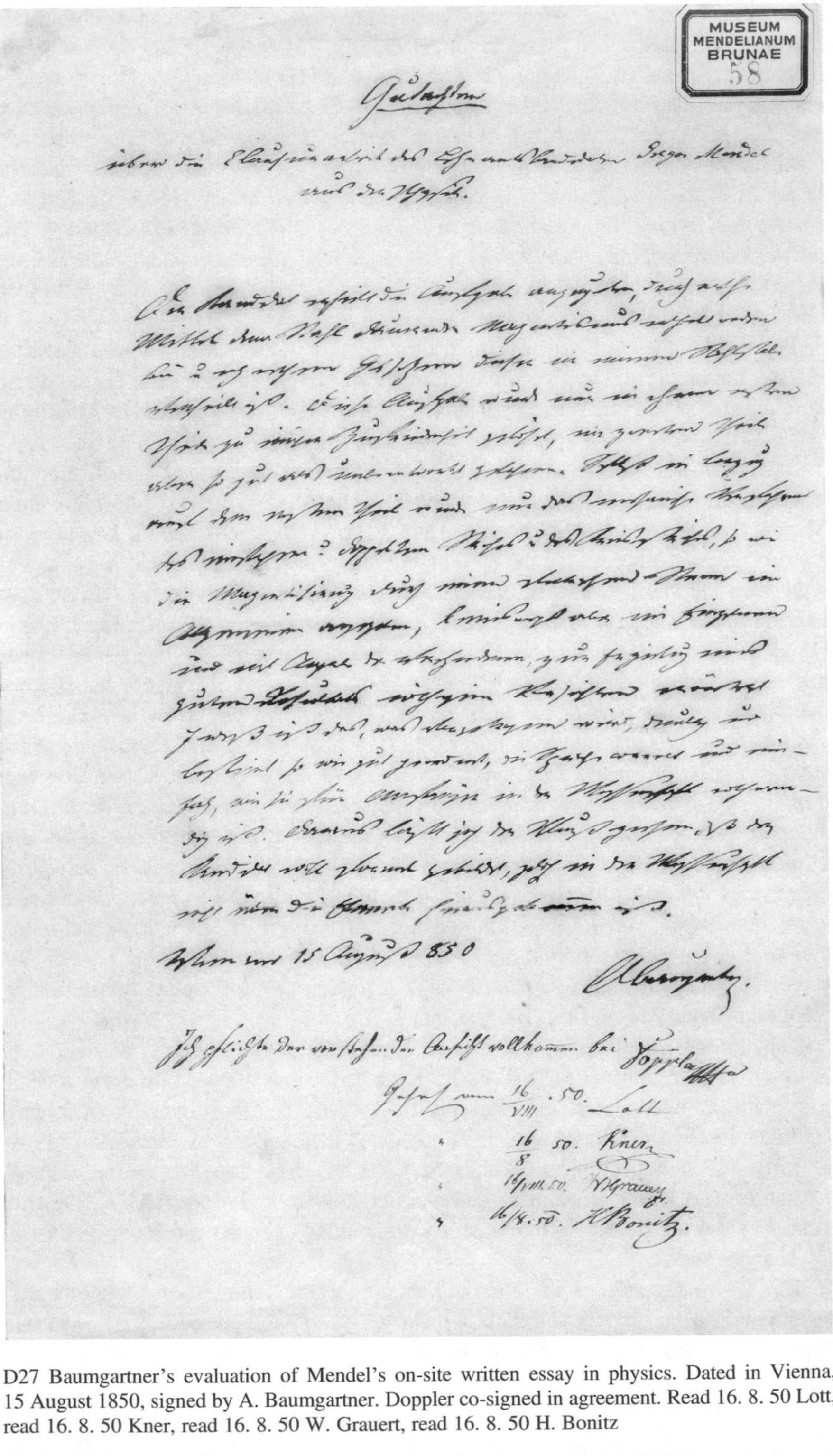

D27 Baumgartner's evaluation of Mendel's on-site written essay in physics. Dated in Vienna, 15 August 1850, signed by A. Baumgartner. Doppler co-signed in agreement. Read 16. 8. 50 Lott, read 16. 8. 50 Kner, read 16. 8. 50 W. Grauert, read 16. 8. 50 H. Bonitz

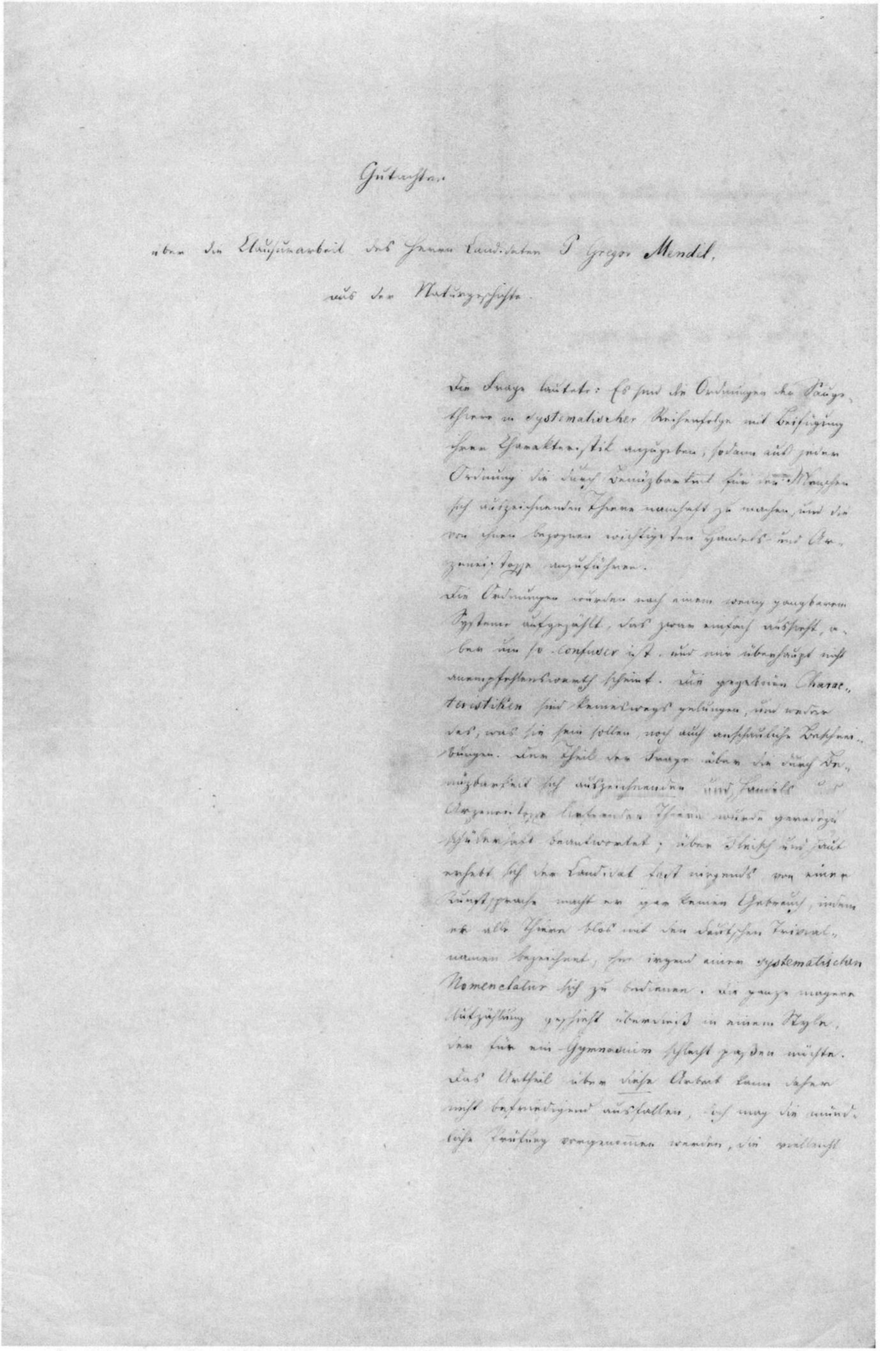

Gutachten

über die Clausurarbeit des Herrn Candidaten P. Gregor Mendel,
aus der Naturgeschichte.

Die Frage lautete: Es sind die Ordnungen der Säugethiere in systematischer Reihenfolge mit Beifügung ihrer Charakteristik anzugeben, sodann aus jeder Ordnung die durch Nutzbarkeit für den Menschen sich auszeichnenden Thiere namhaft zu machen, und die von ihnen bezogenen wichtigeren Handels- und Arzneistoffe anzuführen.

Die Ordnungen wurden nach einem wenig gangbaren Systeme aufgezählt, das zwar einfach aussieht, aber eben so confuser ist, und mir überhaupt nicht [illegible] scheint. Die gegebenen Characteristiken sind keineswegs gelungen, und weder das, was sie sein sollen, noch auch anschauliche Beschreibungen. Der Theil der Frage über die durch Nutzbarkeit sich auszeichnenden und Handels- und Arzneistoffe liefernden Thiere wurde [illegible] beantwortet; aber [illegible] erhebt sich der Candidat fast nirgends von einer Kunstsprache macht er gar keinen Gebrauch, indem er alle Thiere bloß mit den deutschen Trivialnamen bezeichnet, ohne irgend einer systematischen Nomenclatur sich zu bedienen. Die ganze [illegible] Aufzählung geschieht überdieß in einem Style, der für ein Gymnasium schlecht passen möchte. Das Urtheil über diese Arbeit kann daher nicht befriedigend ausfallen, doch mag die mündliche Prüfung vorgenommen werden, die vielleicht

D33 The first page of the evaluation by Kner of Mendel's on-site written essay in natural history on classification of mammals and their utility to man

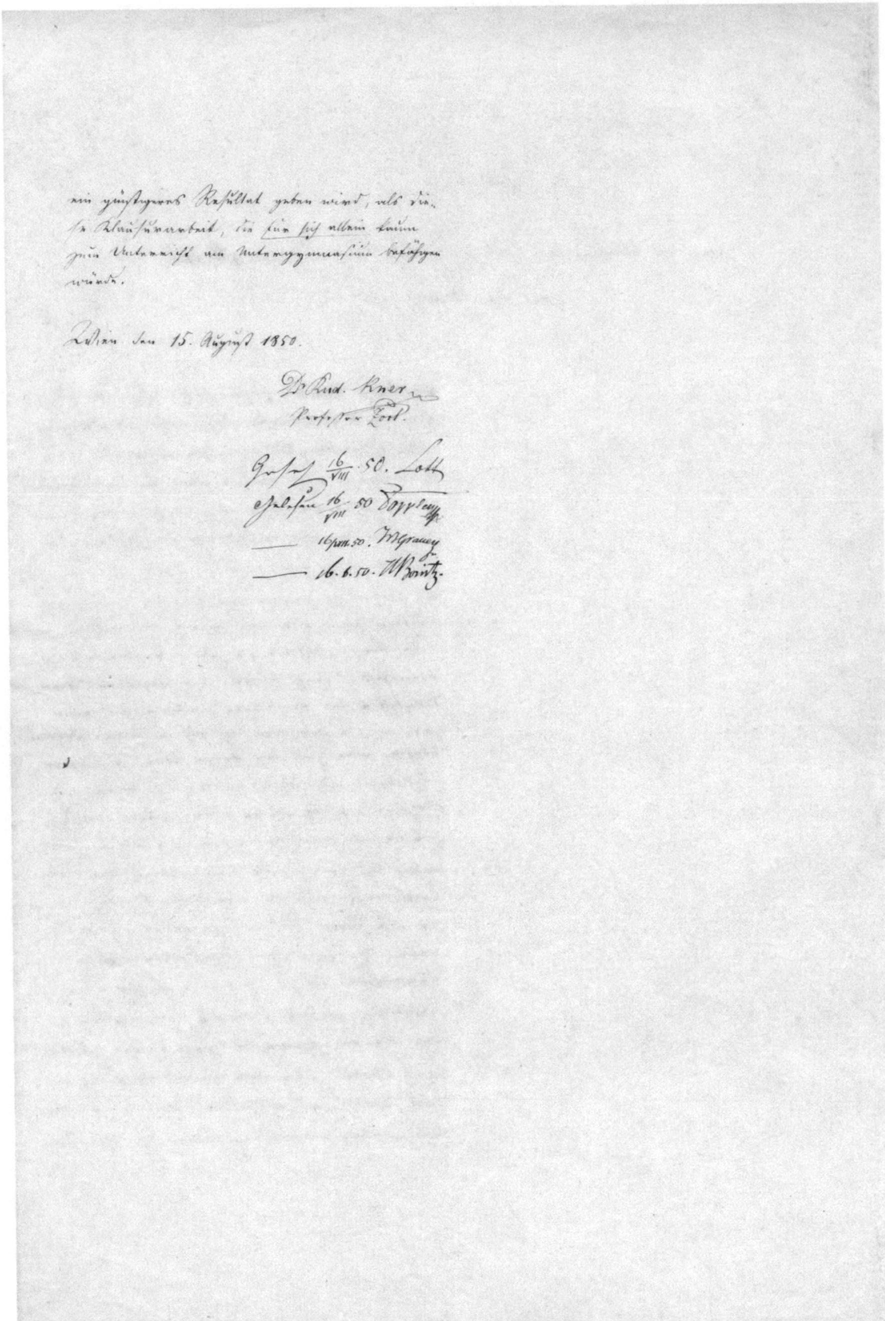

ein günstigeres Resultat geben wird, als diese Klausurarbeit, die für sich allein kaum zum Unterricht am Untergymnasium befähigen würde.

Wien den 15. August 1850.

Dr Rud. Kner
Professor Zool.

Gesehen 16/VIII 50. Lott
Gelesen 16/VIII 50 Doppler
— 16/VIII 50. Grauert
— 16.8.50. Bonitz

D33 The second page of the evaluation by Kner of Mendel's on-site written essay in natural history on classification of mammals and their utility to man. Dated in Wien/Vienna, 15 August 1850, signed by Dr. Rud. Kner, Professor Zool.; vidi 16. 8. 50 Lott, 16. 8. 50 Doppler, 16. 8. 50 Grauert, 16. 8. 50 Bonitz

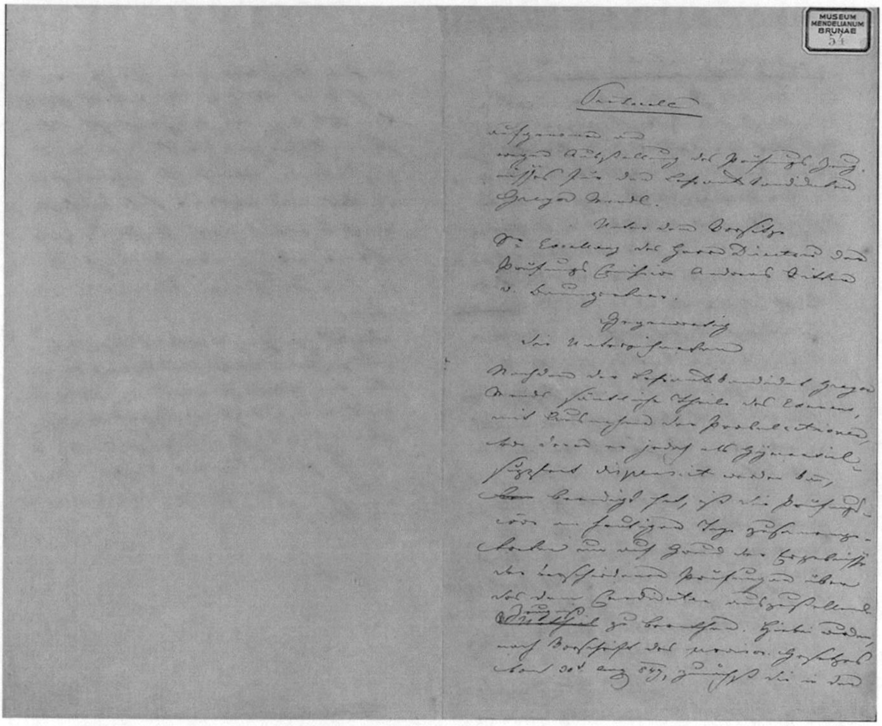

D67 The Brno original copy of the Protocol reporting the minutes of Mendel's oral examination from physics presided over by Baumgartner and from natural history by Kner. Dated in Wien/ Vienna, 16 August 1850, without signatures

One original copy of the Protocol is deposited in the archives of the University of Illinois in Urbana-Champaign. One in the Mendelianum of the Moravian Museum in Brno. It was published in facsimile by Gicklhorn (1969, p. 150) and transcribed on p. 157.

An excerpt from Mendel's Scientific *Gymnasium* Examination Certificate states that Mendel is not qualified to teach physics even in lower Gymnasia. English translation by Eden and Cedar Paul in Iltis (1966, pp. 71–2): *In the viva voce examination the candidate was asked various questions relating to different departments of physics. His answers confirmed the opinion which the examiners had already formed from his written answers. Though he has studied diligently, he lacks insight, and his knowledge is without the requisite clarity, so that the examiners find it necessary, for the time being, to declare him unqualified to teach physics in the lower schools. Since, however, in view of the candidate's unmistakable good will, we feel entitled to suppose that, if carefully guided, he will be enabled by further study to fit himself with the qualifications of a high-school teacher as required by law, we hereby declare that he shall be entitled to present himself for re-examination after the lapse of not less than one year.*

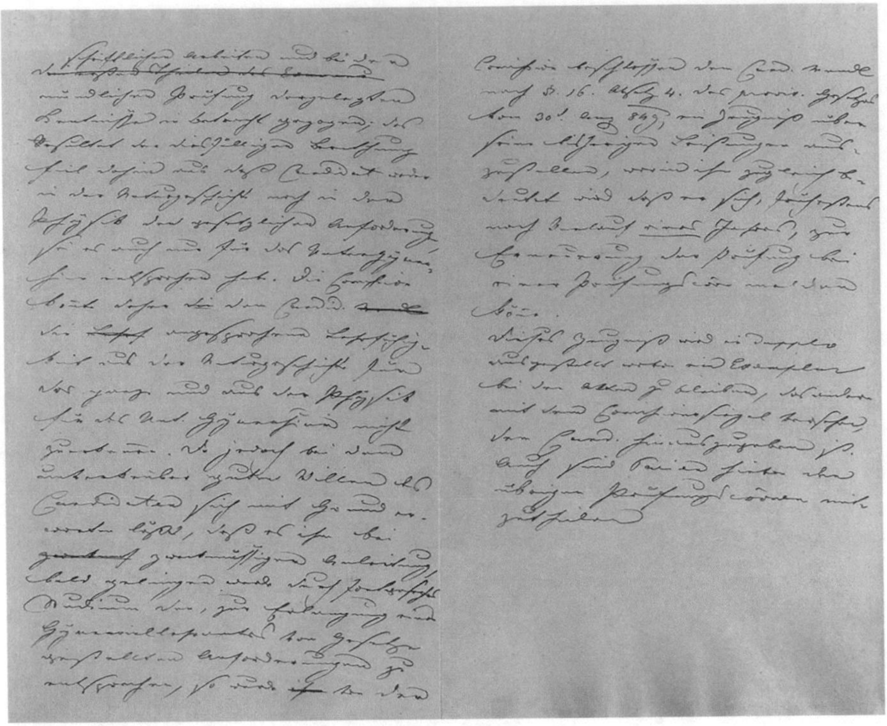

(continued)

Kner concludes that the result of the viva voce examination in natural history was more satisfactory to this extent, that the candidate showed more knowledge and gave evidence of more diligent study than might have been expected from the written papers. It was plain that he was devoid neither of industry nor of talent. It would seem, however, that he can have had no opportunity for acquiring exhaustive knowledge, and must have lacked access to the necessary means of study, so that he is not as yet competent to become a teacher. Still, we may hope if he is given opportunity for more exhaustive study together with access to better sources of information he will soon be able to fit himself, at least for work as a teacher at lower schools.

[English translation by Eden and Cedar Paul in Iltis (1966, p. 72).]

The report about the result of Mendel's university examination became officially known to Napp until 1 year later in the summer of 1851. In the meantime, Mendel began to work in the Agricultural Society. For a short time, he substituted the ill professor of natural history at the Technical Institute in Brno in the spring of 1851.

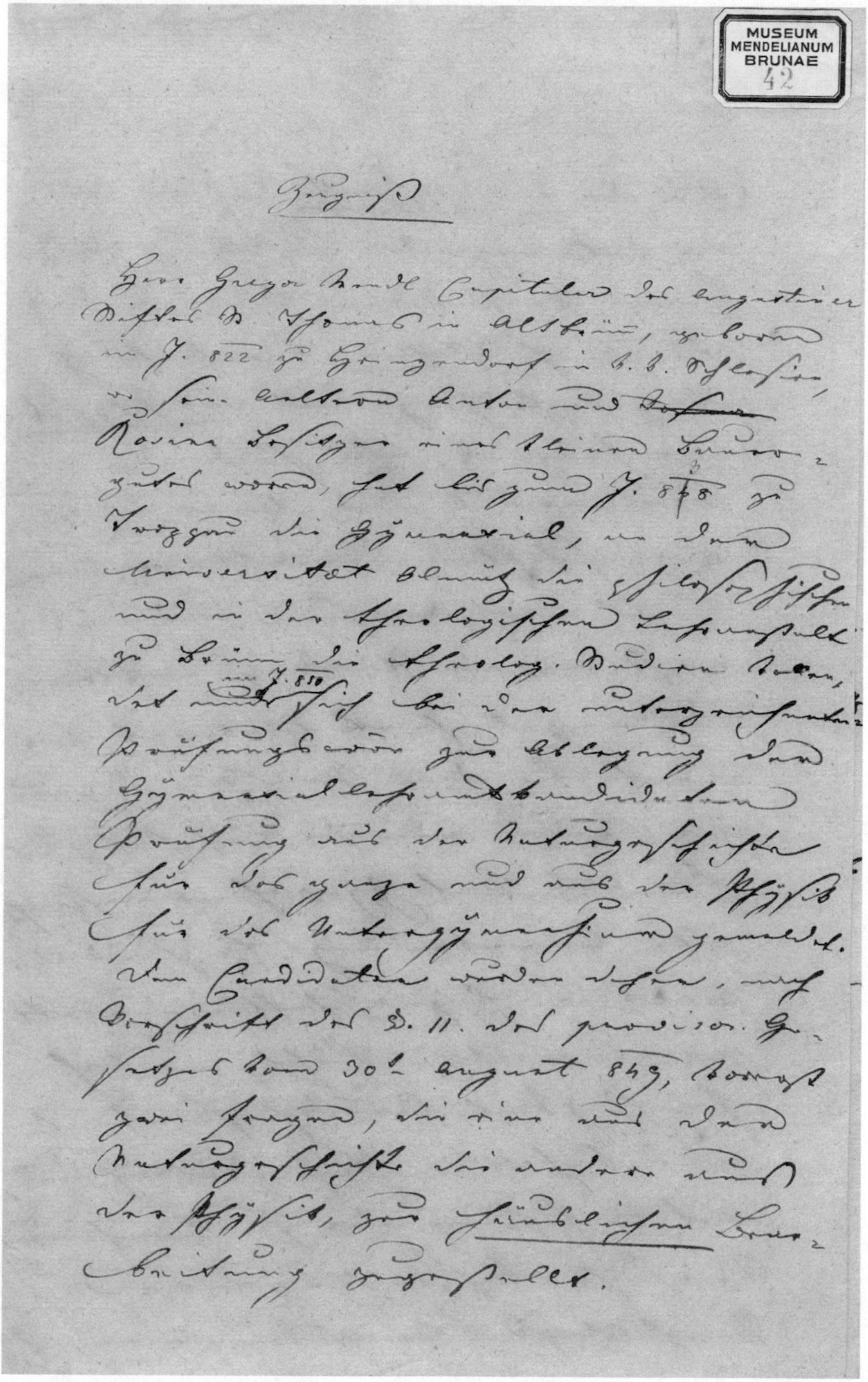

Zeugniß

Herr Gregor Mendel Capitular des Augustiner Stiftes St. Thomas in Altbrünn, geboren im J. 822 zu Heinzendorf in k. k. Schlesien, von seinen Aeltern Anton und Rosine Besitzer eines kleinen Bauerngutes waren, hat bis zum J. 843 zu Troppau das Gymnasium, an der Universität Olmütz die philosophischen und in der theologischen Lehranstalt zu Brünn die theolog. Studien beendet, hat sich im J. 850 bei der unterzeichneten Prüfungscommission zur Ablegung der Gymnasiallehramtscandidaten Prüfung aus der Naturgeschichte für das ganze und aus der Physik für das Untergymnasium gemeldet. Dem Candidaten wurden daher, nach Vorschrift des §. 11. des provisor. Gesetzes vom 30. August 849, vorerst zwei Fragen, die eine aus der Naturgeschichte die andere aus der Physik, zur häuslichen Bearbeitung zugestellt.

D68 Mendel's Final Certificate: The Imperial-Royal Scientific *Gymnasium* Examination Certificate from physics and natural history. Six pages. Dated in Wien/Vienna, 17 October, signed by A. Baumgartner, H. Bonitz, W. Grauert, Theodor Georg von Karajan, Rud. Kner, F. Lott

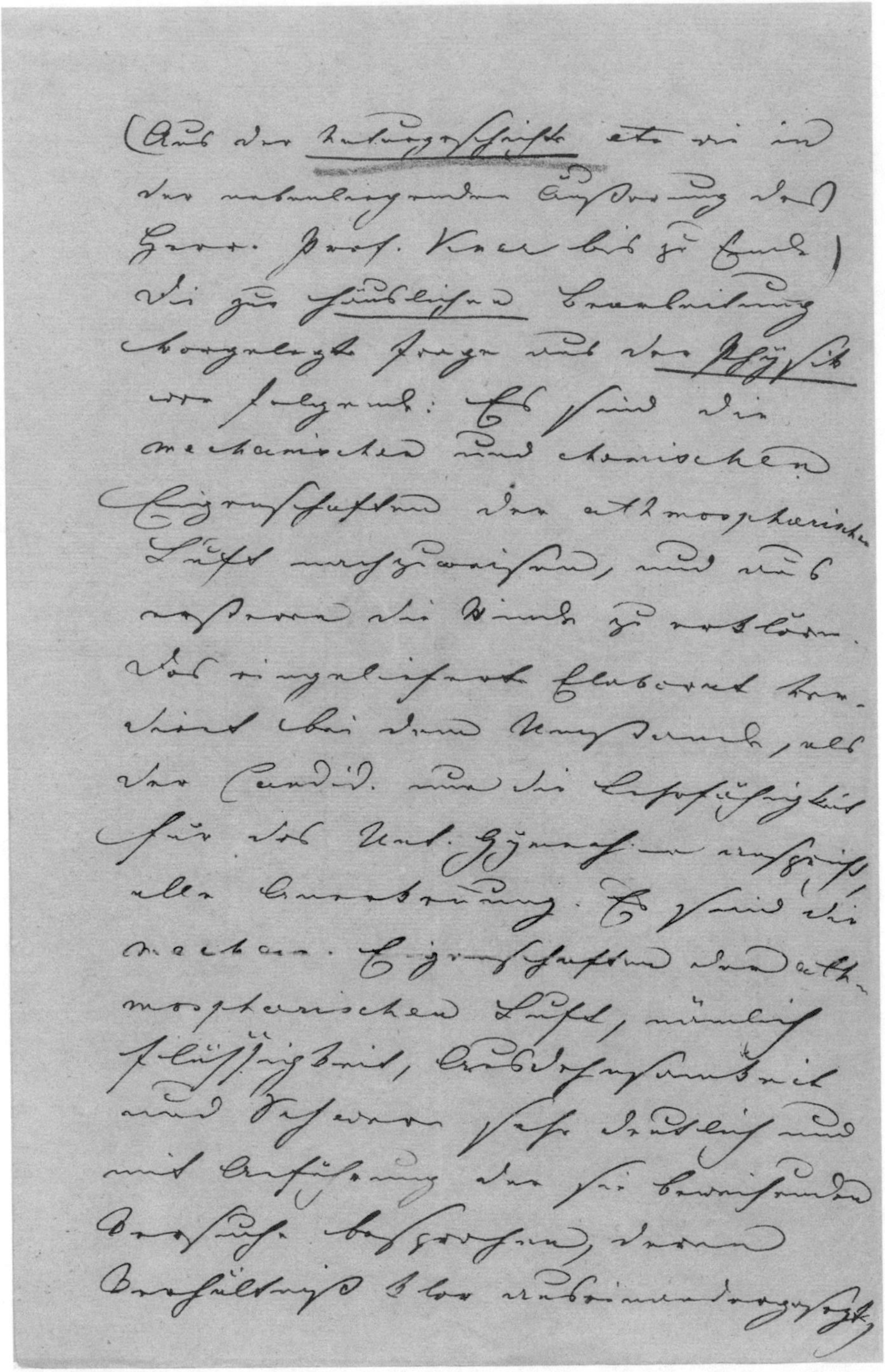

(Aus der Naturgeschichte ist sie in
der nebenliegenden Äusserung des
Herrn Prof. Kner bis zu finden)
Die zur häuslichen Bearbeitung
vorgelegte Frage aus der Physik
war folgende: Es sind die
mechanischen und chemischen
Eigenschaften der atmosphärischen
Luft nachzuweisen, und daraus
die Winde zu erklären.
Das eingelieferte Elaborat ver-
dient bei dem Umstande, als
der Candid. nur die Lehrfähigkeit
für das Unt. Gymnasium anspricht,
alle Anerkennung. Es sind die
mechan. Eigenschaften der atmosphärischen Luft, nämlich
Flüssigkeit, Ausdehnsamkeit
und Schwere sehr deutlich und
mit Anführung der sie beweisenden
Versuche besprochen, deren
Verhältniss klar auseinandergesetzt

(continued)

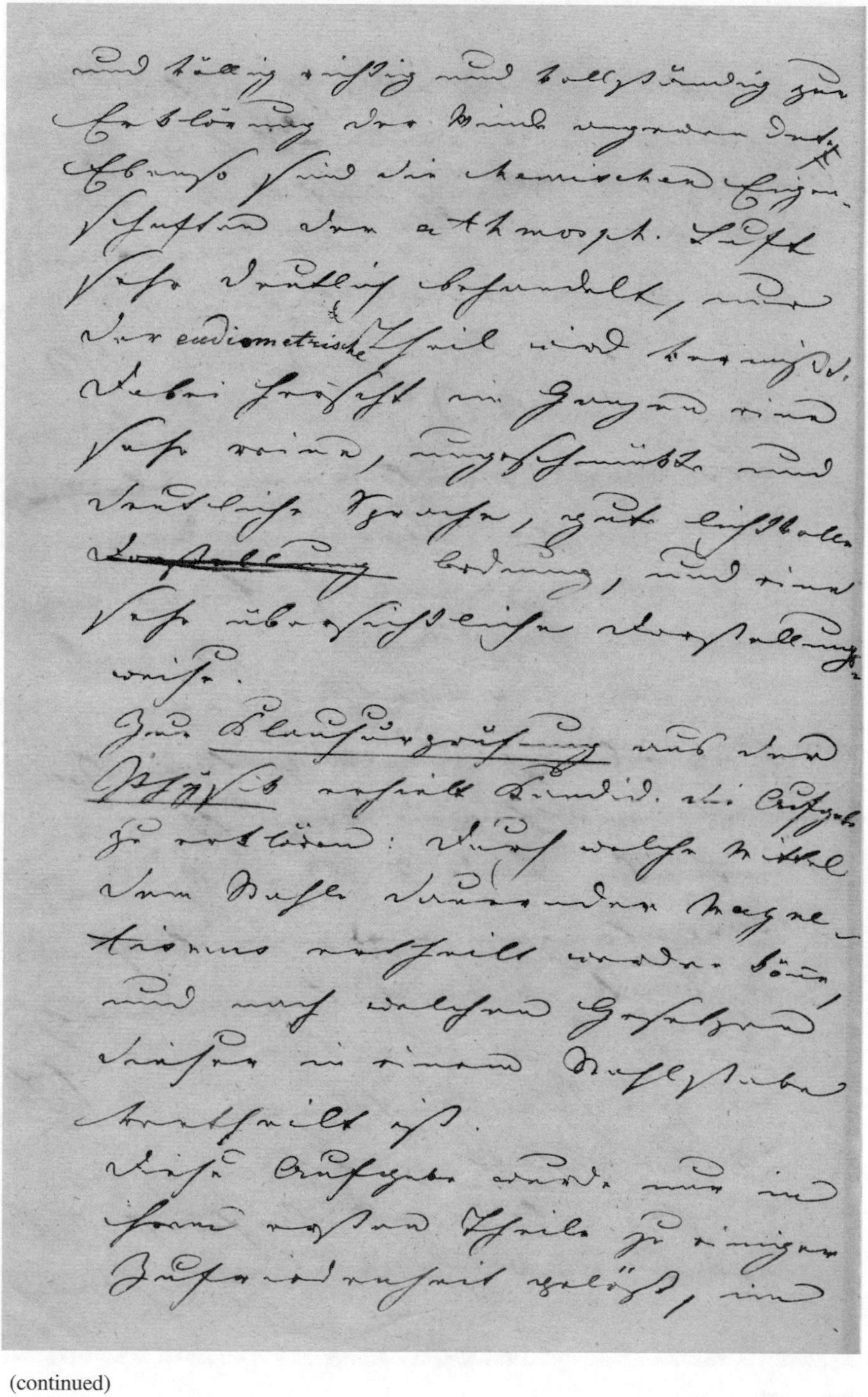

und tüchtig, richtig und vollständig zur
Erklärung der Winde angewendet.
Ebenso sind die chemischen Eigenschaften der atmosph. Luft
sehr deutlich behandelt, nur
der eudiometrische Theil wird vermißt.
Dabei herrscht im Ganzen eine
sehr reine, ungeschmückte und
deutliche Sprache, gute lichtvolle
~~Darstellung~~ Ordnung, und eine
sehr übersichtliche Darstellungs-
weise.
Zur Clausurprüfung aus der
Physik erhielt Candid. die Aufgabe
zu erklären: durch welche Mittel
dem Stahle dauernder Magne-
tismus ertheilt werden kann
und nach welchen Gesetzen
dieser in einem Stahlstabe
vertheilt ist.
Diese Aufgabe wurde nur im
seinem ersten Theile zu einiger
Zufriedenheit gelöst, im

(continued)

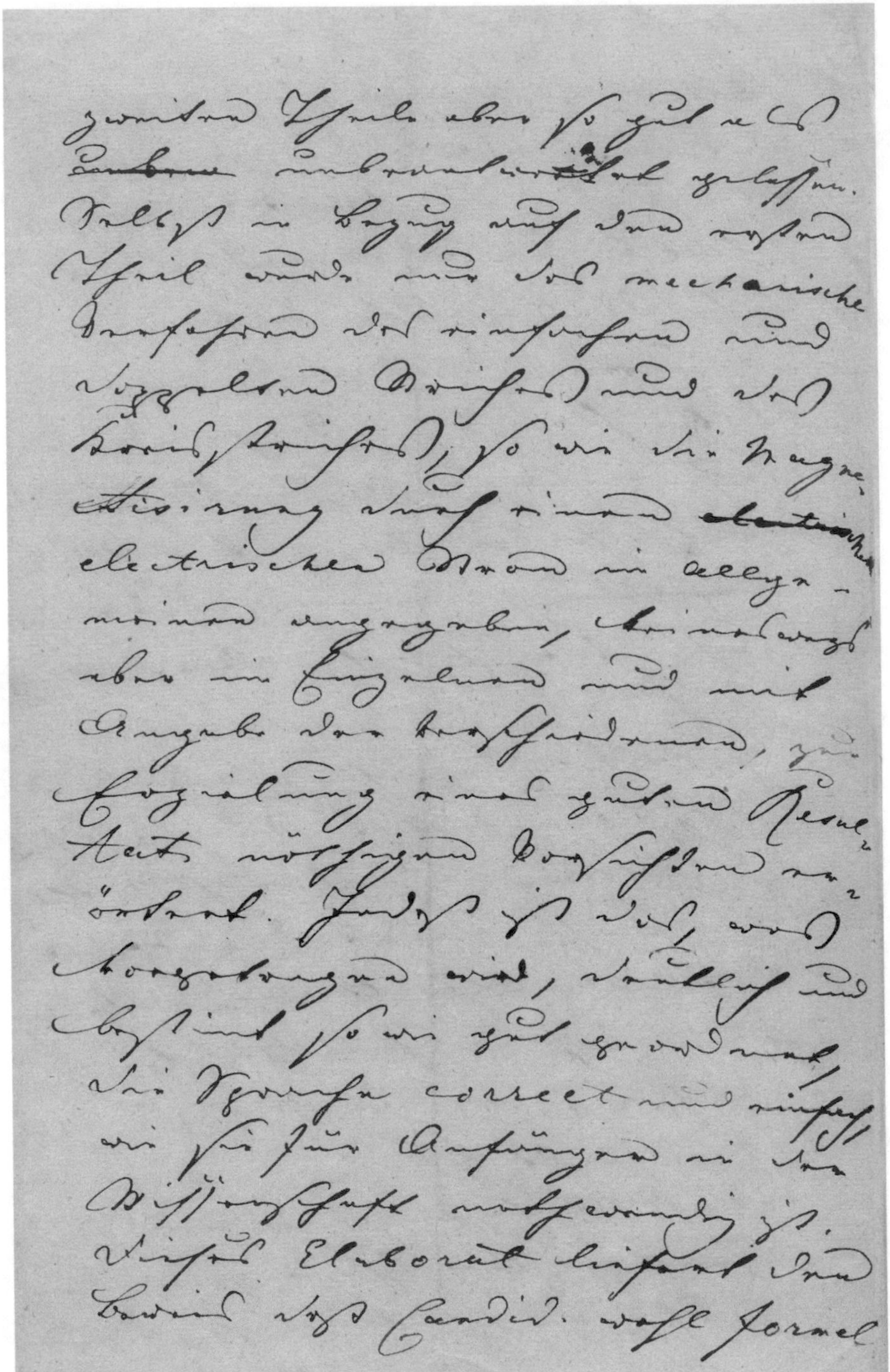
zweiten Theile aber so gut als unberücksichtigt gelassen. Selbst in Bezug auf den ersten Theil würde nur das mechanische Verfahren des einfachen und doppelten Streichens und des Kreisstreichens, so wie die Magnetisirung durch einen electrischen Strom im allgemeinen angegeben, keineswegs aber im Einzelnen und mit Angabe der Verschiedenen, zur Erzielung eines guten Resultats nöthigen Vorsichten erörtert. Jedoch ist das, was vorgetragen wird, deutlich und bestimmt so wie gut geordnet, die Sprache correct und einfach wie sie für Anfänger in der Wissenschaft nothwendig ist. Dieses Elaborat liefert den Beweis daß Candid. wohl formel

(continued)

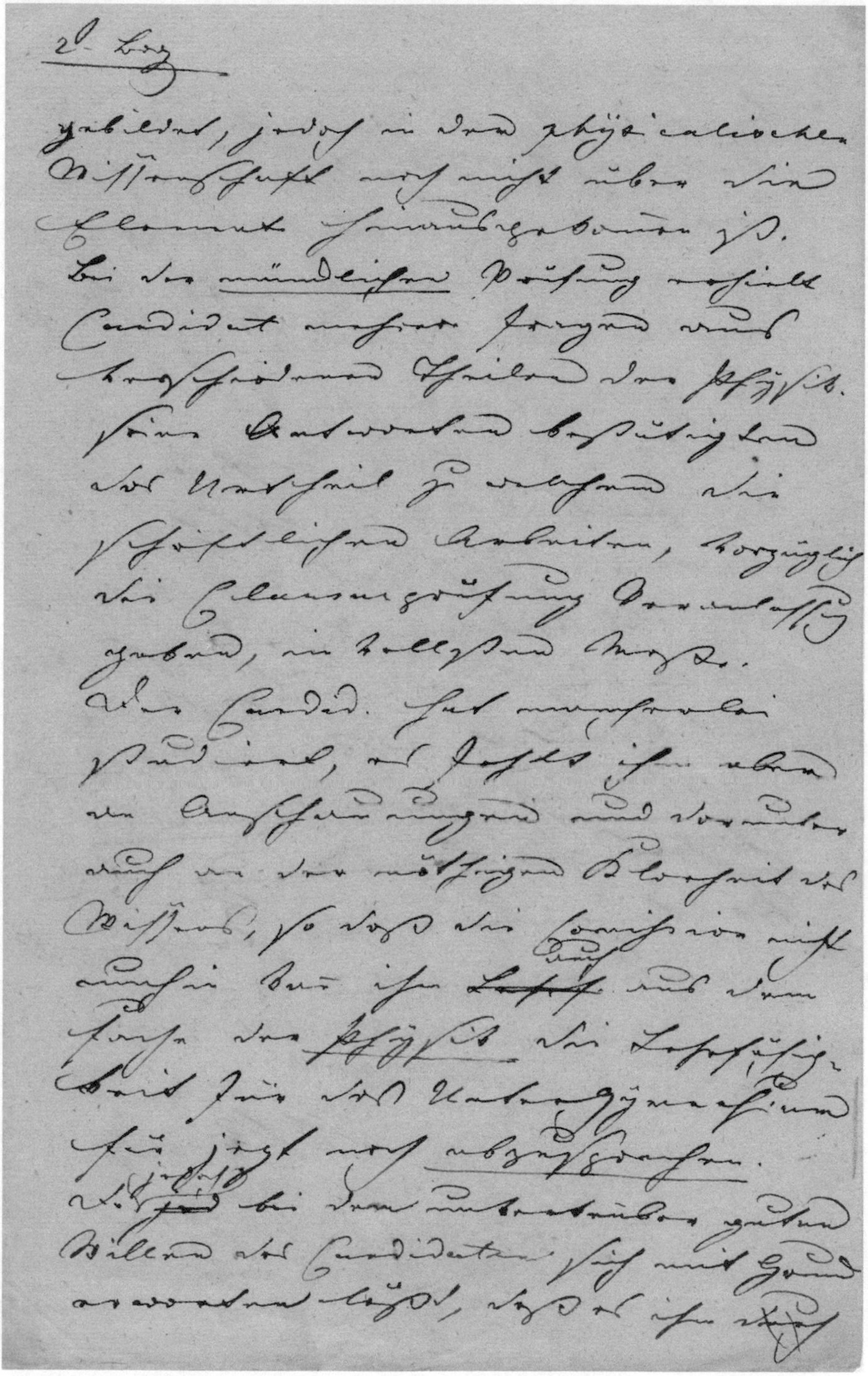

2. Bog.

gebildet, jedoch in den physicalischen
Wissenschaften noch nicht über die
Elemente hinausgekommen ist.
Bei der mündlichen Prüfung erhielt
Candidat mehrere Fragen aus
verschiedenen Theilen der Physik.
Seine Antworten bestätigten
das Urtheil zu welchem die
schriftlichen Arbeiten, vorzüglich
die Clausurprüfung veranlassung
gaben, in vollstem Maße.
Der Candid. hat mancherlei
studirt, es fehlt ihm aber
an Anschauungen und vermuthlich
auch an der nöthigen Klarheit des
Wissens, so daß die Commission nicht
umhin kann ihm ~~Lehrf~~ auch aus dem
Fache der Physik die Lehrfähigkeit für das Untergymnasium
für jetzt noch abzusprechen.
Da jedoch bei dem unverkennbaren guten
Willen des Candidaten sich mit Grund
erwarten läßt, daß es ihm durch

(continued)

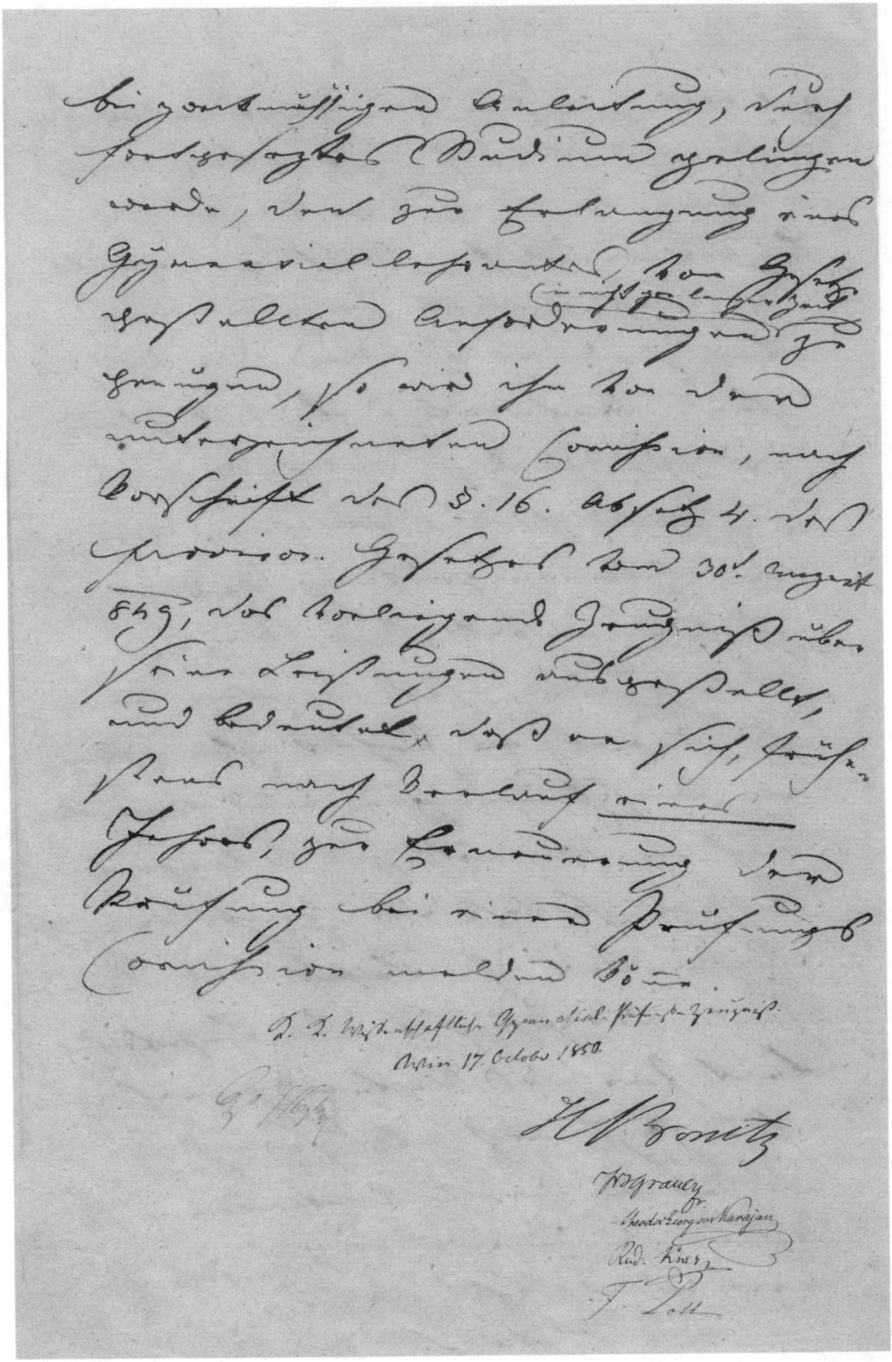

bei zweckmäßiger Anleitung, durch fortgesetzte Studien gelingen werde, den zur Erlangung eines Gymnasiallehramtes vom Gesetze gestellten Anforderungen zu genügen, so wird ihm von der unterzeichneten Commission, nach Vorschrift des §. 16. Absatz 4. des provisor. Gesetzes vom 30t. August 849, das vorliegende Zeugniß über seine Leistungen ausgestellt, und bedeutet, daß er sich, frühestens nach Verlauf eines Jahres, zur Erneuerung der Prüfung bei einer Prüfungs Commission melden könne.

K. K. Wissenschaftliche Gymnasial-Prüfungs-Zeugniß.
Wien 17. October 1850.

H Bonitz
W Grauert
Dr. Theodor Georg von Karajan
Rud. Kner
F. Lott

(continued)

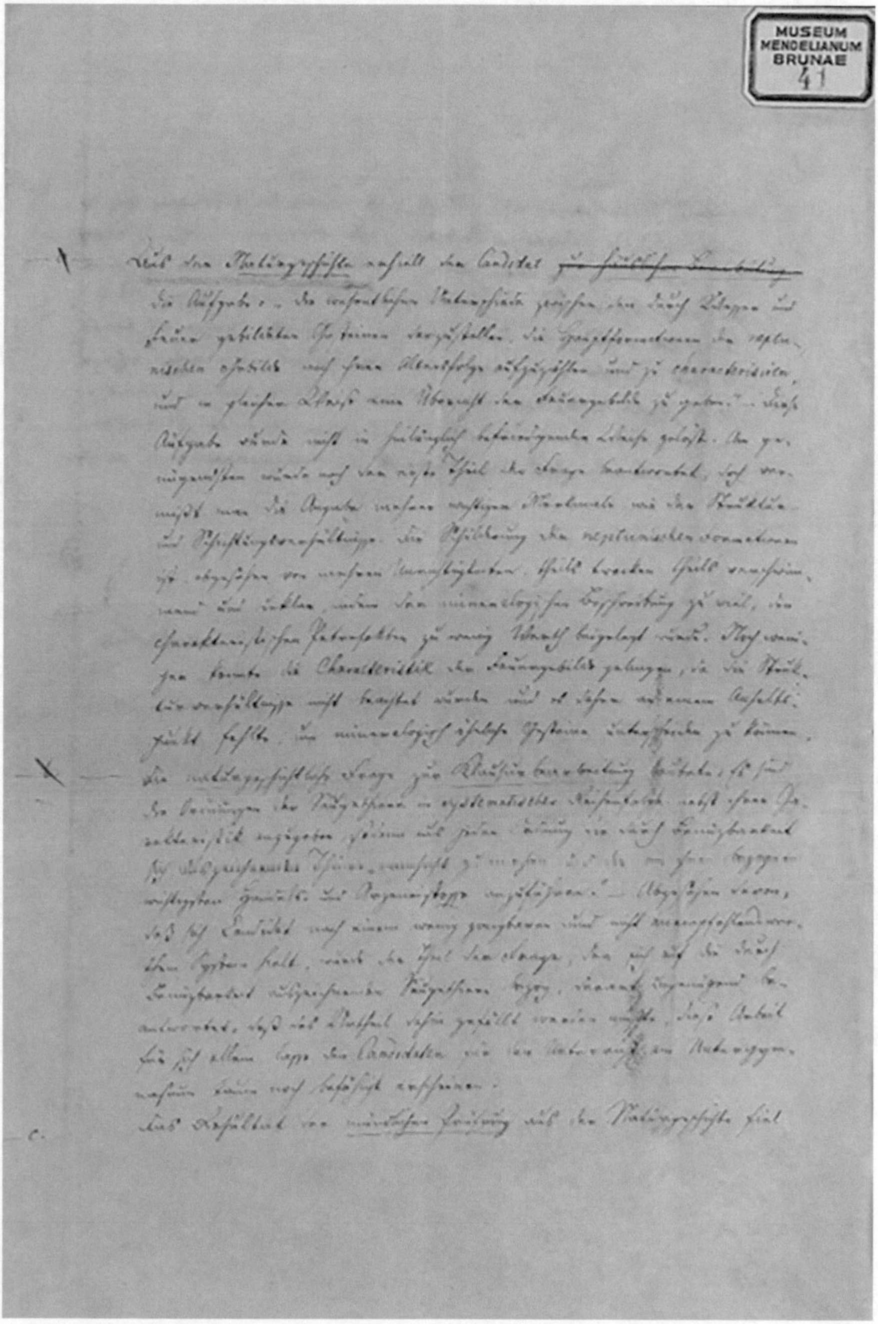
MUSEUM
MENDELIANUM
BRUNAE
41

D66 Kner's complex evaluation of Mendel's homework, on-site written and viva voce examinations in natural history. Two pages. Kner registered Mendel's examination from natural history belatedly most probably *pro domo* for formality reasons. Dated in Wien/Vienna, 11 November 1850, signed by Dr. Rudolf Kner. Transcription published by Gicklhorn (1973, pp. 82–3)

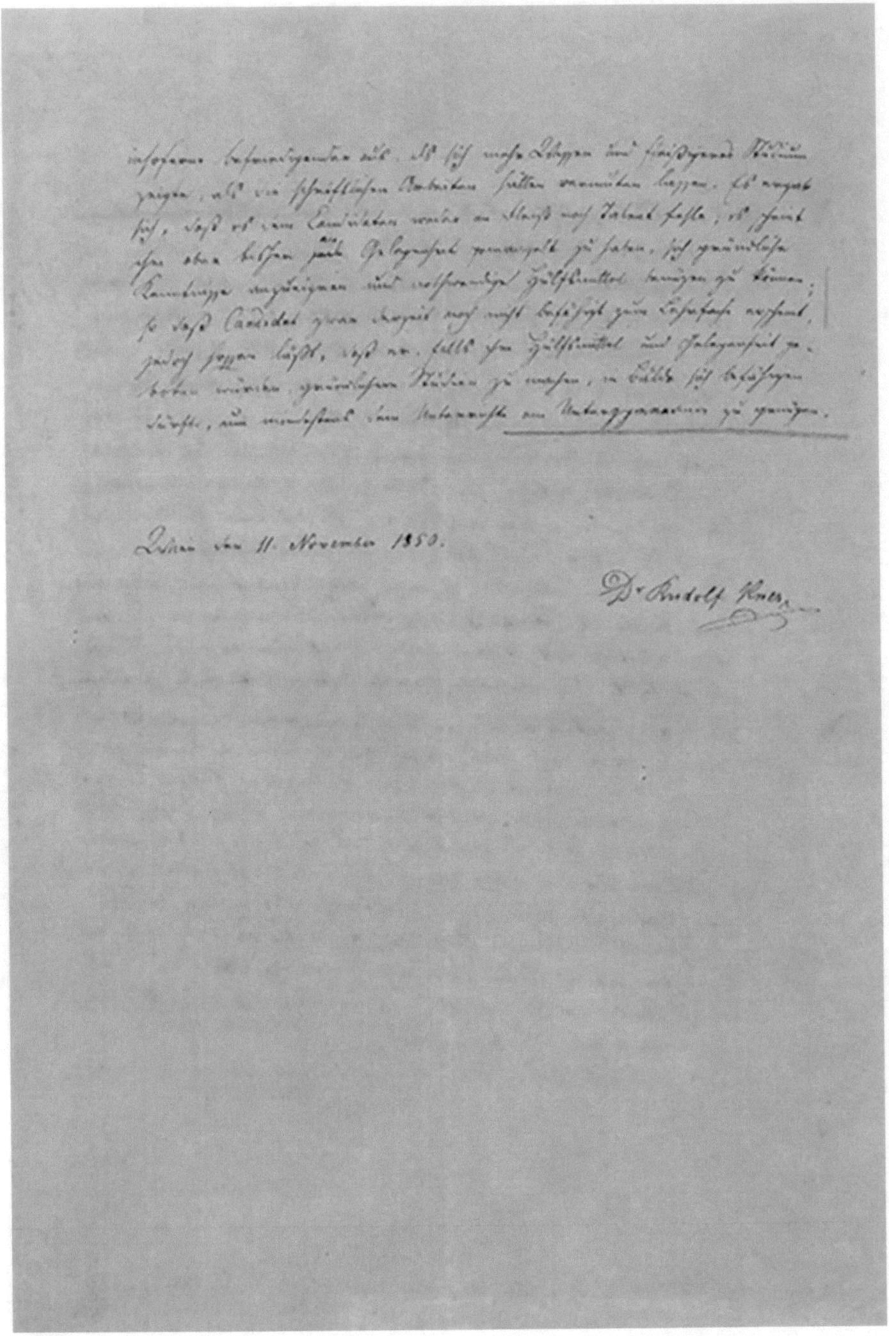

insofern befriedigender als, als sie sich mehr Übung und fleißigeres Studium zeigen, als die schriftlichen Arbeiten hätten vermuten lassen. Es ergab sich, daß es dem Candidaten weder an Fleiß noch Talent fehle, es scheint ihm aber bisher jede Gelegenheit gemangelt zu haben, sich gründliche Kenntnisse anzueignen und vollständige Hülfsmittel benützen zu können; so daß Candidat zwar derzeit noch nicht befähigt zum Lehrfache erscheint, jedoch hoffen läßt, daß er, falls ihm Hülfsmittel und Gelegenheit geboten würden, gründlichen Studien zu machen, er bald sich befähigen dürfte, um mindestens den Unterricht an Untergymnasien zu geben.

Wien den 11. November 1850.

Dr. Rudolf Kner

(continued)

References and Historical Printed Sources

Gicklhorn R (1969) Gregor Mendels Lehramtsprüfung und Studienzeit in Wien. Allgemeine Biologie der Medizinischen Fakultät der Universität Wien 7(4):145–159

Gicklhorn R (1973) Wurde Gregor Mendel bei der Lehramtsprüfung an der Wiener Universität ungerecht behandelt? Biol Rundschau 11:73–84

Iltis H (1966) Life of Mendel. Translated by Eden and Cedar Paul. Hofner Publ. Co, New York

Weiling F (1993/1994) Johann Gregor Mendel. Der Mensch und Forscher. Forscher in der Kontroverse. J. G. Mendel im Urteil der Zeitgenossen. Medizinische Genetik 5:35–51, 208–222, 274–289, 379–393, 6:35–51, 241–255.

Chapter 6
Member of the Agricultural Society in Brno 1851–1884

Mendel worked in the Natural Science Section of the Imperial-Royal Moravian-Silesian Agricultural Society for the Improvement of Agriculture, Knowledge of Nature and the Country (in brief, the Agricultural Society) from 23 July 1851 until the end of his life. The fields of his interest were agriculture, meteorology, apiculture and pomology. From a novice, he gradually grew to become a respected expert, and he eventually occupied the highest executive position as a functionary. The Agricultural Society partially fulfilled the role of an academy in Moravia. Its members founded a museum, as well as schools and institutes of education for the lower and higher levels, and its members aimed to establish a technical university. The Agricultural Society was solving practical problems for Moravia. Its programme was devoted to the building of modern practical schools and institutes of technical learning for the implementation of demonstrative methods into the school curriculum for instruction of natural history, chemistry and physics. In 1872, Mendel was elected to the highest position in the Central Board of the Agricultural Society in the absence of an aristocratic president. Similar to his superior Abbot Napp Mendel was engaged in the development of state prosperity in agricultural economics. He taught natural history and physics for 14 years and was known as a professor and practical experimenter. Of special interest is the newspaper report of 1861 documenting Mendel's attempts to improve vegetables and ornamental flowers (van Dijk et al. 2018). His practical approach was to employ artificial fertilization. A *Moravian Correspondent* newspaper article reported Mendel's attempts at local plant improvement: *The vegetables grown by the professor, such as peas, fisols, cucumbers, and beans, are high towering bushes that are distinguished by a massive production of fruit which, in size and taste, leave nothing to be desired. … Until now, the experiments carried out with potatoes were less successful. … The carnations and fuchsias, of which the Professor grew several 100 pots, stand out by their abundance and full splendor in an astonishing way* (van Dijk et al. 2018, p. 349).

While in exile in America, Klácel recalled the experiments with potatoes they made when studying variants obtained from one piece of potato. The criticism in the Brno newspapers on Mendel's imported vegetables was misinterpreted from the

A. Matalová, E. Matalová, *Gregor Mendel - The Scientist*, Springer Biographies,
https://doi.org/10.1007/978-3-030-98923-1_6

viewpoint of acclimatization, not hybridization. At this point, we may mention Mendel's reluctance to publish his experiments on the acclimatization of tropical stingless bees, which were reported by his friend A. Tomaschek (1879). Mendel wrote about his acclimatization experiment only to A. Kühne in a private letter in 1881. Acclimatization was an accidental activity for Mendel. His long-term research was focused on the study of plant forms. His Pisum developmental series made clear to him how to make informed choices of plants as parents for hybridization, considering all possible combinations of plant traits in the progeny. He studied both actual and possible means for plant improvement. The Agricultural Society in Brno was an organization supporting increased agricultural production, as well as exploration and exploitation of natural resources, documentation of plants and animals in the region, control of drinking water in Brno's wells, and the dissemination of knowledge about Moravian and Silesian traditions. For classification and systemization of data, scientific terms were created dependent on the European context, statistics were implemented in research, and education was reorganized to cope with the new industrial and commercial trends. The Agricultural Society successfully tackled economic problems with a flexible structure. Its members were open to discourse on urgent practical problems arising from scientific progress and the consequent intensified specialization and application of progressive methods.

The Agricultural Society included educated individuals from aristocratic and feudal houses, freemason lodges, private and secret societies, clergymen, and other people interested in the development of science and growth of prosperity in Moravia and Silesia. In 1806 Christian Carl André (1763–1831), a protestant pedagogue of Saxon origin, author and editor of journals (*Patriotisches Tagblatt, Hesperus, Oekonomische Neuigkeiten*), managed to merge the Moravian Agricultural Society for the Improvement of Field Economy, and the private Society for Natural Sciences and Knowledge of the Country. The united organization co-opted the Silesian Agricultural Society, resulting in the Moravia-Silesian Society for the Improvement of Agriculture, Knowledge of Nature and the Country. The Agricultural Society cultivated both agriculture and natural sciences (physics including meteorology, natural science including geology, mineralogy, mathematics, botany, zoology, entomology, technologies) and social sciences (traditions of Moravian and Silesian history, archaeology, statistics etc.). The Agricultural Society was Utraquist, and its official language was German. The printed organs of the Agricultural Society were the *Mittheilungen der kaiserlich-königlichen Mährisch-Schlesischen Gesellschaft zur Beförderung des Ackerbaues, der Natur- und Landeskunde in Brünn*. The *Mittheilungen* were printed in German; advice and instructions for practical work were also published in Czech. The implementation of new technologies and discoveries appeared in the form of leaflets and calendars for landholders.

In 1817, the Agricultural Society established and operated a museum in the Bishop's Court, the former residence of the Bishop of Olomouc. (Brno had its own bishop from 1777.) The Agricultural Society built an open library and a picture gallery in the museum. (The museum, the library and the gallery have since become significant self-sustaining scientific and cultural institutions in Moravia.) A school system of technical education was established and developed in Brno. Traditional Gymnasia were unable to expand their curricula to include the latest achievements of technical and scientific development. The *Realschule* and the Technical Institute (*Technische Lehranstalt*) opened the way for industrial growth, trade and commerce. The statue of Merkur, God of Trade, standing at the entrance of the Agricultural Society Museum since 1865, symbolizes a dynamic trade policy, and Brno as a modern industrial town in Mendel's time.

The excerpt from the resolution by the emperor of Austria reads in Brodesser et al. (2002, p. 12), translated by Lucie Sedláčková: *I allow the museum to merge with the Moravian and Silesian Society for Improvement of Agriculture, Natural Sciences and Knowledge of the Country and to be called the Franz Museum. At the same time, I confirm that the Bishop's Court should be lent to the famous society to use for free. However, the building is explicitly dedicated for the museum to use, with a reservation that after the end of this use the court will be returned to the Olomouc Bishopric. It is necessary to convey my satisfaction to those people who have already provided considerable donations to the museum.*

Transcription of Mendel's review of the book that should be commented in the Mittheilungen: *Als naturphilosophische Abhandlung für die Zwecke der Gesellschaft von geringerem Interesse. Gr. Mendel.*

Transcription of Mendel's comment on the Annual Reports of the Zoological-Botanical Society in Vienna: *Enthällt nebst mehreren Aufsätzen von rein wissenschaftlichem Interesse pag. 235 eine Notiz von. G. V. Frauenfeld über Vertilgung des Raps Käfers. Wird der Redaktion der Mittheilungen empfohlen. Gr. Mendel.*

In the Natural Science Section of the Agriculture Society Mendel was named responsible for meteorology. Dr. Zawadski was an expert for botany, zoology, physics; Dr. Schwippel for physics, mineralogy; Dr. Allé for medicine, chemistry, physiology and general natural history; Gottlieb for mechanics and crafts; Prof. Heinrich for mineralogy; Gärtner for entomology, Kolenati for zoology; Dr. Melion for mineralogy; Nave for botany; Jul Müller for entomology; Prof. Niessl for astronomy, botany; Nowotný for crafts; Spens for astronomy; Prof. Mendel for meteorology.

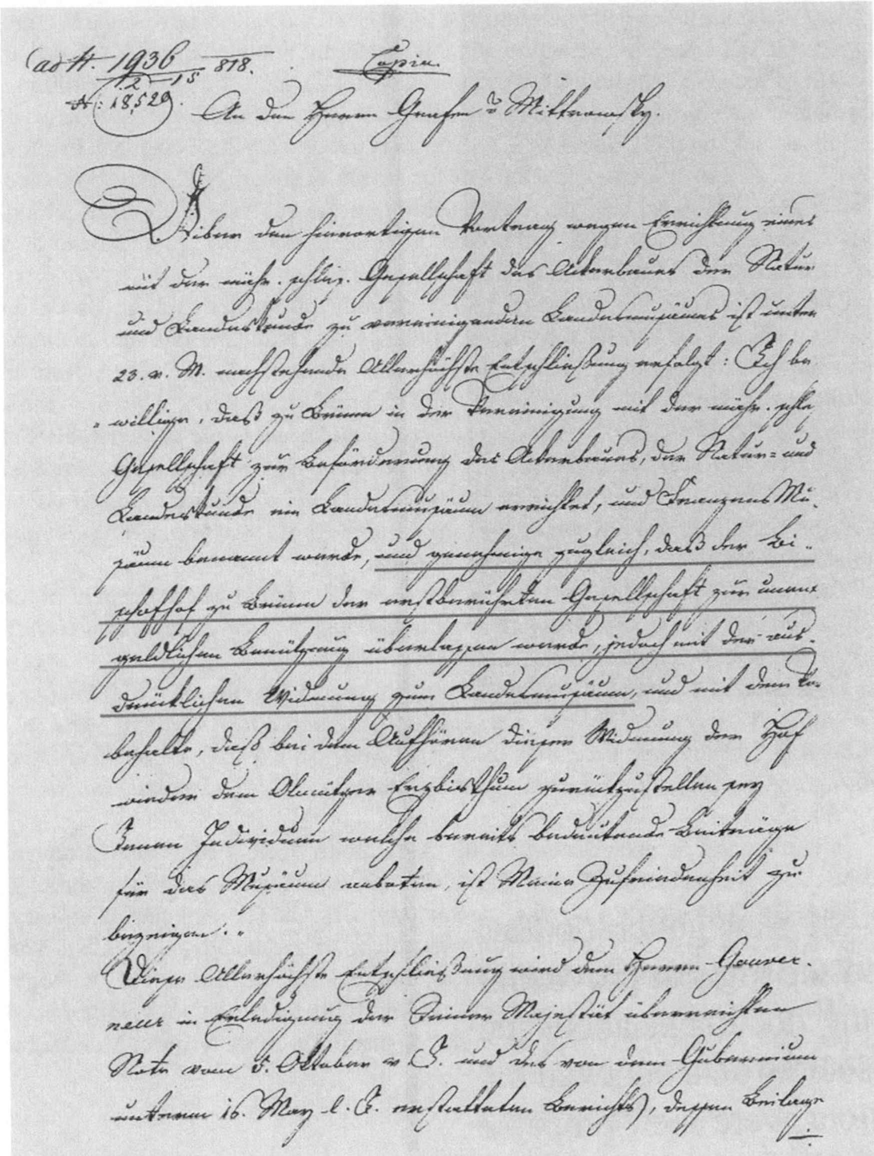

(ad N. 1936 818. Copia
12 15
N: 18529

An den Herrn Grafen v. Mittrowsky.

Über den einverständigen Antrag wegen Errichtung eines mit der mähr. schles. Gesellschaft des Ackerbaues der Natur und Landeskunde zu vereinigenden Landesmuseums ist unterm 23. v. M. nachstehende Allerhöchste Entschließung erfolgt: Ich bewillige, daß zu Brünn in der Vereinigung mit der mähr. schles. Gesellschaft zur Beförderung des Ackerbaues, der Natur- und Landeskunde ein Landesmuseum errichtet, und Franzens Museum benannt werde, und genehmige zugleich, daß der Wirtschaftshof zu Brünn der erstbenannten Gesellschaft zur unentgeltlichen Benützung überlassen werde, jedoch mit der ausdrücklichen Widmung zum Landesmuseum, und mit dem Vorbehalte, daß bei dem Aufhören dieser Widmung der Hof wieder dem Aerarium [illegible] zurückzustellen sey. Jenen Individuen welche bereits bedeutende Beiträge für das Museum erbaten, ist Meine Zufriedenheit zu bezeigen.

Diese Allerhöchste Entschließung wird dem Herrn Gouverneur in Erledigung der Seiner Majestät überreichten Note vom 8. Oktober v. J. und des von dem Gubernium unterm 16. May l. J. erstatteten Berichts, dessen Beilage

MM Earl Geisslern's letter to Earl Mittrowsky on the resolution by Franz I of Austria of 23 July 1817, allowing establishment of the museum, and the Moravian and Silesian Society for Support of Agriculture, Knowledge of Nature and the Land to merge with the museum. Dated in Wien/Vienna, 04 August 1817

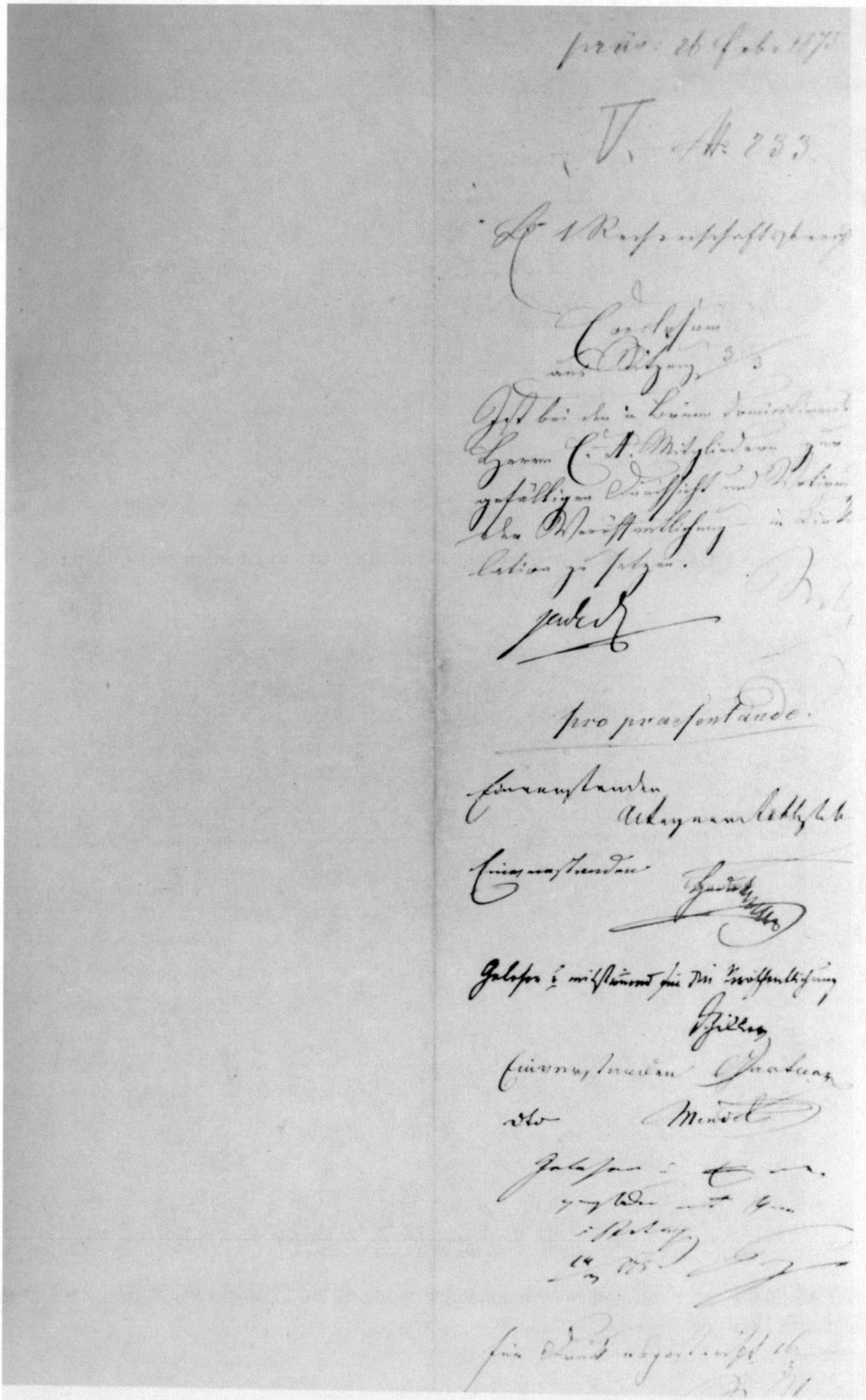

D679 Mendel's agreement with the museum's annual report by Moritz Wilhelm Trapp (1825–1895), who was curator of the museum of the Agricultural Society following Adolf Meinecke, Franz Diebl and Adolf Orcony

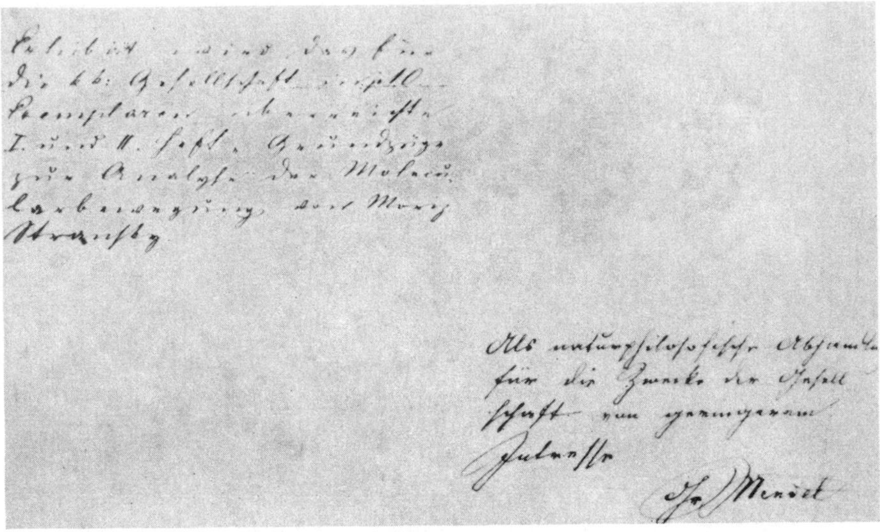
Exhibit wird das für die k.k. Gesellschaft eingesandte Exemplar ... I. und II. Heft. Grundzüge zur Analyse der Molecularbewegung von Moriz Strakosch

Als naturphilosophische Abhandlung für die Zwecke der Gesellschaft von geringerem Interesse

Gr. Mendel

SA Brno Mendel's expert opinion on the books intended for advertisement in the Mittheilungen

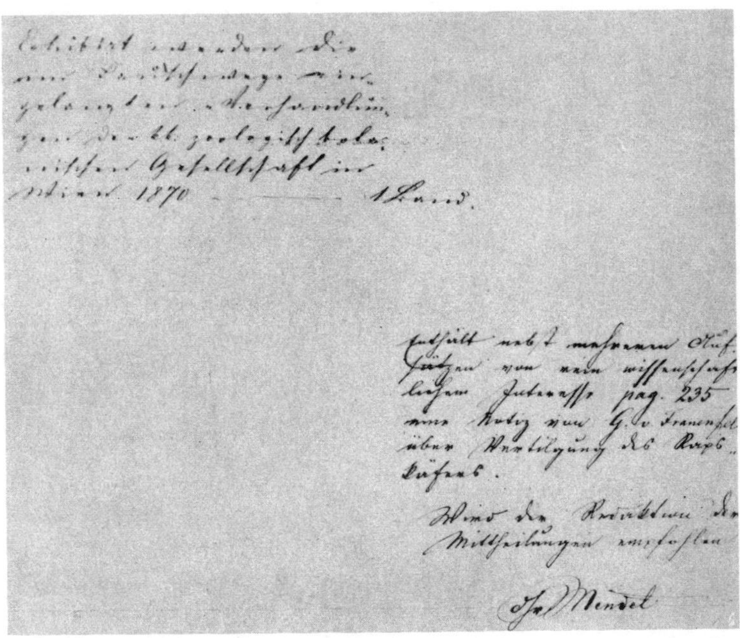
Exhibit werden die vom Verfasser eingelangten Verhandlungen der k.k. zoologisch-botanischen Gesellschaft in Wien 1870 — 1 Band.

Enthält nebst mehreren Aufsätzen von rein wissenschaftlichem Interesse pag. 235 eine Notiz von G. v. Frauenfeld über Vertilgung des Rapskäfers.

Wird der Redaction der Mittheilungen empfohlen

Gr. Mendel

D681 Mendel's expert opinion on the Annual Reports of the Zoological-Botanical Society in Vienna for 1870

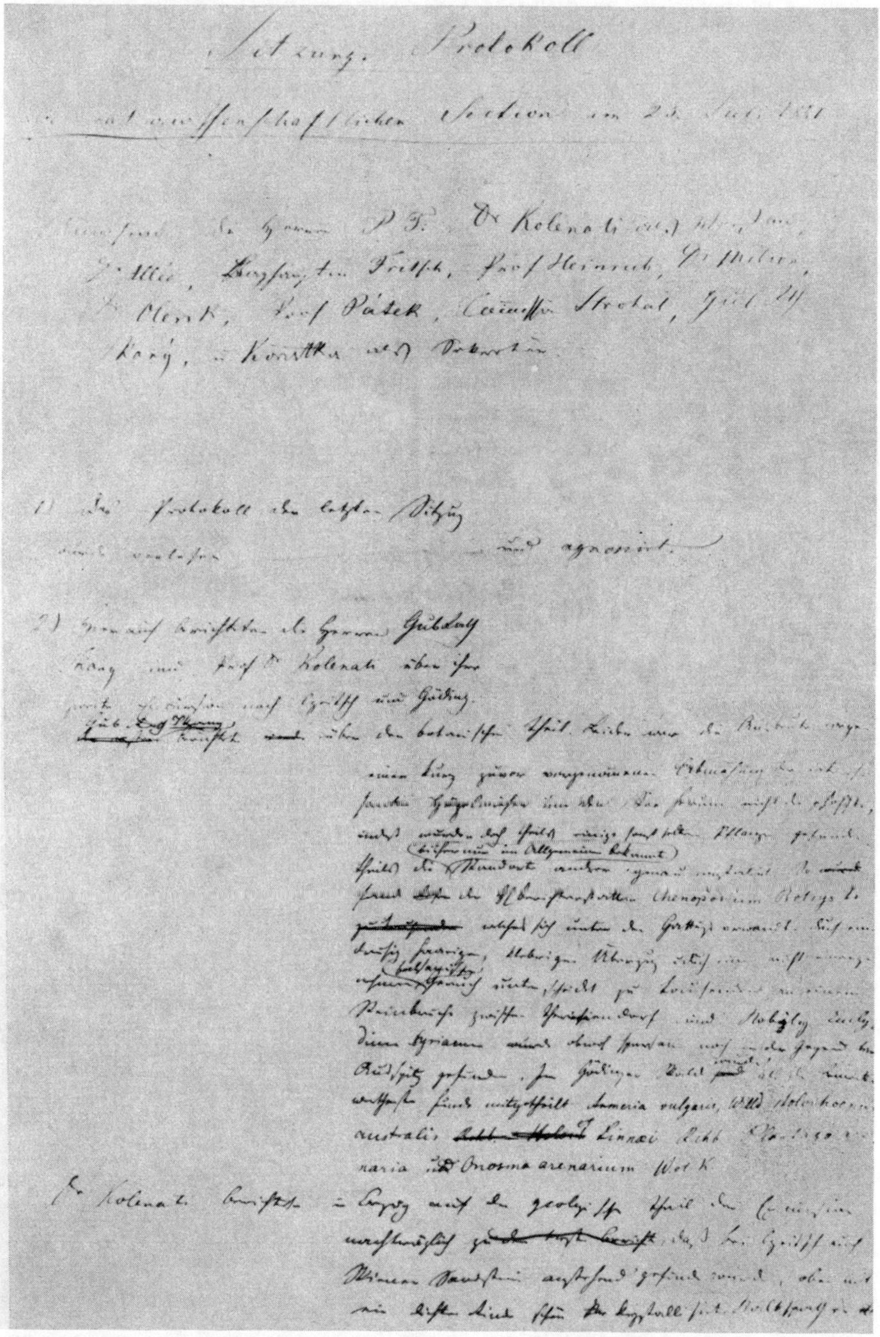

FM3 The first page of the minutes of the meeting of the Natural Science Section of the Agricultural Society, held on 23 July 1851, when Mendel was elected as a member of the Natural Science Section. Dated in Brünn/Brno 23 July 1851, signed by Kořistka. Two pages

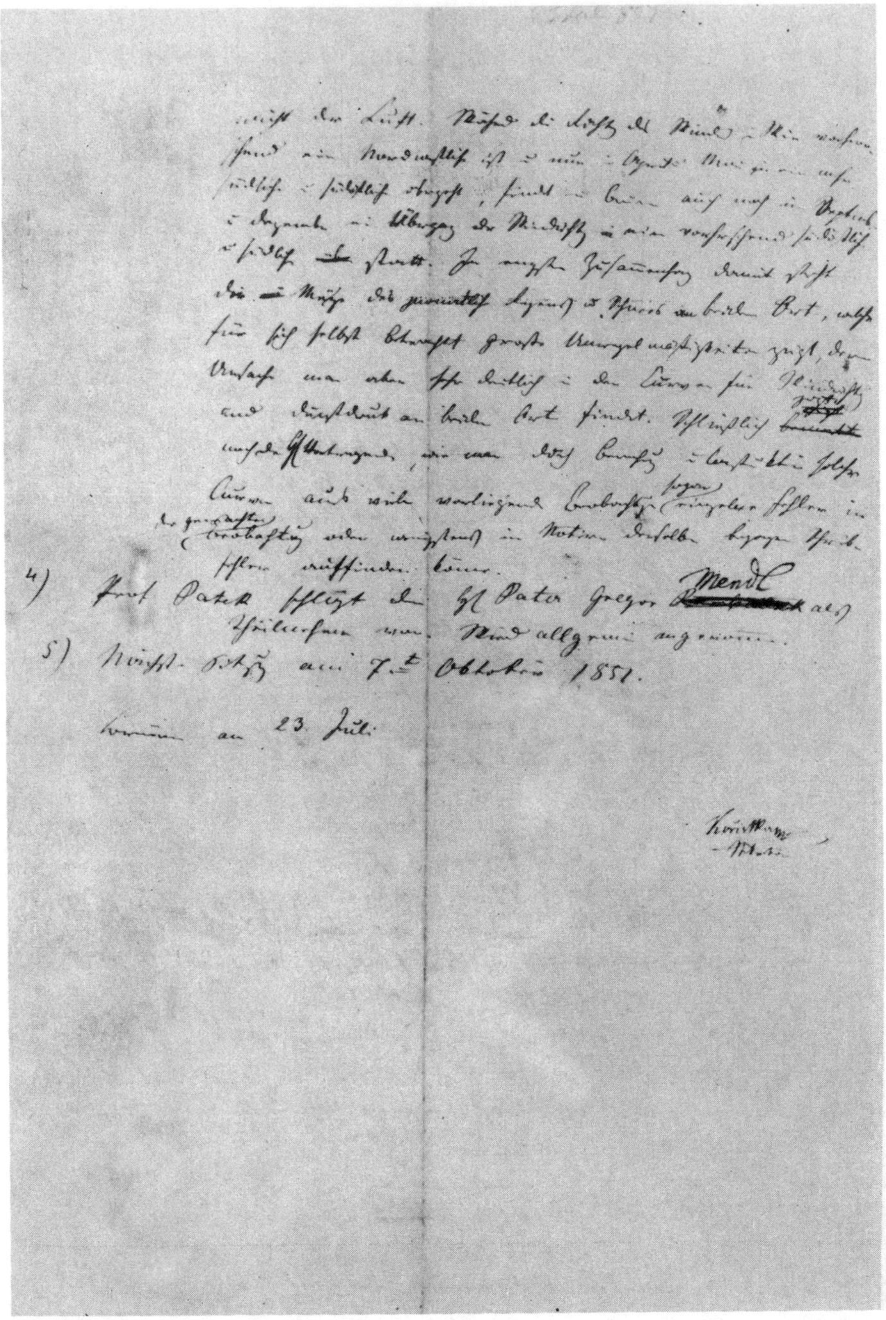
4) Prof Patek schlägt den H. Pater Gregor Mendel

5) Nächste Sitzung am 7. Oktober 1851.

Brünn am 23. Juli

Kořistka

FM3 The second page of the minutes of the meeting of the Natural Science Section of the Agricultural Society. Item number 4 is a proposal by Prof. Pátek to appoint Pater Gregor Mendel as a member of the Natural Science Section of the Agricultural Society. Dated in Brno 23 July 1851, signed by Kořistka

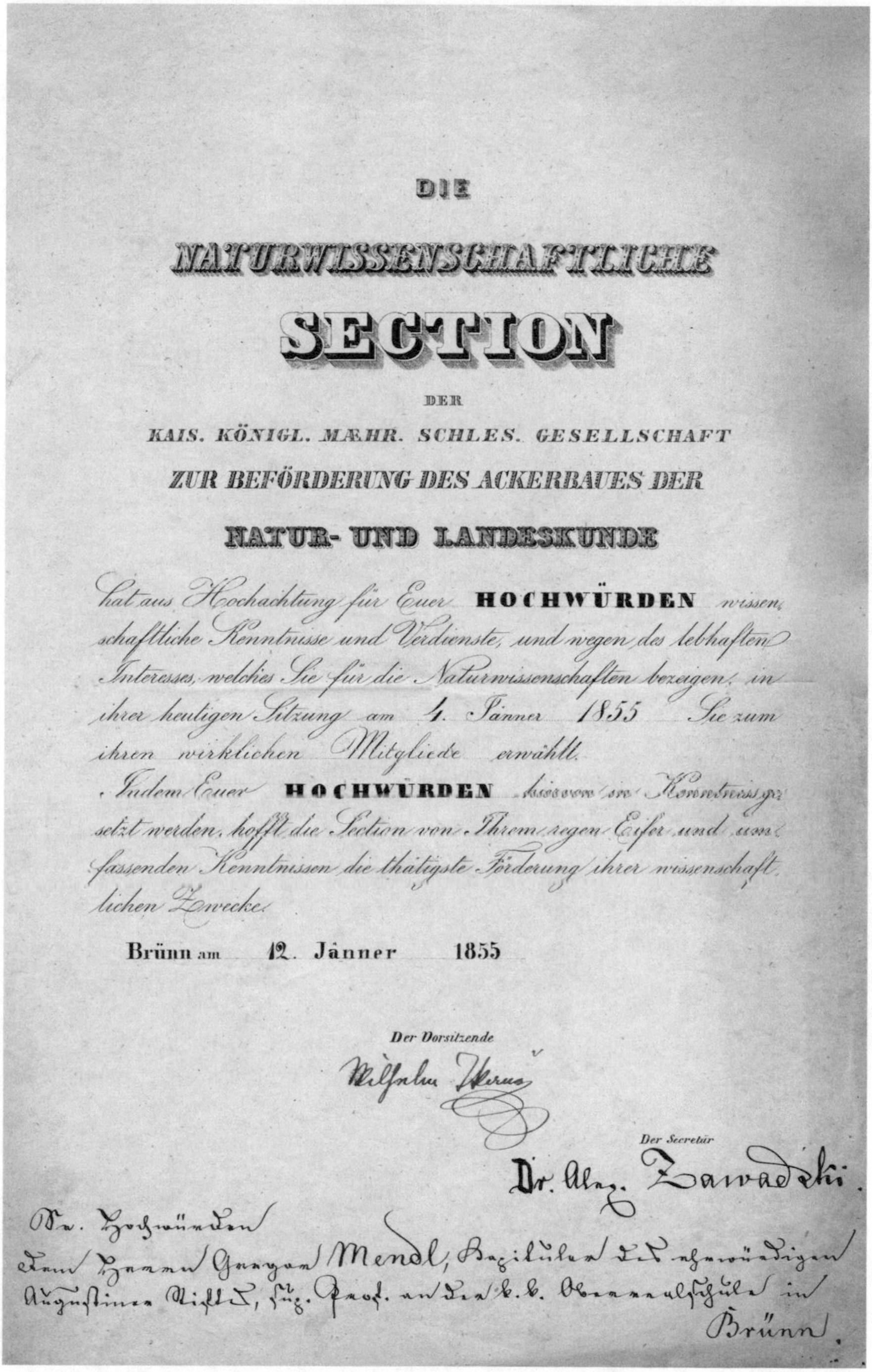

DIE

NATURWISSENSCHAFTLICHE

SECTION

DER

KAIS. KÖNIGL. MÆHR. SCHLES. GESELLSCHAFT

ZUR BEFÖRDERUNG DES ACKERBAUES DER

NATUR- UND LANDESKUNDE

hat aus Hochachtung für Euer **HOCHWÜRDEN** wissenschaftliche Kenntnisse und Verdienste, und wegen des lebhaften Interesses, welches Sie für die Naturwissenschaften bezeigen, in ihrer heutigen Sitzung am 4. Jänner 1855 Sie zum ihren wirklichen Mitgliede erwählt.

Indem Euer **HOCHWÜRDEN** hievon in Kenntniss gesetzt werden, hofft die Section von Ihrem regen Eifer und umfassenden Kenntnissen die thätigste Förderung ihrer wissenschaftlichen Zwecke.

Brünn am 12. Jänner 1855

Der Vorsitzende

Wilhelm Tkany

Der Secretär

Dr. Alex. Zawadski

Sr. Hochwürden
Herrn Herrn Gregor Mendel, Capitular des hochwürdigen Augustiner Stiftes, suppl. Prof. an der k.k. Oberrealschule in Brünn.

D80 Mendel's decree from the Natural Science Section of the Agricultural Society. Dated in Brno, 12 January 1855, signed by Wilhelm Tkaný, Dr. Alex. Zawadski

Regulativ

der naturwissenschaftlichen Section

der k. k. mähr. schles. Gesellschaft zur Beförderung des Ackerbaues, der Natur- und Landeskunde in Brünn.

I. Von der Section, ihren Zwecken und Mitteln.

§. 1. Die naturwissenschaftliche Section der k. k. Gesellschaft zur Beförderung des Ackerbaues, der Natur- und Landeskunde besteht aus der Vereinigung von Freunden der Naturwissenschaften zur Erörterung wissenschaftlicher Gegenstände aus dem Gebiete der Naturgeschichte und Naturlehre und speciellen Förderung dieser Fächer.

§. 2. Dieselben stellen sich zur Erreichung dieses Zwekkes die Aufgabe:

Die botanische, zoologische, mineralogische und geologische Beschaffenheit der Kronländer Mähren und Schlesien, deren metereologische und physikalische Verhältnisse zu erforschen und bekannt zu machen, wissenschaftliche Entdeckungen in diesen Fächern, welche für die Zwecke der k. k. Gesellschaft von Belang sind, zu prüfen und derselben mitzutheilen, über die von derselben der Section zugewisenen einschlägigen Geschäftsgegenstände Berichte zu erstatten, die naturhistorischen Sammlungen des Franzensmuseums zu bereichern und für diese Zwecke sachkundige Männer in- und außerhalb des Landes zu gewinnen.

D373 The first page of the Regulations of the Natural Science Section of the Agricultural Society. Part One defines the Section, its Goals and Instruments. A leaflet of four pages

§. 3. Die Mittel zur Erreichung dieses Zweckes sind:

a) Wissenschaftliche Arbeiten der Sectionsglieder, welche theils freiwillige, theils aufgetragene sind.

Die Mittheilungen eigener Forschungen im Gebiete der Naturwissenschaften bilden den Gegenstand der Erstern; außerdem aber trägt die Section durch Beschluß Arbeiten und Berichterstattung einzelnen Mitgliedern auf; beide Arten von Arbeiten werden der Section vorgelegt, von ihr besprochen und entweder der k. k. Gesellschaft übermittelt oder von der Section selbst weiter benützt.

b) Verbindung mit andern Vereinen gleicher Wirkungsart Behufs gegenseitiger Mittheilungen und Arbeiten-Austausches, welche von der Section beschlossen und dem Central-Ausschuße der Gesellschaft angezeigt wird.

c) Stellung geeigneter Anträge an die k. k. Gesellschaft, sowohl auf Erwerbung wissenschaftlicher Hilfsmittel, als auch erforderlichen Falls auf Abordnung von Mitgliedern zu Beobachtungen, Forschungen oder andern nur an Ort und Stelle zu erzielenden wissenschaftlichen Zwecken.

II. Von den Mitgliedern, ihren Rechten und Pflichten.

§. 4. Die Sectionsglieder sind entweder zugleich Mitglieder der k. k. mähr. schles. Gesellschaft oder andere Theilnehmer. Erstere treten in Gemäßheit der Gesellschaftsstatuten über einfache Anmeldung beim Sections-Vorstande ein, alle übrigen Theilnehmer werden über Vorschlag eines Sectionsgliedes durch Wahl der Section aufgenommen.

§. 5. Jeder Eintretende verpflichtet sich zu einem jährlichen Beitrage von 2 fl. CM., der im Vorhinein im ersten Vierteljahre zu leisten ist, zur Uebernahme der ihm aufgetragenen Arbeiten und zur möglichsten Förderung der im §. 2 angeführten Aufgaben.

D373 The second page of the Regulations of the Natural Science Section of the Agricultural Society. Page two defines the rights and duties of its members

§. 6. Der Austritt aus der Section steht Jedermann frei; jedoch hat der Austretende den Beitrag für das Jahr, in welchem der Austritt erfolgt, zu leisten.

§. 7. Die Section wählt aus ihrer Mitte durch absolute Stimmenmehrheit der Anwesenden einen Vorstand und einen Secretär (Schrift- oder Geschäftsführer) auf die Dauer eines Jahres. Ersterer führt den Vorsitz, leitet die Verhandlungen und unterzeichnet die Protokolle so wie die Correspondenzen nach den Vormerkungen und Entwürfen des Letztern.

Correspondenzen mit Staatsbehörden und andern Corporationen haben immer nur durch den Centralausschuß der k. k. Gesellschaft zu geschehen.

§. 8. Die Functionäre verwalten ihr Amt stets vom 1. Jänner bis zum letzten Dezember und sind nach Ablauf ihres Jahres wieder wählbar. Tritt während desselben ein oder der andere ab, so wird bis zum Ablauf der Zeit ein Stellvertreter gewählt. In Verhinderungsfällen des Vorstandes wird von Fall zu Fall ein Sectionsglied zum jeweiligen Vorsitz gewählt.

§. 9. Alle Functionen und Arbeiten werden ohne Anspruch auf Emolumente geleistet. Es werden aber in Uebereinstimmung mit den Gesellschaftsstatuten jene Auslagen auf Verlangen vergütet, welche aus Anlaß übertragener Geschäfte erwachsen.

§. 10. Alle Sectionsglieder haben in den Versammlungen gleichen Rang und gleiches Stimmrecht. Diese Versammlungen sind öffentlich und finden in der Regel wenigstens einmal im Monat Statt. Außerdem hält die Section am Tage vor der allgemeinen Gesellschaftsversammlung eine vorberathende Sitzung.

§. 11. Außer den Berichtserstattungen über aufgetragene Arbeiten werden in den Versammlungen wissenschaftliche Mittheilungen, einzelne Vorträge über Fachgegenstände gehalten. Auch Nichtmitgliedern kann über Anmeldung beim Vorstande die Abhaltung von wissenschaftlichen Mittheilungen gestattet werden. Fortgesetzte Vorlesungen aber können nur unter Beobachtung der für das Unterrichtswesen bestehenden Gesetze öffentlich Statt finden.

D373 The third page of the Regulations of the Natural Science Section of the Agricultural Society defines the organizational position of the Chairman and of the Secretary, who were elected for a period of 1 year

§. 12. Außerordentliche Versammlungen zur Berathung von Sections-Angelegenheiten können vom Vorstande auch außer der gewöhnlichen Zeit, jedoch mit Angabe der zu verhandelnden Gegenstände und unter Vorladung jedes einzelnen Sectionsgliedes berufen werden.

§. 13. Eine eigene von dem Centralausschuße bestätigte Geschäftsordnung regelt die Art und Weise sowohl der berathenden als der wissenschaftlichen Versammlungen.

§. 14. Alle Sectionsglieder haben das Recht, unter Beobachtung der von der Gesellschaft festgesetzten Vorschriften die wissenschaftlichen Sammlungen des Franzensmuseums zu benützen. Die Versammlungen werden gleichfalls in den Räumlichkeiten desselben abgehalten.

§. 15. Eine Aenderung gegenwärtigen Regulativs kann mit zweidrittel Stimmem aller auf specielle Vorladung Anwesenden beschlossen werden, und ist in Gemäßheit der Gesellschaftsstatuten an die Bestätigung der k. k. Gesellschaft gebunden.

Gedruckt bei R. Rohrer's Erben.

D373 The fourth and closing page of the Regulations of the Natural Science Section of the Agricultural Society. Printed by R. Rohrer in Brno, undated

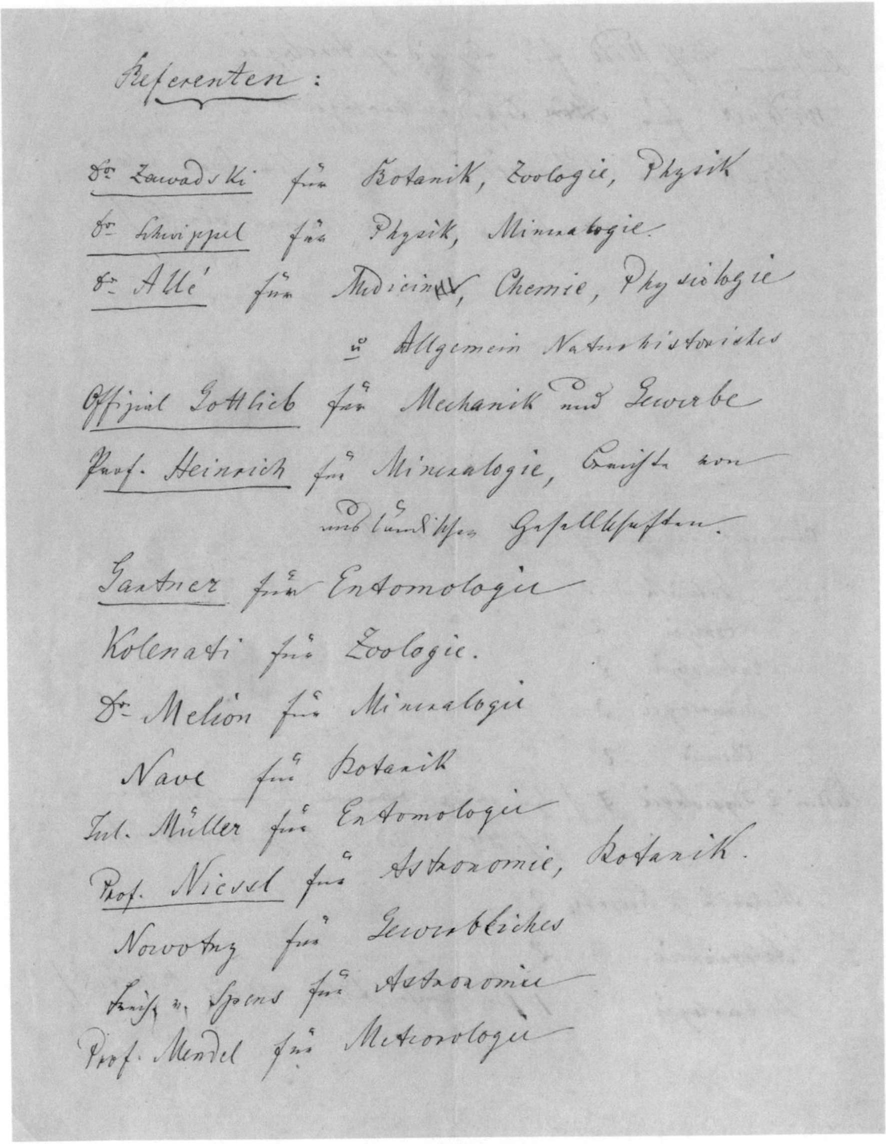

Referenten:

Dr. Zawadski für Botanik, Zoologie, Physik
Dr. Schwippel für Physik, Mineralogie
Dr. Allé für Medicin, Chemie, Physiologie
u. Allgemein Naturhistorisches
Offizial Gottlieb für Mechanik und Gewerbe
Prof. Heinrich für Mineralogie, Schriften von
ausländischen Gesellschaften.
Gartner für Entomologie
Kolenati für Zoologie.
Dr. Melion für Mineralogie
Nave für Botanik
Jul. Müller für Entomologie
Prof. Niessl für Astronomie, Botanik.
Nowotny für Gewerbliches
Freih. v. Spens für Astronomie
Prof. Mendel für Meteorologie

D674 The first page of experts of the Natural Science Section of the Agricultural Society shows Mendel as the only expert in meteorology

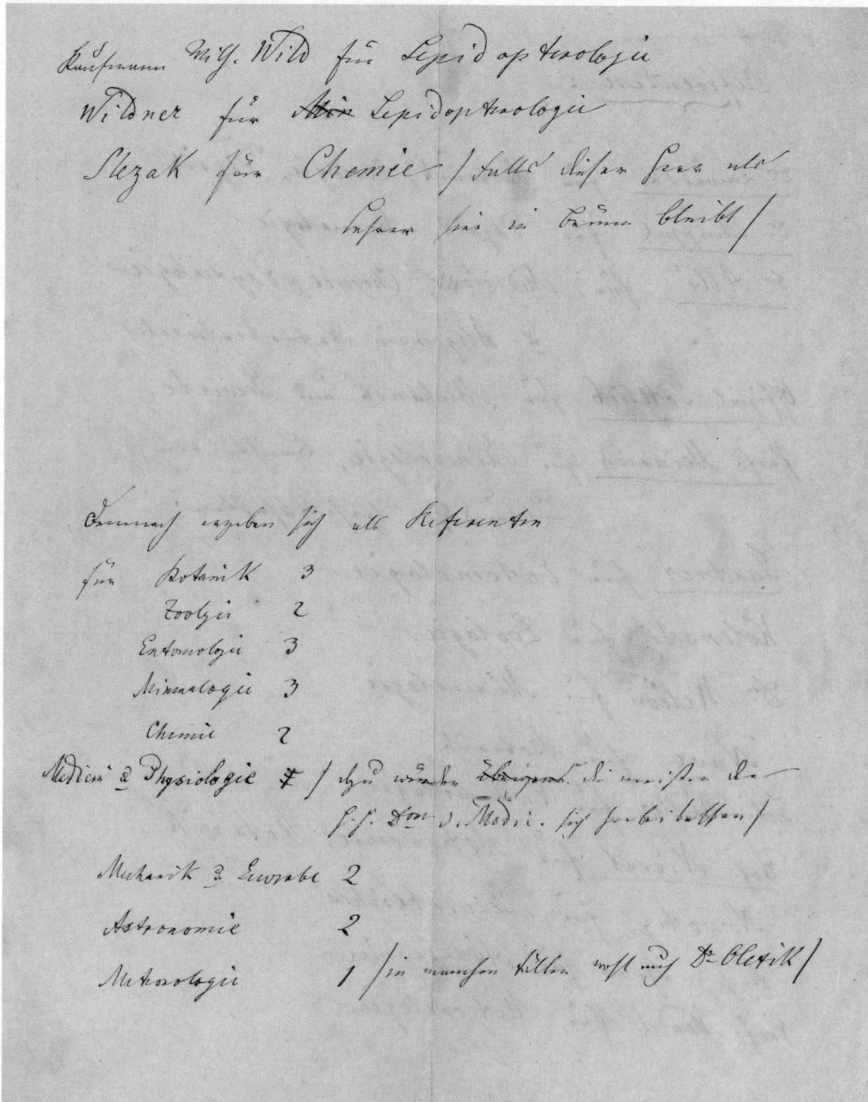

Kaufmann Wilh. Wild für Lepidopterologie
Wildner für ~~Mi~~ Lepidopterologie
Slezak für Chemie (sollte dieser Herr als Lehrer hier in Brünn bleibt)

Demnach ergeben sich als Referenten

für Botanik	3	
Zoologie	2	
Entomologie	3	
Mineralogie	3	
Chemie	2	
Medicin & Physiologie	7	(… die meisten der h.h. Drn d. Medic. sich herbeilassen)
Mechanik & Gewerbe	2	
Astronomie	2	
Meteorologie	1	(in manchen Fällen wohl auch Dr. Olexik)

D674 The second page of experts of the Natural Science Section of the Agricultural Society. The last line gives one meteorologist for Brno. The added note in brackets to number one specifies that in some cases also Dr. Olexik (*in manchen Fällen wohl auch Dr. Olexik*) may participate in the expertise

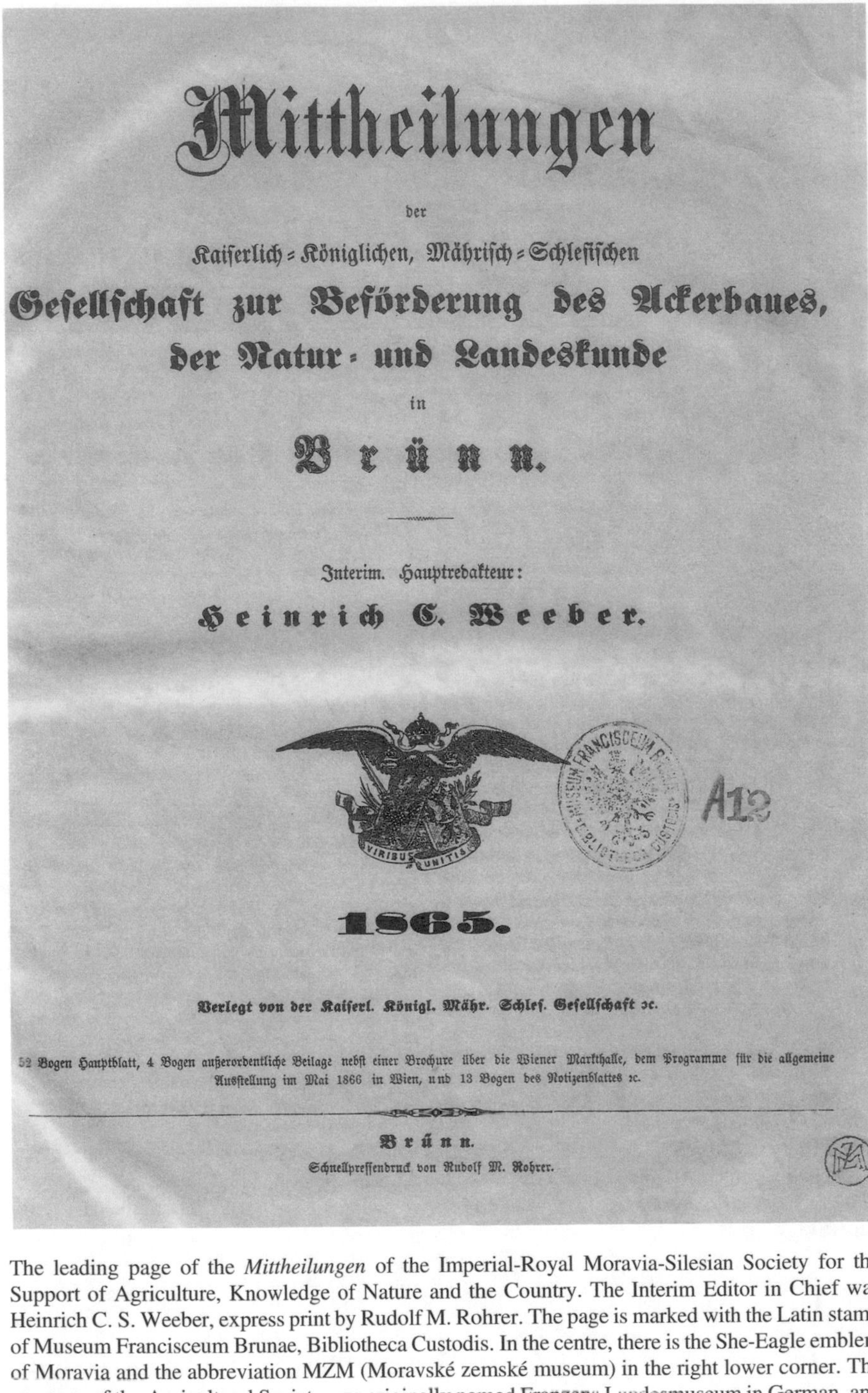

Mittheilungen

der

Kaiserlich-Königlichen, Mährisch-Schlesischen

Gesellschaft zur Beförderung des Ackerbaues, der Natur- und Landeskunde

in

Brünn.

Interim. Hauptredakteur:

Heinrich C. Weeber.

1865.

Verlegt von der Kaiserl. Königl. Mähr. Schles. Gesellschaft ꝛc.

52 Bogen Hauptblatt, 4 Bogen außerordentliche Beilage nebst einer Brochure über die Wiener Markthalle, dem Programme für die allgemeine Ausstellung im Mai 1866 in Wien, und 13 Bogen des Notizenblattes ꝛc.

Brünn.

Schnellpressendruck von Rudolf M. Rohrer.

The leading page of the *Mittheilungen* of the Imperial-Royal Moravia-Silesian Society for the Support of Agriculture, Knowledge of Nature and the Country. The Interim Editor in Chief was Heinrich C. S. Weeber, express print by Rudolf M. Rohrer. The page is marked with the Latin stamp of Museum Francisceum Brunae, Bibliotheca Custodis. In the centre, there is the She-Eagle emblem of Moravia and the abbreviation MZM (Moravské zemské museum) in the right lower corner. The museum of the Agricultural Society was originally named Franzens Landesmuseum in German, and was renamed to Moravské zemské museum in Czech after the First World War

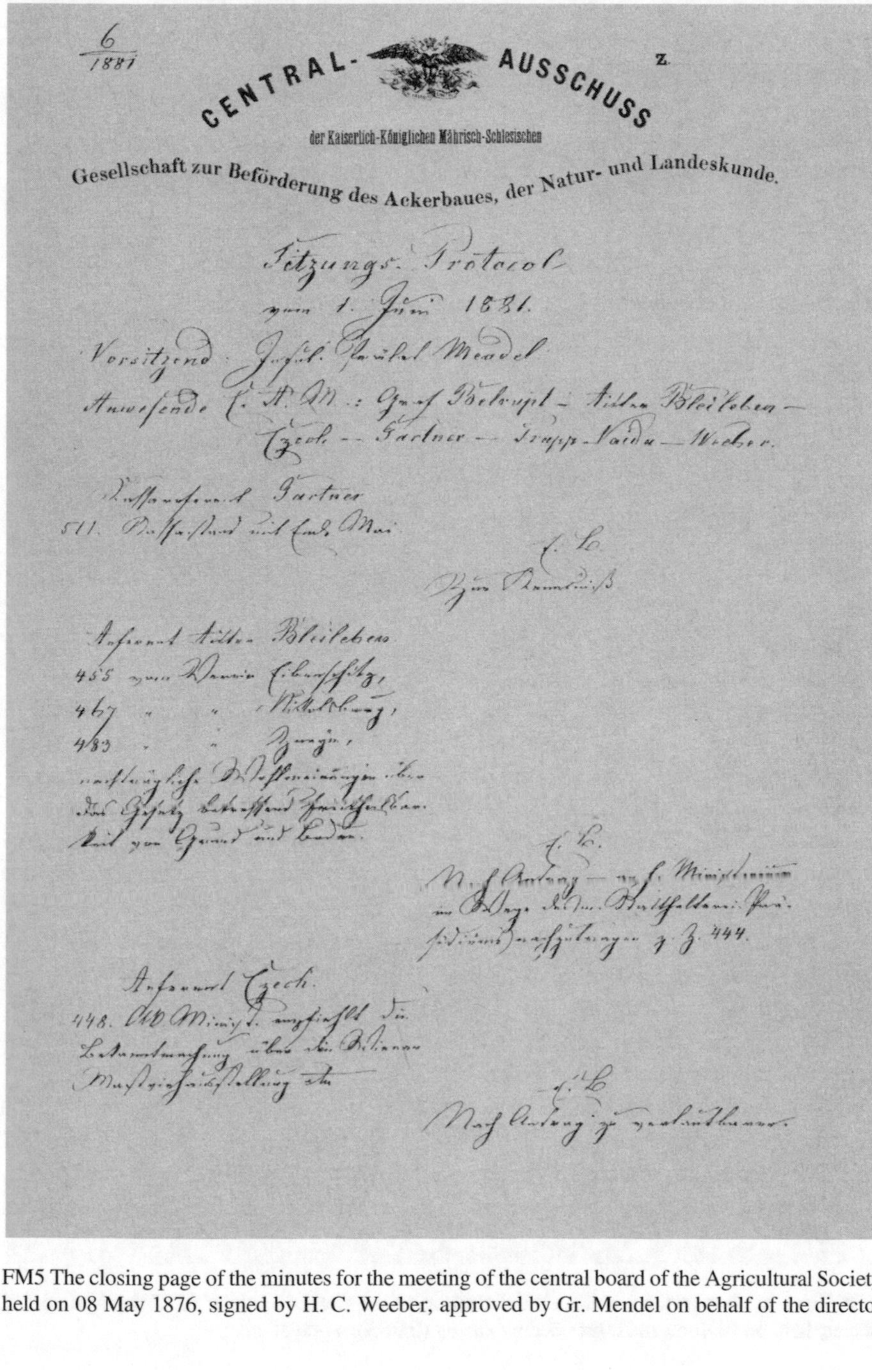

6/1881

CENTRAL-AUSSCHUSS

Z.

der Kaiserlich-Königlichen Mährisch-Schlesischen

Gesellschaft zur Beförderung des Ackerbaues, der Natur- und Landeskunde.

Sitzungs-Protocol

vom 1. Juni 1881.

FM5 The closing page of the minutes for the meeting of the central board of the Agricultural Society held on 08 May 1876, signed by H. C. Weeber, approved by Gr. Mendel on behalf of the director

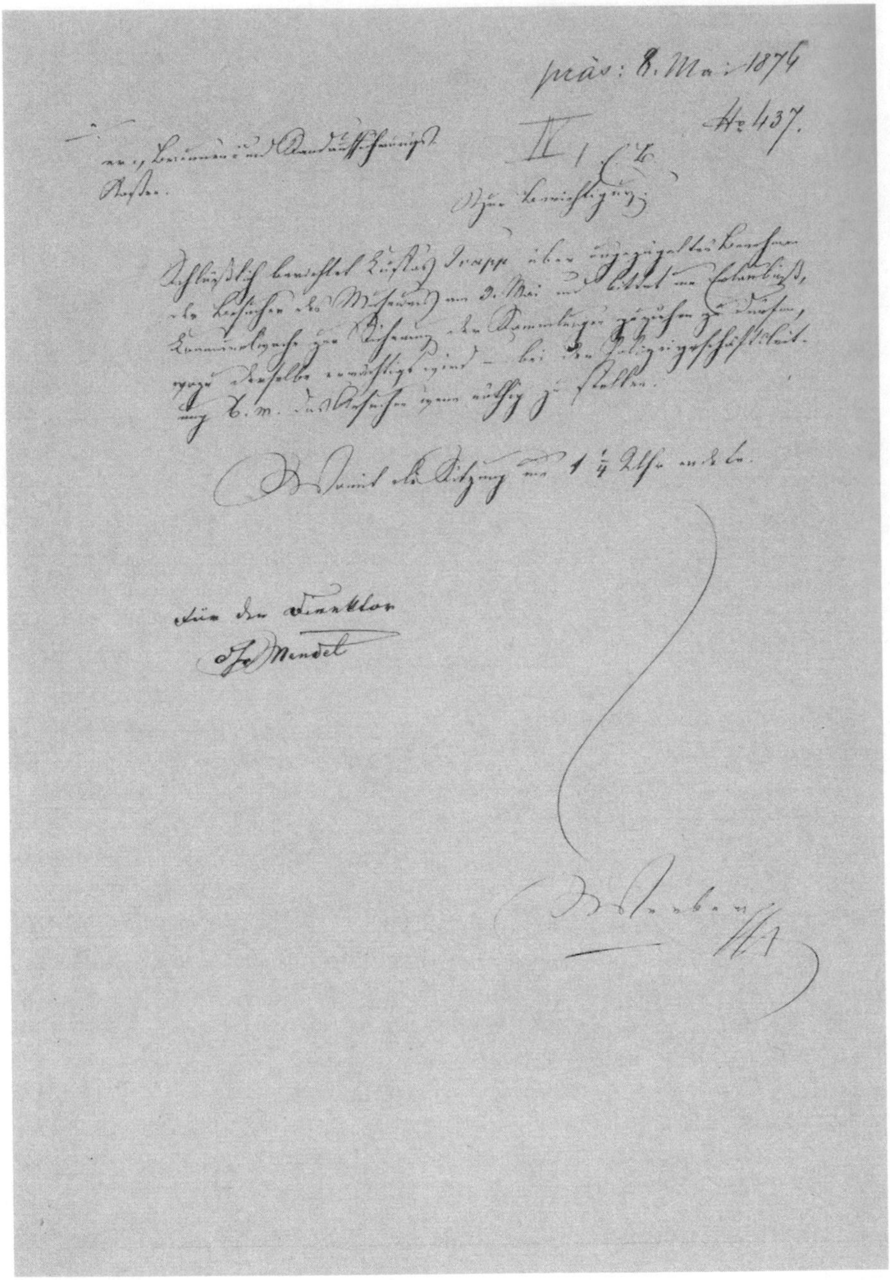
präs: 8. Mai 1876

No 437.

Mendel

FM5 The first page of the minutes on the meeting of the Central Committee of the Agricultural Society held on 01 June 1881, presided by *mitred* (infu l.) *prelate Mendel*

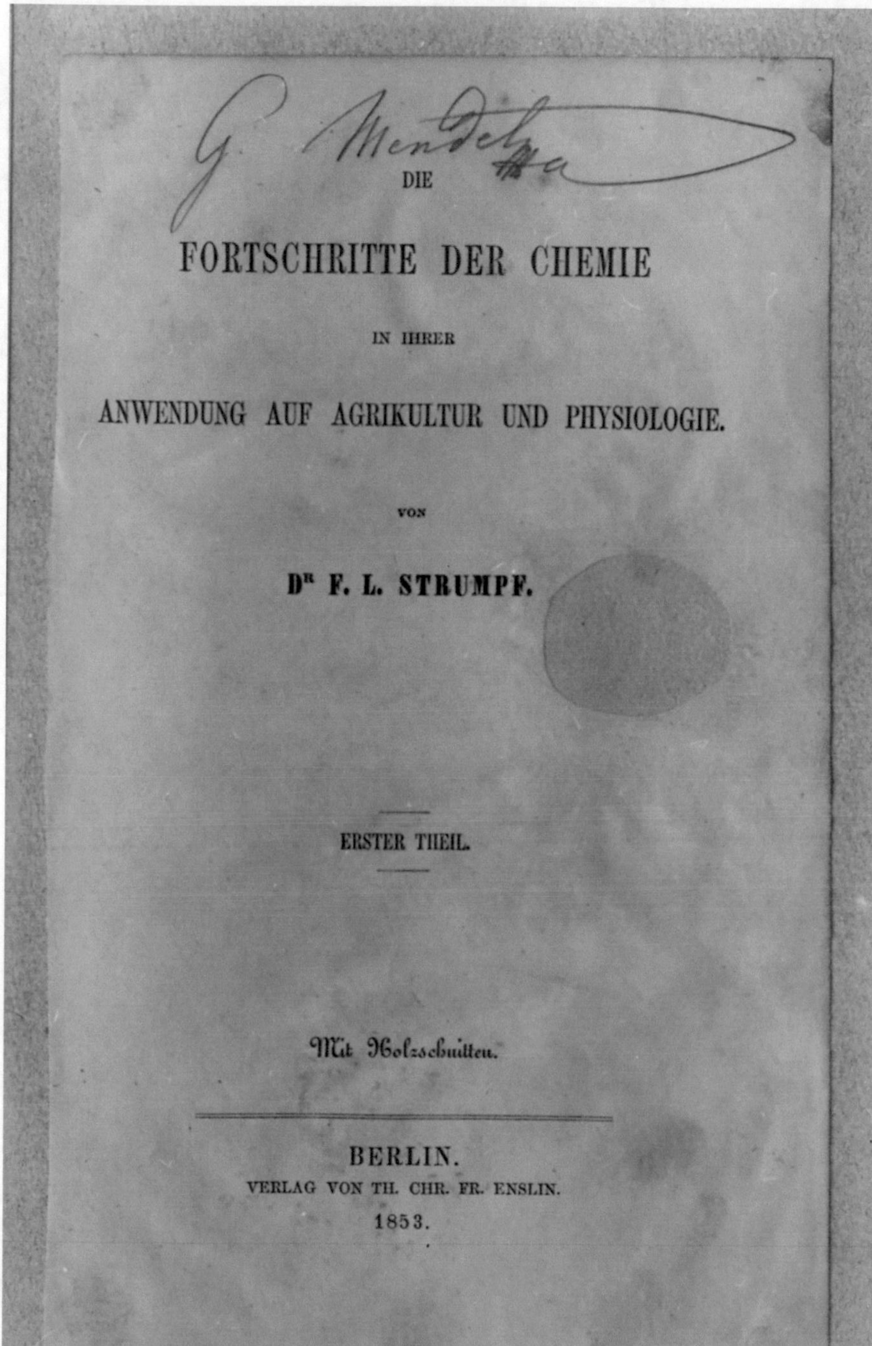

G. Mendel

DIE

FORTSCHRITTE DER CHEMIE

IN IHRER

ANWENDUNG AUF AGRIKULTUR UND PHYSIOLOGIE.

VON

DR F. L. STRUMPF.

ERSTER THEIL.

Mit Holzschnitten.

BERLIN.

VERLAG VON TH. CHR. FR. ENSLIN.

1853.

D582 Mendel's signature on the cover of The Progress of Chemistry and Its Implementation in Agriculture and Physiology by Dr. F. L. Strumpf. Vol. 1, Berlin, Published by Til. Chr. Fr. Enslin, 1853. The original is physically held at Villanova University in Pennsylvania

References and Historical Printed Sources

Brodesser S, Břečka J, Mikulka J (2002) Serving understanding and the glory of the land. History of the Moravian Museum, Brno

Tomaschek A (1879-80) Ein Schwarm der amerikanischen Bienenart Trigona lineata Lep. lebend in Europa. Part one in Zoologischer Anzeiger 2:782-7, part two in Zoologischer Anzeiger 3:60-5

van Dijk PJ, Weissing FJ, Noel Ellis TH (2018) How Mendel's interest in inheritance grew out of plant improvement. Genetics 210:347–355

Chapter 7
Substitute Teacher at the Technical Institute 1851

Transcription: *Loebliches Directorat!—Nachdem der Gefertigte durch die Zuschrift des loebl. Directorates dto. 6. Juni F. 3. Z. 211, an der Supplirung der allgemeinen Naturgeschichte in dem Vorbereitungsjahrgange der hierortigen k. k. technischen Lehranstalt in Folge der Wiederherstellung des krank gewesenen Professors Dr. J. Helcelet, enthoben wurde: so ersucht derselbe, Ein Loebl. Directorat wolle die Einleitung treffen, dass ihm für seine Supplirung von 7. April d. J. bis 6. Juni d. J. eine entsprechende Remuneration angewiesen werde.—Brünn am 15. Juni 851—Gregor Mendel—Capitular d. Stiftes St. Thomas.*

At the Technical Institute, students were taught physics, chemistry, geometry, mechanics, commerce and national economy. Mendel was appointed substitute teacher at the proposal of Professor Kolenatý, whose specialty was natural history. The staff meeting of the Technical Institute was presided over by Director Fl. Schindler in the presence of Dr. Teireich, Dr. Hruby, Dr. Kolenaty, Kořistka, Marin, Ringhofer, Quadrat, Regner, Ritter von Bleileben and Joseph Auspitz. Auspitz was the scribe for the minutes. The minutes of the Technical Institute teaching staff of 03 April 1851 state: *Various professors, especially Professor Kolenati, having borne witness to the extensive studies of Pater Gregor Mendel in natural history, affirm that there can be no doubt that he is fully competent in respect of both scientific knowledge and as a teacher* (Richter 1943, p. 65).

Dr. Florian Schindler, who came to Brno from the Technical Academy in Lviv, was the Director of the Technical Institute. He appreciated Mendel's substitute teaching: *The Directorate takes pleasure in the opportunity to express to Your Reverence its utmost commendation for the zeal shown during the time of your employment here, the profitable nature of your teaching, the judicious treatment of your students and your engaging behaviour toward all members of the institute, and to offer its most fundamental gratitude for your self-sacrificing effort and your active promotion of the school's objectives.* Translation by Scott Abbott and Daniel J. Fairbanks, 2021.

The Technical Institute in Brno accepted university professors who were exiled from the Austrian monarchy, having lost their posts for involvement in the

A. Matalová, E. Matalová, *Gregor Mendel - The Scientist*, Springer Biographies,
https://doi.org/10.1007/978-3-030-98923-1_7

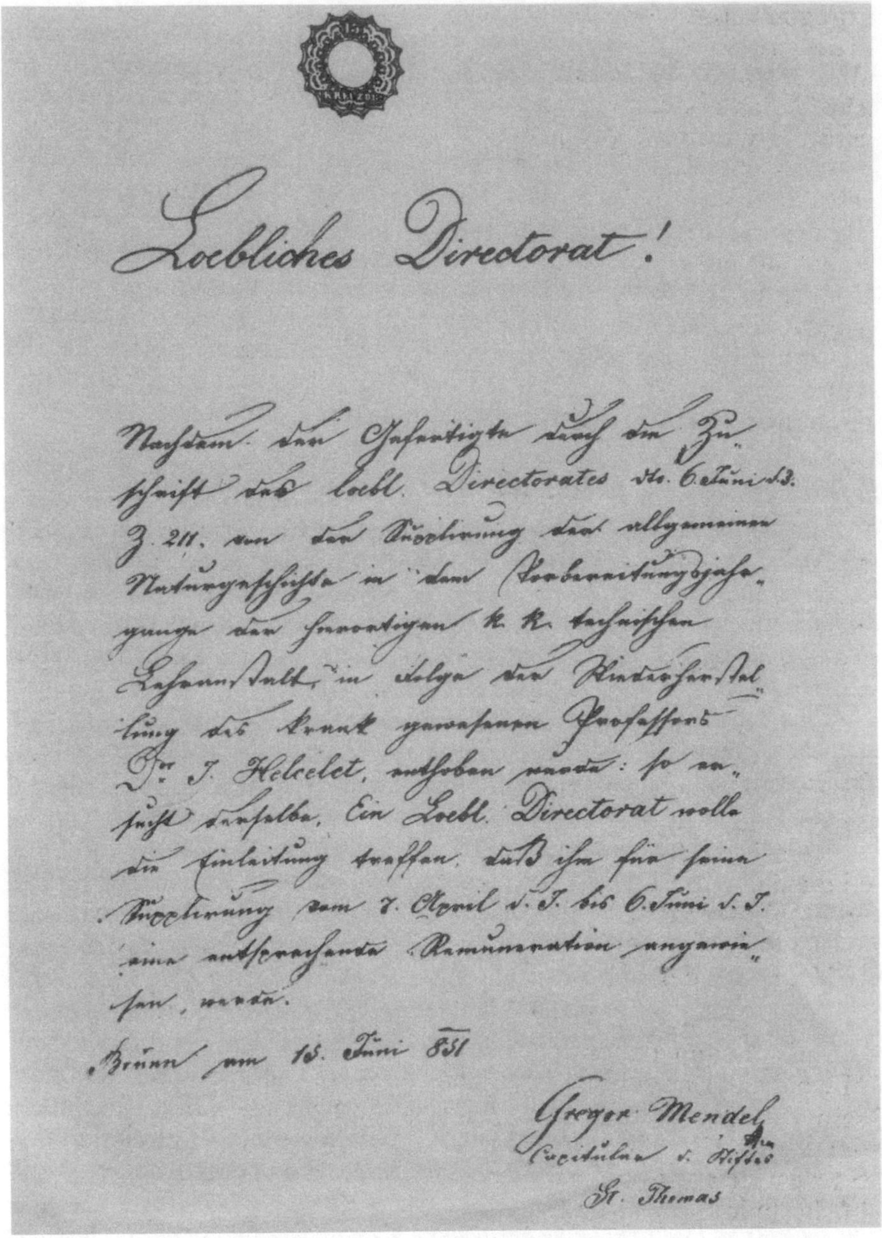

Loebliches Directorat!

Nachdem der Gefertigte durch die Zuschrift des loebl. Directorates ddo. 6. Juni d. J. Z. 211. von der Supplierung der allgemeinen Naturgeschichte in dem Vorbereitungsjahrgange der hierortigen k. k. technischen Lehranstalt, in Folge der Wiederherstellung des krank gewesenen Professors Dr. J. Helcelet, enthoben wurde: so ersucht derselbe, Ein Loebl. Directorat wolle die Einleitung treffen, daß ihm für seine Supplierung vom 7. April d. J. bis 6. Juni d. J. eine entsprechende Remuneration angewiesen werde.

Brünn am 15. Juni 851

Gregor Mendel
Capitular des Stiftes
St. Thomas

Richter 1943. Letter by Mendel to the Directorate of the Technical Institute in Brno asking for remuneration for his substitute teaching in natural history for Professor Johann Helcelet, teacher of general natural history, who was ill, from 07 April to 06 June 1851. Dated in Brno, 15 June 1851, signed by Gregor Mendel

revolutionary events of 1848. In Brno, they organized the establishment of the Realschule, laying the foundation of a new type of technical high school education. Joseph Auspitz (Professor of Commerce, who was forced to leave Vienna) was appointed the Director of the Realschule. Alexander Zawadski (Professor of Physics and Natural History, who was expelled from the University of Lviv) promoted new methods of teaching natural history at both the Technical Institute and the Realschule. Brno profited from the organizational skills and progressive methods of teaching and research. It was Zawadski who founded the association of researchers in Brno, known as the *Naturforschender Verein* a self-contained organization of the Agricultural Society that published the proceedings in which Mendel's lectures on his experiments with plant hybrids were printed. It would have been impossible for Mendel to publish this paper in any other publication in Brno.

References and Historical Printed Sources

Richter O (1943) Johann Gregor Mendel wie er wirklich war. Neue Beiträge zur Biographie des berühmten Biologen aus Brünns Archiven mit 31 Abbildungen im Texte. Verhandlungen des Naturforschenden Vereines in Brünn, Abteilung für Naturforschung der Deutschen Gesellschaft für Wissenschaft und Volkstumforschung in Mähren 74 (für das Jahr 1942):1-262. Druck von Josef Klär, Brno

Chapter 8
Extraordinary Student at the Viennese University 1851–1853

Mendel's certificate by the Scientific Examination Commission of the University of Vienna was issued on 17 October 1850. The protocol of Mendel's failed examination had to be registered by the Moravian Educational Council and by the Moravian Directorate of Gymnasia, which was a complicated and time-consuming process in the traditional Austrian state bureaucracy. It reached the Old Brno Monastery as late as 09 August 1851.

Napp learned that the examiners acknowledged Mendel's talent and industry, and that they were of the opinion that Mendel needed better guidance. Napp consulted Baumgartner and determined to send Mendel to the University of Vienna to study physics with Christian Doppler. The start of Doppler's physics course in 1851 was delayed due to the relocation of his Physical Institute to a private house in Erdberg, enabling Mendel to join Doppler's lectures after the official beginning of the winter semester at the Viennese University in 1851.

English translation of Napp's letter to Schaaffgotsche by Eden and Cedar Paul appeared in Iltis (1966, pp. 75–6): *In view of the fact that Father Gregor Mendel has proved unsuitable for work as a parish priest, but has on the other hand shown evidence of exceptional intellectual capacity and remarkable industry in the study of natural sciences, and seeing that his praiseworthy knowledge in this field has been recognized by Count Baumgartner, the minister for commerce, though for the full practical development of his powers in this respect it would seem necessary and desirable to send him to Vienna University where he will have full opportunities for study—I propose, with this end in view, to send him thither in the course of the present month, and to arrange that during his stay in Vienna he shall reside and board in the monastery of the Brothers of Mercy, where he will comply with the rule of the house and will perform his religious duties.*

In his reply Schaaffgotsche gave his approval, *provided that in Vienna the above-named priest shall lead the life proper to a member of a religious order, and shall not become estranged from his profession* [English translation by Eden and Cedar Paul in Iltis (1966, p. 76)].

A. Matalová, E. Matalová, *Gregor Mendel - The Scientist*, Springer Biographies,
https://doi.org/10.1007/978-3-030-98923-1_8

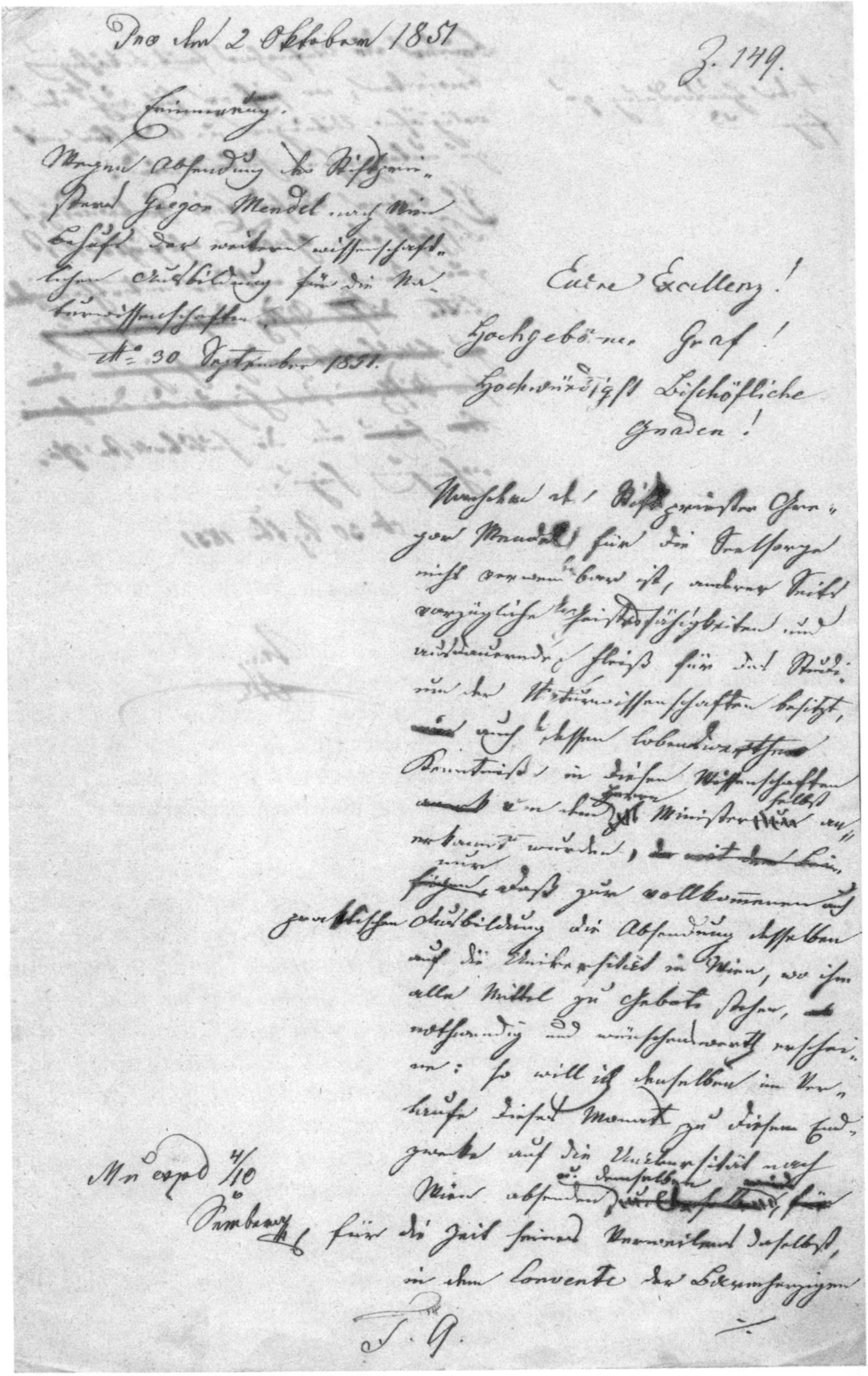
Z. 149.

Euere Excellenz!

Hochgeborner Graf!

Hochwürdigst Bischöfliche Gnaden!

D88 The first page of the draft of a letter by Napp to Schaaffgotsche, informing him that he was sending Mendel to the University of Vienna. 30 September 1851

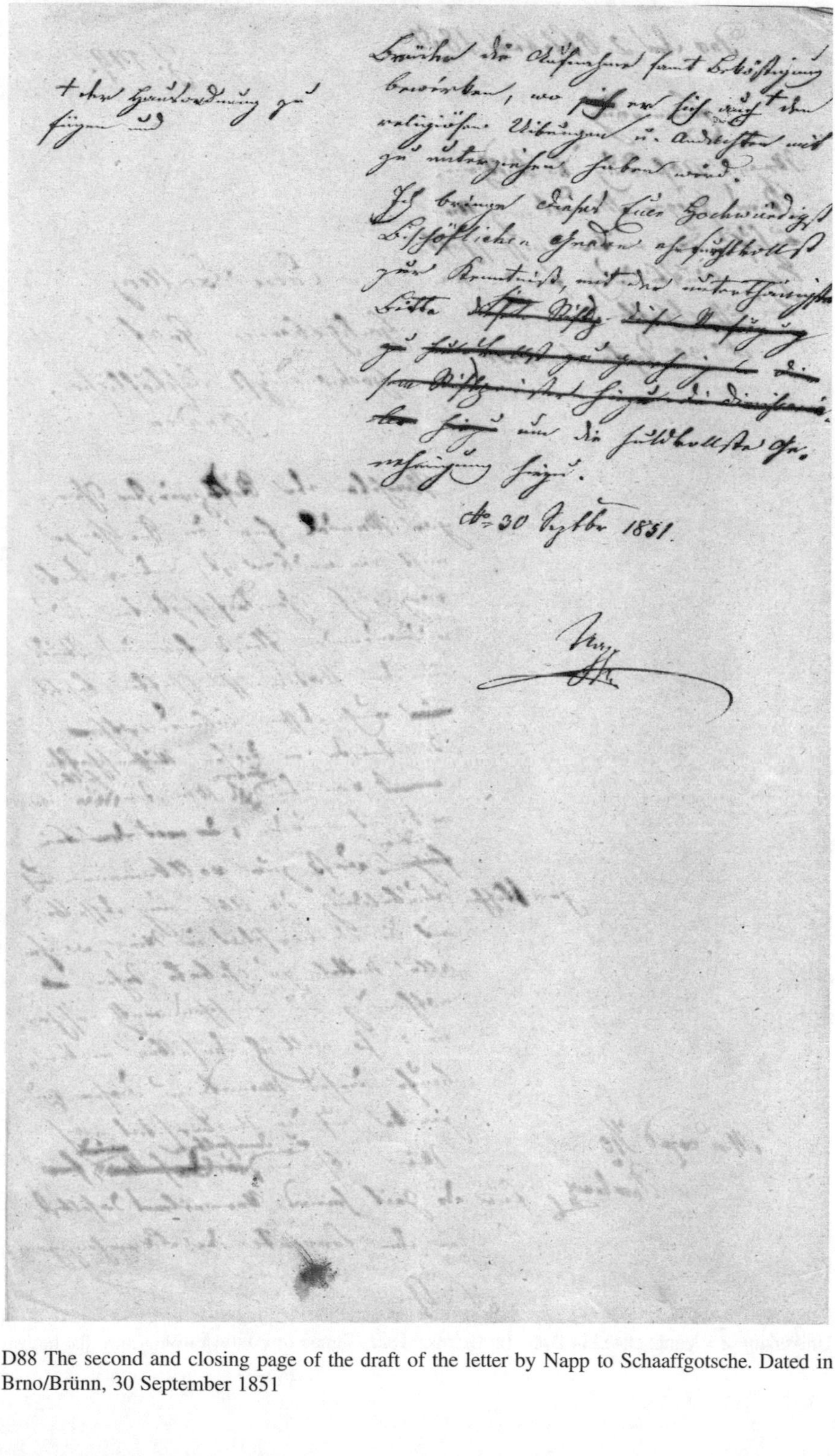

D88 The second and closing page of the draft of the letter by Napp to Schaaffgotsche. Dated in Brno/Brünn, 30 September 1851

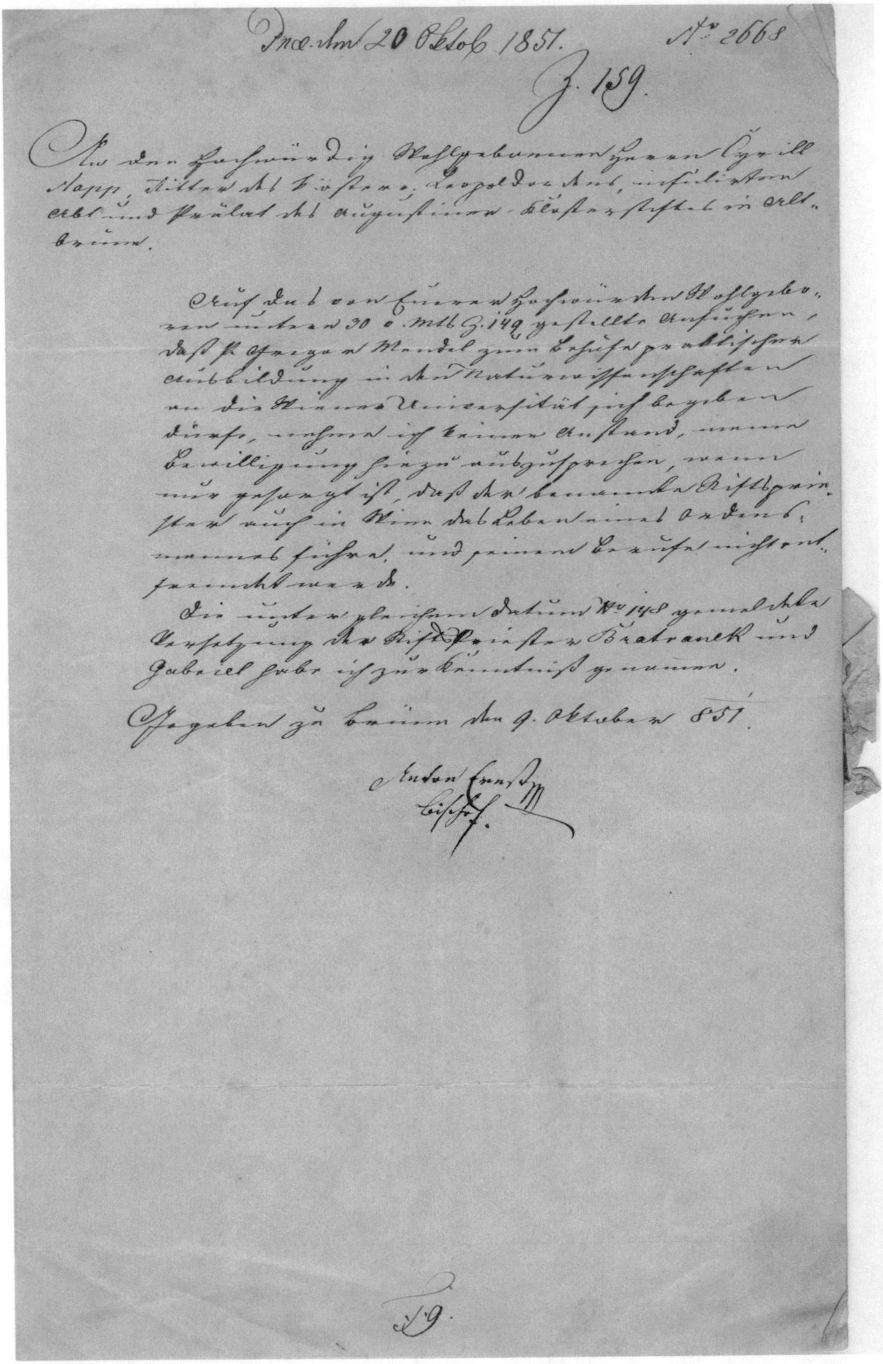

Praes. am 20 Octob 1851. No 2668

Z. 159.

An den Hochwürdig Wohlgebornen Herrn Cyrill Napp, Ritter des k. österr. Leopoldordens, infulirten Abt und Prälat des Augustiner-Klosterstiftes in Alt-Brünn.

Auf das von Euer Hochwürden Wohlgeboren unterm 30 v. Mts Z. 149 gestellte Ansuchen, daß P. Gregor Mendel zum Behufe praktischer Ausbildung in den Naturwissenschaften an die Wiener Universität sich begeben dürfe, nehme ich keinen Anstand, meine Bewilligung hiezu auszusprechen, wenn nur gesorgt ist, daß der benannte Stiftspriester auch in Wien das Leben eines Ordensmannes führe, und seinem Berufe nicht entfremdet werde.

Die unter gleichem Datum No 148 gemeldete Versetzung der Stiftspriester Bratranek und Gabriel habe ich zur Kenntniß genommen.

Gegeben zu Brünn den 9. Oktober 851.

Anton Ernst
Bischof.

159.

D90 Schaaffgotsche's letter to Napp giving his approval of Mendel's study of natural sciences at the University of Vienna. Dated in Brno, 09 October 1851, signed by Anton Ernst

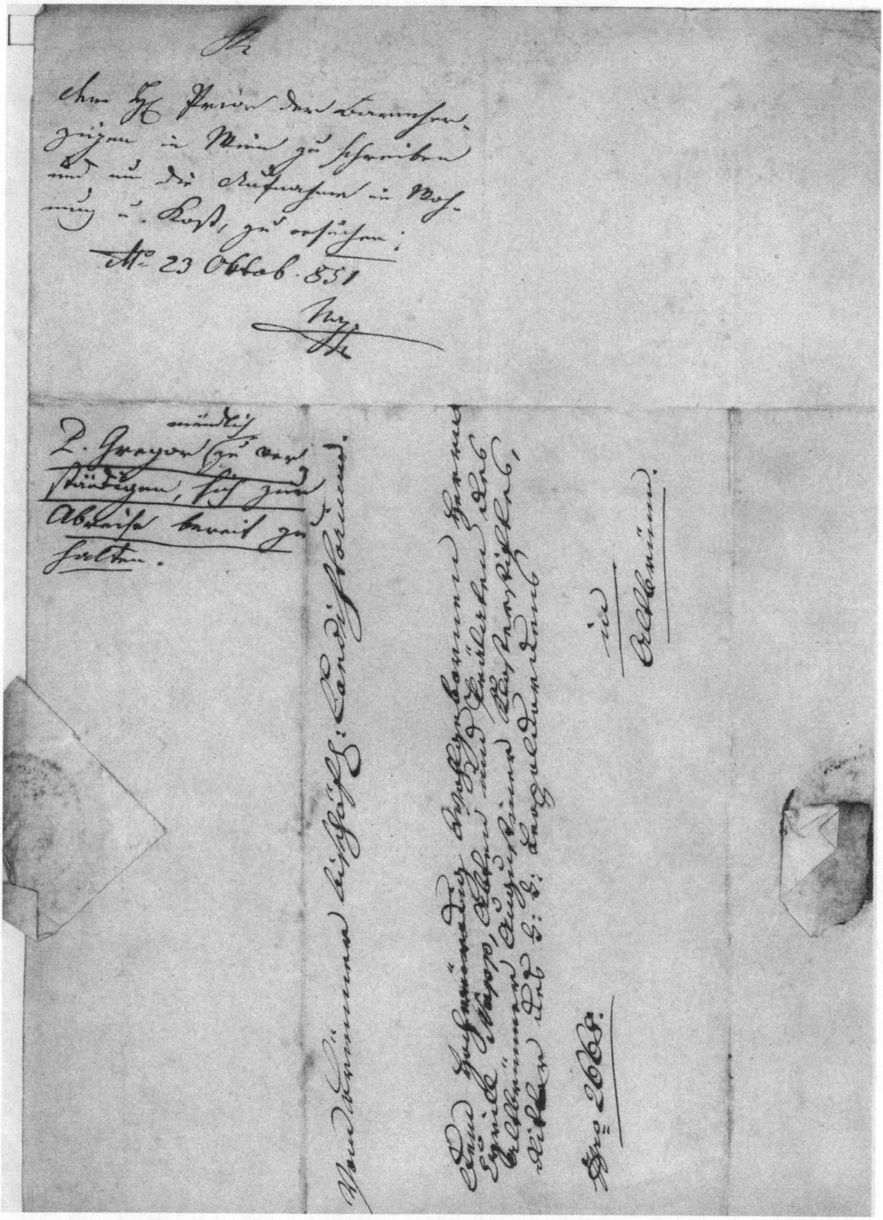

D90 The envelope of Schaaffgotsche's letter on 09 October 1851 with two notes in Napp's hand. Transcription of note 1: *P. Gregor Mendel mündilich zu verständigen, sich zur Abreise bereit zu halten.* Transcription of note 2: *Dem H. Prior der Barmherzigen in Wien zu schreiben und um die Aufnahme in Wohnung u. Kost, zu ersuchen. 23 Oktob. 851 Napp.* Translation by Scott Abbott and Daniel J. Fairbanks (2021): Note 1: *Notify Pater Gregor Mendel to be ready for departure.* Note 2: *Write to the Prior of the Brothers of Mercy in Vienna to request admission for lodging and board. 23 October 1851 Napp*

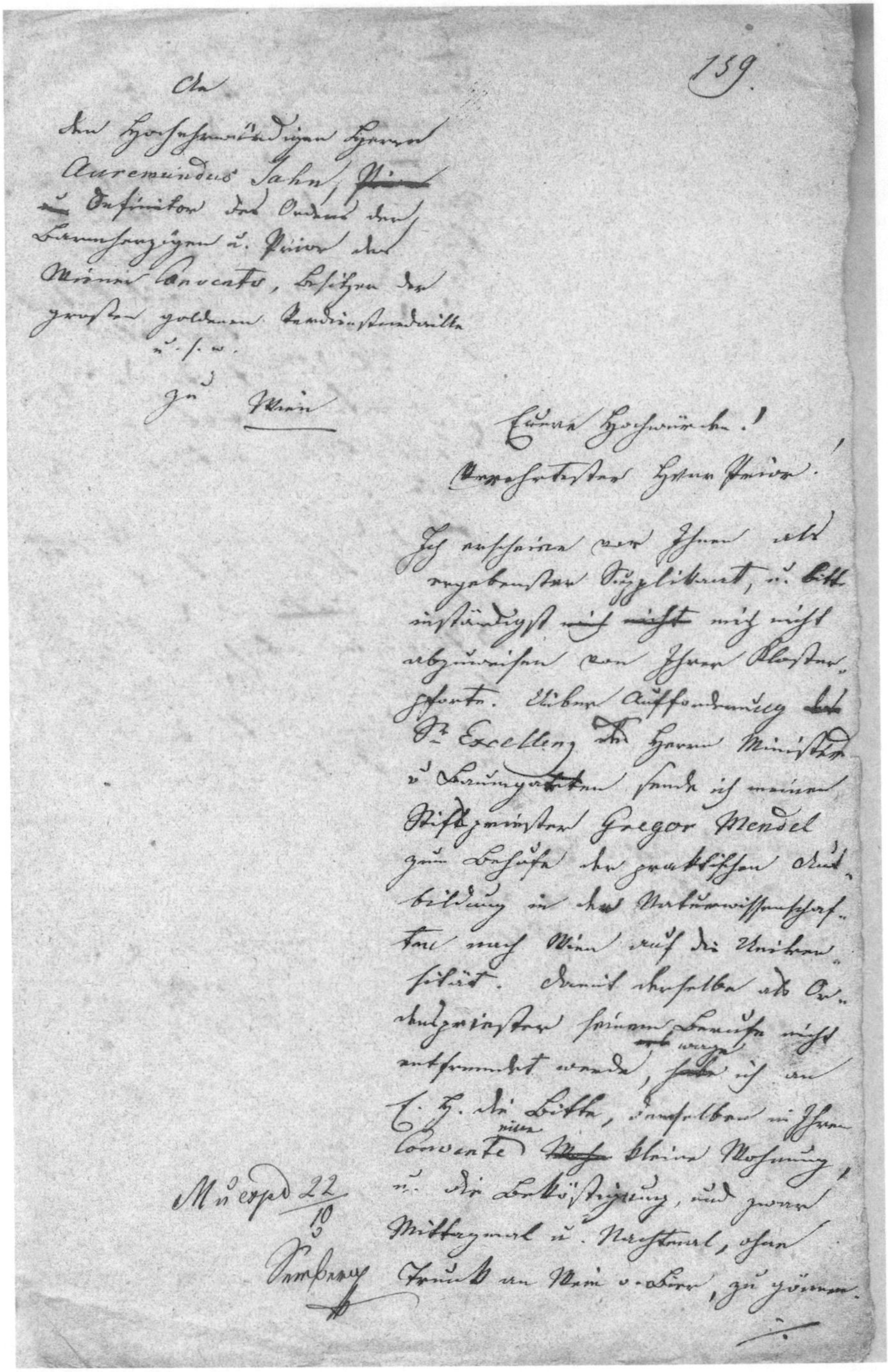

D92 The first page of the draft of a letter by Abbot Napp to Prior Auremundus Jahn, requesting admission for Mendel into the monastery of the Brothers of Mercy in Vienna during his university studies in Vienna

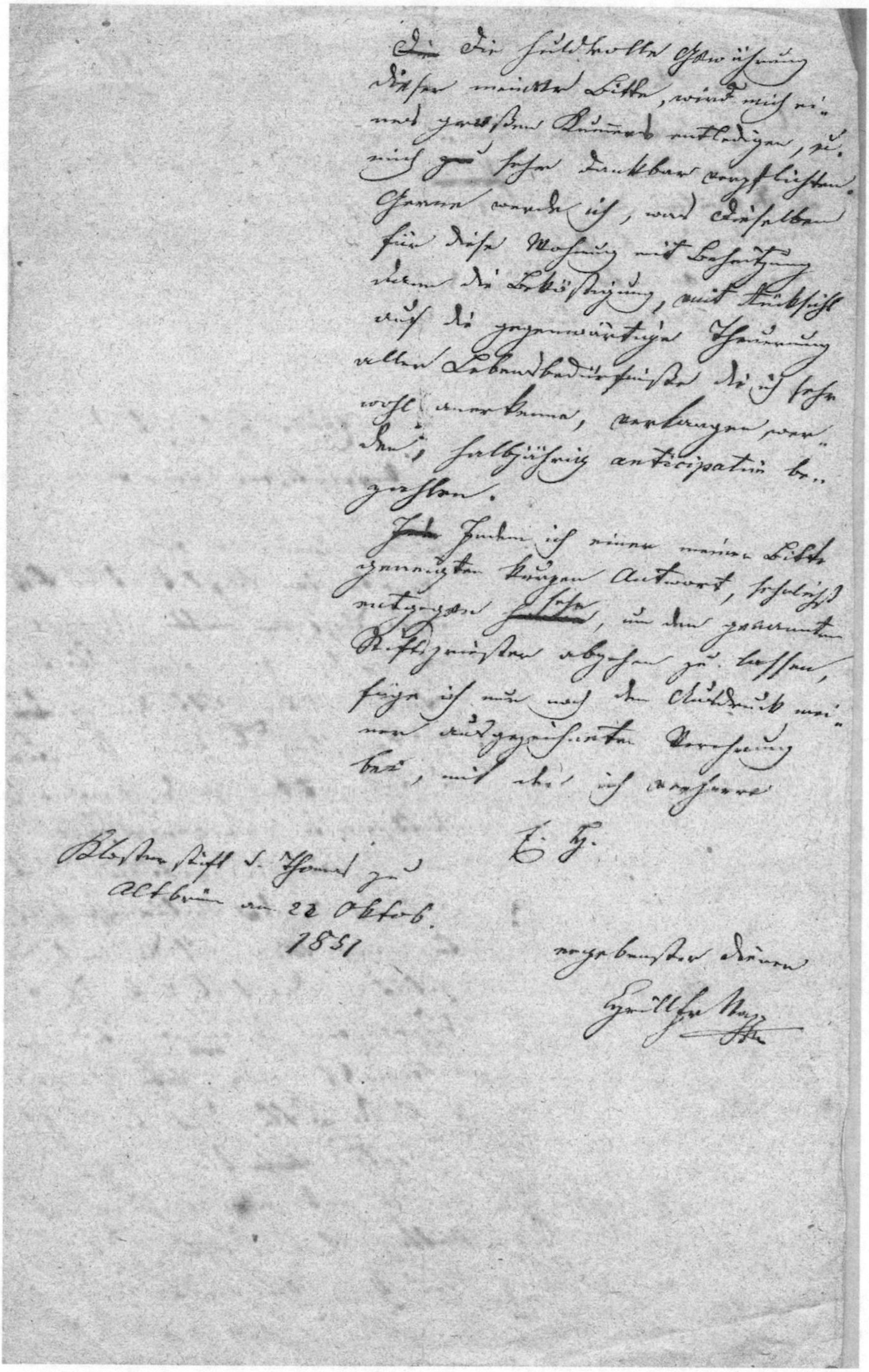

D92 The second page of the draft of Napp's letter to Prior Jahn requesting admission for Mendel into the monastery of the Brothers of Mercy in Vienna. Dated in St. Thomas Monastery in Old Brno, 22 October 1851, signed by Cyrill Fr. Napp

Translation of the draft of the letter by Napp to Jahn, 22 October 1851 by Eden and Cedar Paul in Iltis (1966, pp. 76–7): *Very Reverend and Most Honourable Herr Prior, I present myself before you as a most humble petitioner, and earnestly beg you not to turn me away from the door of your monastery. At the instigation of His Excellency Minister von Baumgartner I am sending Gregor Mendel, one of the priests of my monastery, to Vienna University for a course of practical training in the natural sciences. Since it is essential that during this course of training Pater Mendel should not be alienated from his profession as a member of the Augustinian order, I venture to ask you to be good enough to provide him at your monastery with modest quarters and board, the latter comprising dinner and supper, but without wine and beer. Your gracious compliance with this request will save me much anxiety and trouble, and will leave me greatly in your debt. It will be a pleasure to me to pay half-yearly in advance whatever sum is requisite for this lodging with heating and for the board, taking into account the recent rise in the cost of the necessaries of life, of which I am well aware.*

Transcription of Napp's entry in the monastery journal: *In Folge dieser Antwort blieb mir nichts übrig, als dem P. Gregor nach Wien mit der Aufgabe abzusenden, um in einem anderen Kloster oder geistlichen Hause eine Wohnung u. die Beköstigung zu finden.—Er reiste demnach am 27ten d. M. mit dem Nachttrain nach Wien ab.—28/10 51—Napp.*

English translation of Napp's entry in the monastic journal by Scott Abbott and Daniel J. Fairbanks (2021): *As a consequence of this answer, I had no choice but to send Pater Gregor to Vienna with the task of finding an apartment and boarding in another monastery or religious house.—He left on 27 of this month on the night train to Vienna.—October 28, 1851—Napp.*

A note in the margin of the draft states: Mundirt u. dem P. Gregor mit nach Wien übergeben am 27/10 Napp. English translation by Scott Abbott and Daniel J. Fairbanks, 2021: Corrected and delivered to P. Gregor to take with him to Vienna on 27 October Napp.

The beginning of the winter semester (from October to April) at the Physics Institute in 1851 was delayed. The Physics Institute had to arrange for provisional lecture halls and laboratories in a hired private house in Erdberg.

On the reverse side of the application form, lectures are given in which Mendel wanted to enrol as an extraordinary student at the University of Vienna.

Nationale registers the lectures which Mendel intended to enrol in the third semester at the University of Vienna from October 1852 to April 1853: Doppler's Demonstrational Experimental Physics, Kner's Zoology, Redtenbacher's General and Medical-Pharmaceutical Chemistry, Redtenbachers's Methods of Analytical Chemistry, Moth's Exercises with Logarithmic-Trigonometric Tables, Unger's Anatomy and Physiology of Plants and Practical Exercises in the Use of the Microscope.

On the reverse side of Meldungsbogen, the lectures are given in which Mendel enrolled in his fourth semester at the University of Vienna, from April 1853 to August 1853 that include Ettingshausen's Preparation and Use of Physical Instruments and Higher Mathematical Physics, and Redtenbacher's Organic Chemistry.

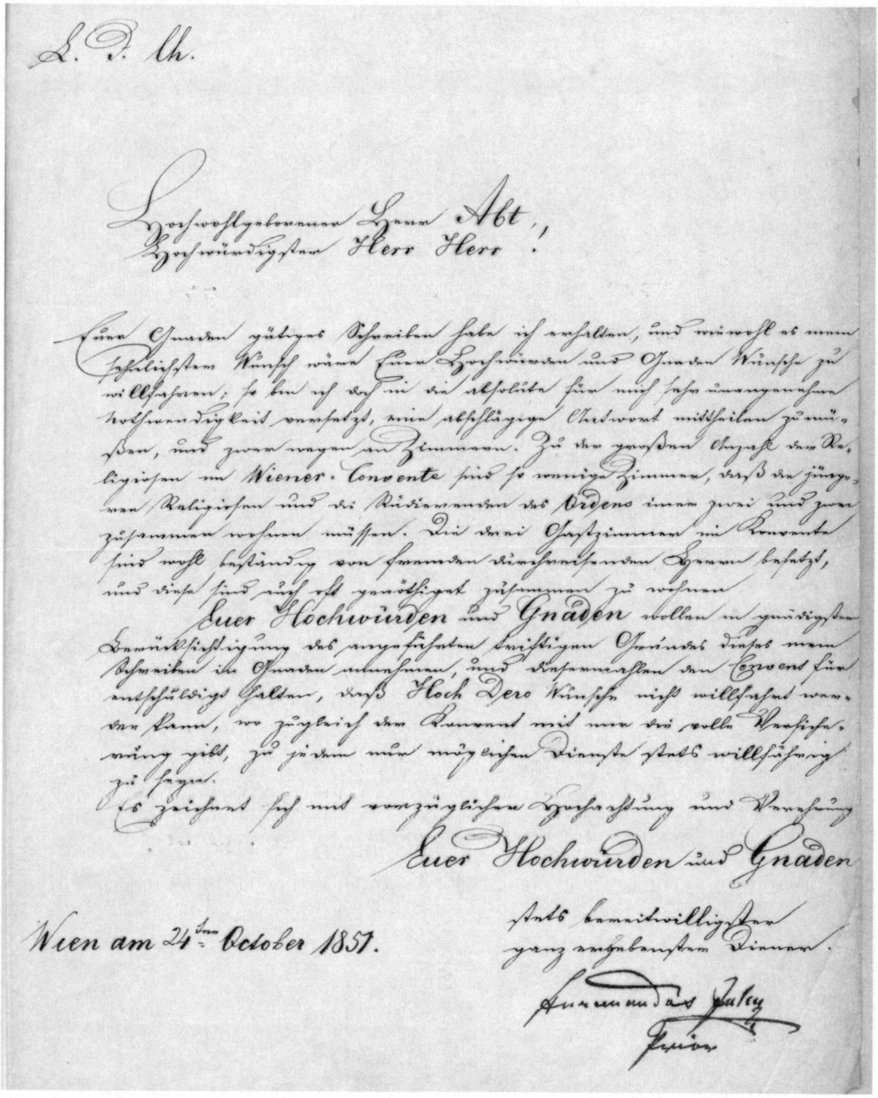

L. J. Ch.

Hochwohlgeborner Herr Abt,
Hochwürdigster Herr Herr!

Euer Gnaden gütiges Schreiben habe ich erhalten, und wie wohl es mein sehnlichster Wunsch wäre Euer Hochwürden und Gnaden Wünsche zu willfahren; so bin ich doch in die absolute für mich sehr unangenehme Nothwendigkeit versetzt, eine abschlägige Antwort mittheilen zu müssen, und zwar wegen der Zimmer. Zu der grossen Anzahl der Religiosen im Wiener-Convente sind so wenige Zimmer, dass die jüngeren Religiosen und die Studierenden des Ordens immer zwei und zwei zusammen wohnen müssen. Die drei Gastzimmer im Convente sind wohl beständig von fremden durchreisenden Herren besetzt, und diese sind auch oft genöthiget zusammen zu wohnen.

Euer Hochwürden und Gnaden wollen in gnädigster Berücksichtigung der angeführten triftigen Gründe dieses mein Schreiben in Gnaden aufnehmen, und dessen ungeachtet den Convent für entschuldigt halten, dass Hoch Dero Wünsche nicht willfahrt werden kann, wo zugleich der Convent mit mir die volle Versicherung gibt, zu jedem nur möglichen Dienste stets willfährig zu seyn.

Es zeichnet sich mit vorzüglicher Hochachtung und Verehrung

Euer Hochwürden und Gnaden

Wien am 24ten October 1851.

stets bereitwilligster
ganz ergebenster Diener.

Auremundus Jahn
Prior

D91 Letter by Prior Jahn informing Napp that he cannot accommodate Mendel in the Viennese Convent because they are full. Dated in Vienna, 24 October 1851, signed by Auremundus Jahn

Mendel also attended the lectures in the Court Museum in Vienna, the meeting place of the Imperial-Royal Zoological-Botanical Society in Vienna. Vincenz Kollar (1797–1860) was an entomologist and the Head of the Zoological Botanical Society. He encouraged Mendel to publish his first two papers, both on insect predation on plants, in the society's proceedings (Mendel 1853, 1854).

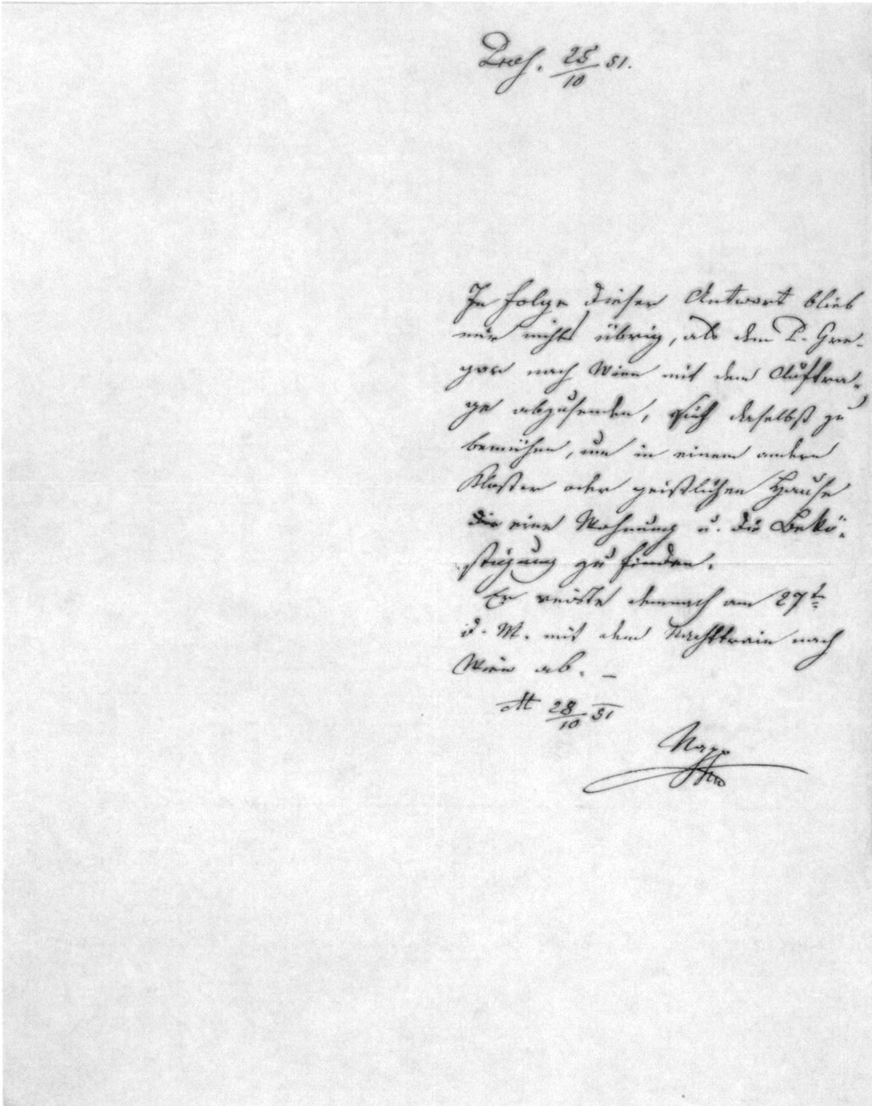

Praes. 25/10 51.

In Folge dieser Antwort blieb mir nichts übrig, als den P. Gregor nach Wien mit dem Auftrage zu entsenden, sich daselbst zu bemühen, um in einem andern Kloster oder geistlichen Hause eine Wohnung u. die Verköstigung zu finden.

Er reiste demnach am 27t. d. M. mit dem Nachttrain nach Wien ab. –

Alt 28/10 51

Napp Abt

Napp's entry in the monastery agenda that Mendel departed for Vienna on the night train on 27 October 1851, with the task of finding himself accommodation in another monastery or religious house. State Archives Brno G84.Dated in Brno, 28 October 1851, signed by Napp

From the lectures in which Mendel enroled, it is evident that physics was his principal interest. Doppler, in his demonstrative physics course, stressed precision of the initial conditions for each experiment. In his experiments with plant hybrids, Mendel devoted 2 years to testing the stability of plant traits. He did not include the

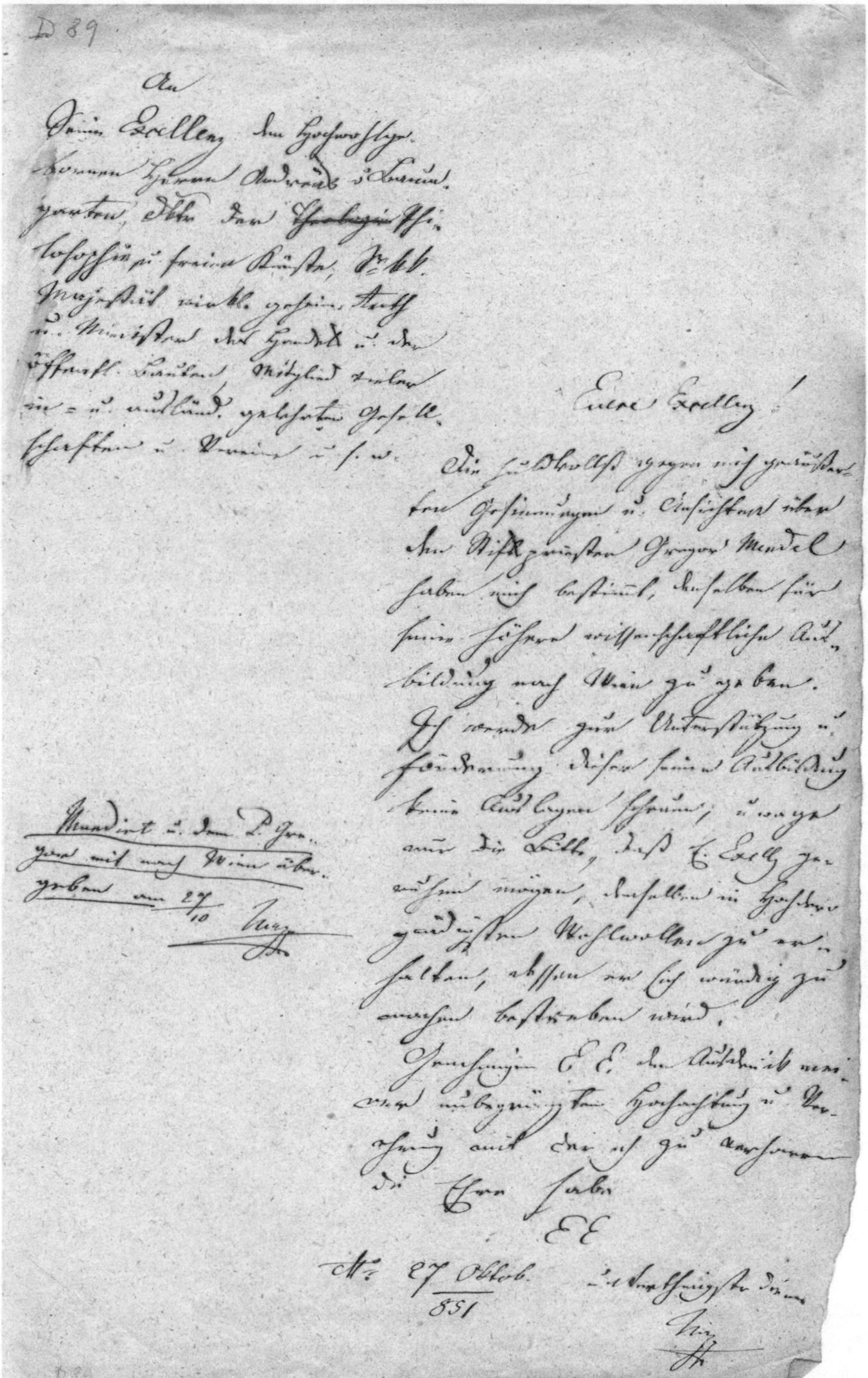

D89 Draft of a letter by Napp to Baumgartner. Mendel took the letter with him to Vienna to present himself personally to Professor Baumgartner. Napp informs Baumgartner that he is sending Mendel

2 preparatory years of testing when he recounted the time required for his experiments. According to Mendel, his experiments lasted 8 years, from 1856 to 1863 (Mendel to Nägeli 1867).

The influence of Doppler and Unger on Mendel as a researcher was fundamental. Unger taught Mendel the anatomy and physiology of plants and trained him in the use of microscopes. At the University of Vienna, Mendel studied simultaneously with J. Nave (1829–1864) from Brno, an expert in microscopy, whom Unger taught to master the use of microscopes. After his return from Vienna, Nave presented a report on microscopes in the Natural Science Section of the Agricultural Society. In 1864, Nave published an *Introduction to Collection, Preparation and Analysis of Higher Plants*. In 1867, he published a handbook for the collection and preparation of freshwater and marine algae, diatoms, desmids, fungi, lichens, mosses and other cryptogamic plants, with instructions for the creation of an herbarium. Mendel's collection of microscopic preparations of green and red algae appears to have been prepared with Nave's participation (Milovidov 1968). The dendrological microscopic preparations of Mendel's are taken from a collection of 100 microtomes of trees appended to the book by Nördlicher (1852) which is preserved in the Augustinian library in Old Brno (Orel and Čunderlík 1985). Mendel cut off the edges of some microtomes for his microscopic preparations. Mendel had three microscopes. The oldest microscope was type No. 1 by Simon Plössl from the years 1828–1831. The second is also the product of the Viennese firm Plössl, type No. 4 from 1863. The third and best microscope was produced by the Viennese firm Carl Reichert in about 1877 (Procházka 1985).

(continued) to study at the University of Vienna, his decision being influenced by Baumgartner's opinion on Mendel's talent and industry. Dated in Brno, 27 October 1851, signed by Napp

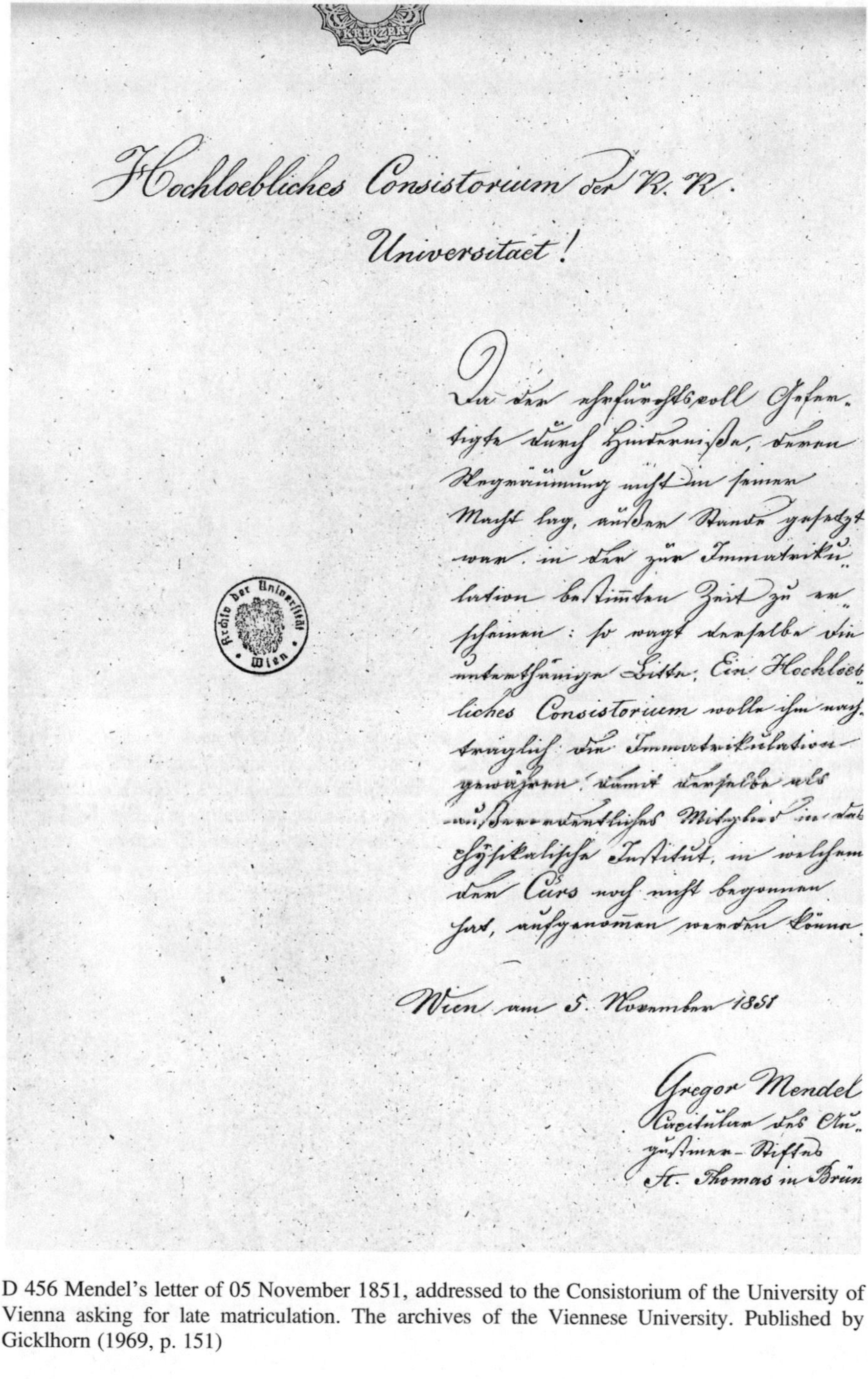

Hochloebliches Consistorium der K. K.
Universitaet!

Da der ehrfurchtsvoll Gefertigte durch Hindernisse, deren Wegräumung nicht in seiner Macht lag, außer Stande gesetzt war, in der zur Immatriculation bestimmten Zeit zu erscheinen: so wagt derselbe die unterthänige Bitte: Ein Hochloebliches Consistorium wolle ihm nachträglich die Immatriculation gewähren, damit derselbe als außerordentliches Mitglied in das physikalische Institut, in welchem der Curs noch nicht begonnen hat, aufgenommen werden könne.

Wien am 5. November 1851

Gregor Mendel
Capitular des Augustiner-Stiftes
St. Thomas in Brünn

D 456 Mendel's letter of 05 November 1851, addressed to the Consistorium of the University of Vienna asking for late matriculation. The archives of the Viennese University. Published by Gicklhorn (1969, p. 151)

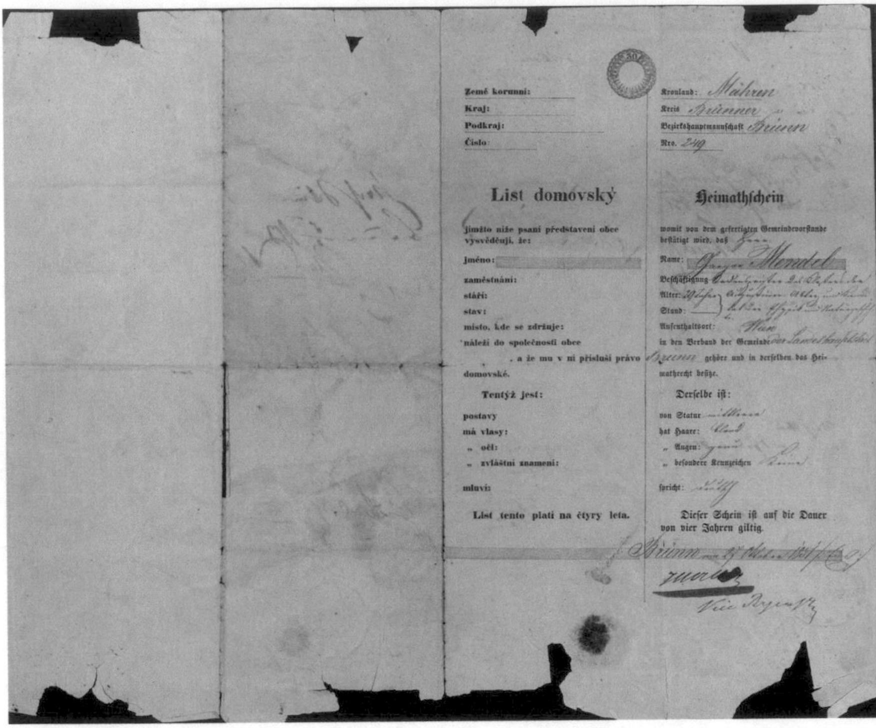

Země korunní:
Kraj:
Podkraj:
Číslo:

Kronland: Mähren
Kreis Brünner
Bezirkshauptmannschaft Brünn
Nro. 249

List domovský

jímžto níže psaní představení obce vysvědčují, že:
jméno:
zaměstnání:
stáří:
stav:
místo, kde se zdržuje:
náleží do společnosti obce . a že mu v ní přísluší právo domovské.

Tentýž jest:
postavy
má vlasy:
„ oči:
„ zvláštní znamení:
mluví:

List tento platí na čtyry leta.

Heimathschein

womit von dem gefertigten Gemeindevorstande bestätigt wird, daß Herr
Name: Gregor Mendel
Beschäftigung
Alter:
Stand:
Aufenthaltsort: Wien
in den Verband der Gemeinde Brünn gehöre und in derselben das Heimathrecht besitze.

Derselbe ist:
von Statur
hat Haare:
„ Augen:
„ besondere Kennzeichen
spricht:

Dieser Schein ist auf die Dauer von vier Jahren giltig.

Brünn am 27 Okt. 1851

D41 Czech-German *Heimatschein* (identity paper, passport), valid for 4 years, giving the dates of Mendel's arrivals in Vienna and Brno, which contains the following personal data on Mendel written in Mendel's hand in the German half of the document of three pages: *Crown land Moravia, district Brünn, township Brünn, ID No. 249 Mr. Gregor Mendel, occupation priest of the Augustinian Abbey and candidate of physics and natural history, stays in Vienna, 29 years old, religion Catholic, the place of residence Brünn, capital of the province. Mean stature, fair hair, grey eyes, other peculiarities none, speaks German.—Dated in Brno, 27 October 1851, signed by J. Hertl*

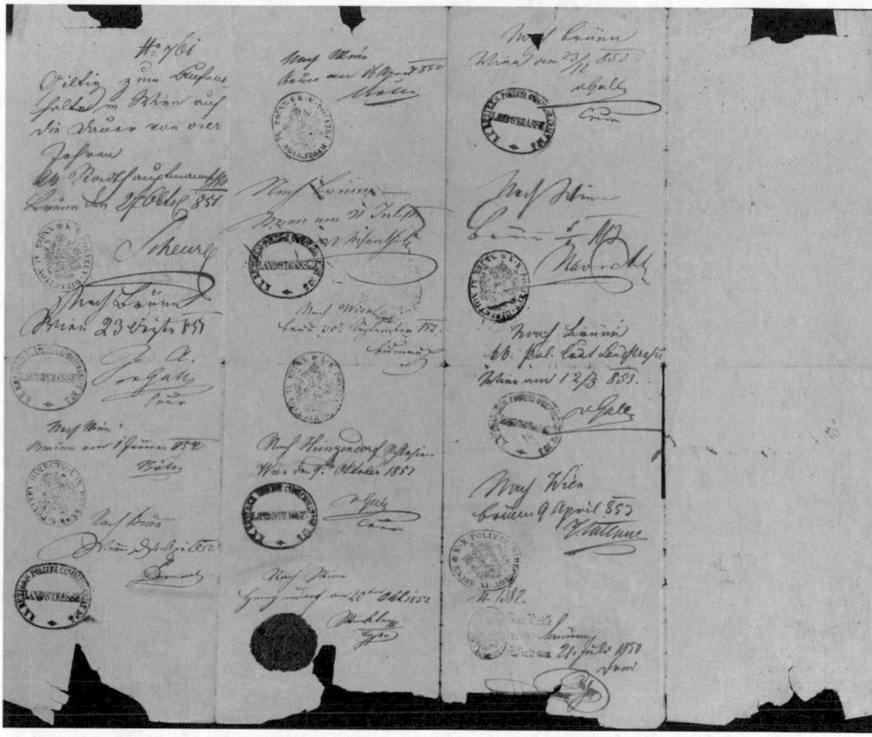

(continued)

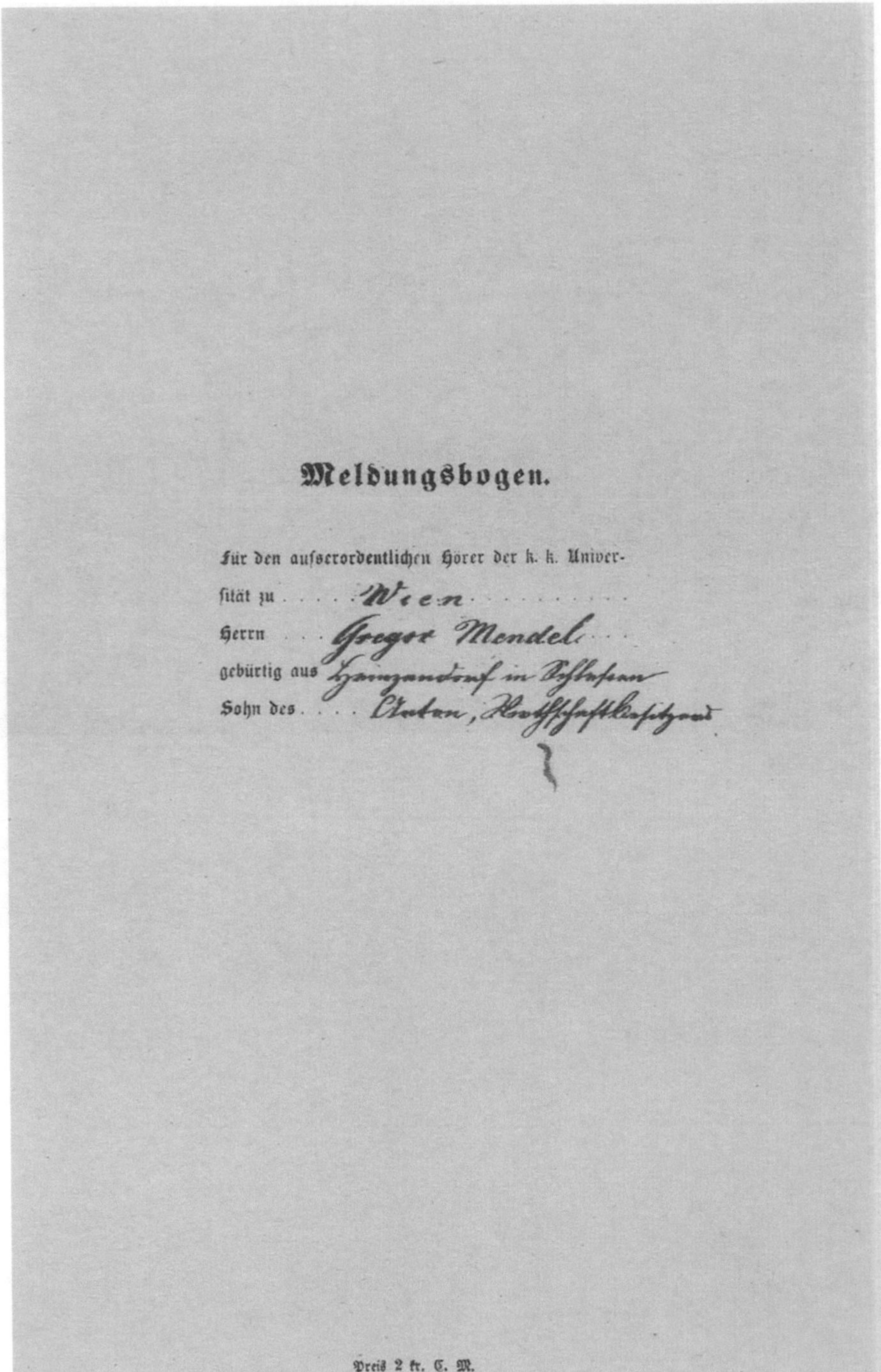

Meldungsbogen.

für den aufserordentlichen Hörer der k. k. Universität zu . . . . Wien . . . .

Herrn . . . Gregor Mendel . . .

gebürtig aus Heinzendorf in Schlesien

Sohn des . . . . Anton, Wirtschaftsbesitzer

Preis 2 kr. C. M.

D56 Mendel's application form (*Meldungsbogen*). As an extraordinary (auditing student) attendee of lectures held at the Imperial-Royal University of Vienna Mendel was not taking examinations. Transcription: *Meldungsbogen für den ausserordentlichen Hörer der k. k. Universität zu Wien—Herr Gregor Mendel—gebürtig aus Heinzendorf in Schlesien—Sohn des Anton Wirtschaftsbesitzer*

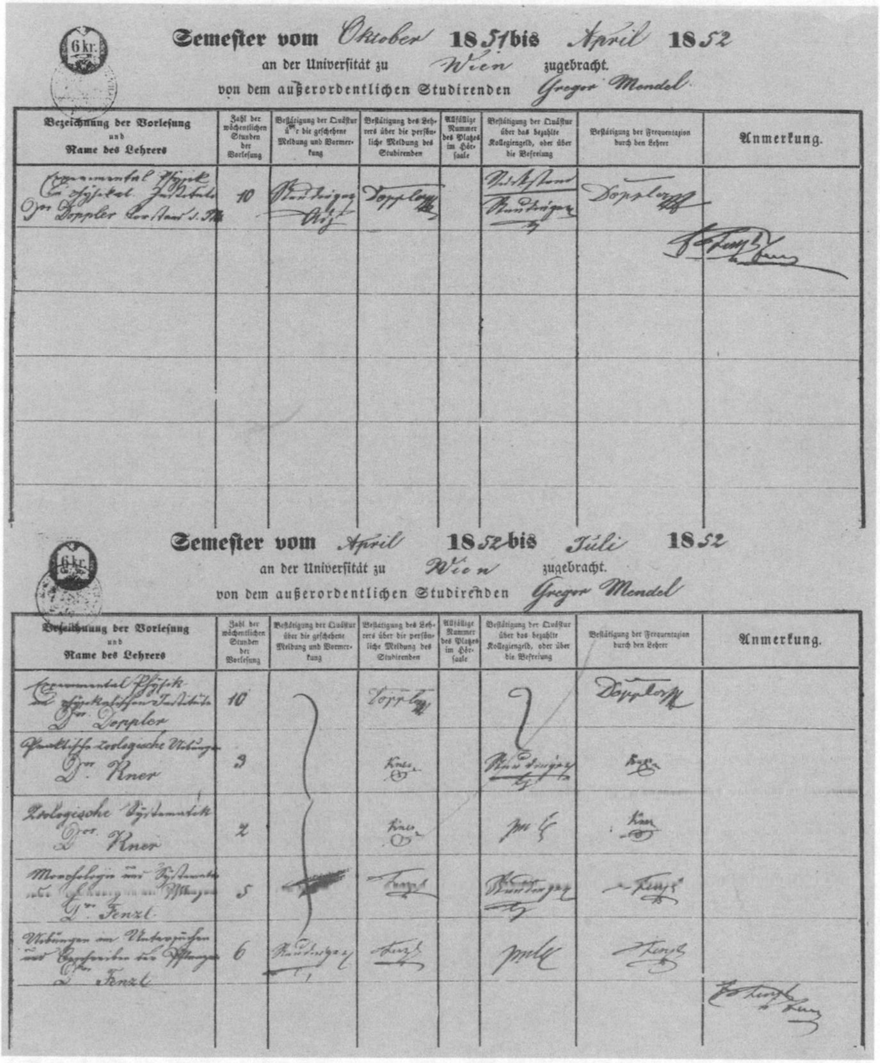

Semester vom Oktober 1851 bis April 1852
an der Universität zu Wien zugebracht.
von dem außerordentlichen Studirenden Gregor Mendel

Bezeichnung der Vorlesung und Name des Lehrers	Zahl der wöchentlichen Stunden der Vorlesung	Bestätigung der Quästur über die geschehene Meldung und Vormerkung	Bestätigung des Lehrers über die persönliche Meldung des Studirenden	Allfällige Nummer des Platzes im Hörsaale	Bestätigung der Quästur über das bezahlte Kollegiengeld, oder über die Befreiung	Bestätigung der Frequentazion durch den Lehrer	Anmerkung.
Experimental Physik im physikal. Institute Dr. Doppler	10	[illegible]	Doppler		[illegible]	Doppler	[illegible]

Semester vom April 1852 bis Juli 1852
an der Universität zu Wien zugebracht.
von dem außerordentlichen Studirenden Gregor Mendel

Bezeichnung der Vorlesung und Name des Lehrers	Zahl der wöchentlichen Stunden der Vorlesung	Bestätigung der Quästur über die geschehene Meldung und Vormerkung	Bestätigung des Lehrers über die persönliche Meldung des Studirenden	Allfällige Nummer des Platzes im Hörsaale	Bestätigung der Quästur über das bezahlte Kollegiengeld, oder über die Befreiung	Bestätigung der Frequentazion durch den Lehrer	Anmerkung.
Experimental Physik im physikalischen Institute Dr. Doppler	10		Doppler			Doppler	
Praktische zoologische Übungen Dr. Kner	3		Kner		[illegible]	Kner	
Zoologische Systematik Dr. Kner	2		Kner		[illegible]	Kner	
Morphologie und Systematik der Phanerogamen Dr. Fenzl	5		Fenzl		[illegible]	Fenzl	
Übungen im Untersuchen und Beschreiben der Pflanzen Dr. Fenzl	6	[illegible]	Fenzl		[illegible]	Fenzl	[illegible]

D56 On the reverse side of *Meldungsbogen* there is Mendel's index of lectures at the University of Vienna in the semester of October 1851 to April 1852 and the second semester from April 1852 to July 1852. Mendel's record of lectures in the first term from October 1851 to April 1852 shows 10 h weekly of Doppler's Experimental Physics. Mendel's index of lectures in the second semester from April 1852 to June 1852 shows ten hours of Doppler's Experimental Physics, three hours of Kner's Practical Exercises in Zoology, two hours of Kner's Zoological Systematics, five hours of Fenzl's Morphology and Systematics of Phanerogamic Plants and 6 h of Fenzl's Exercises in Analysis and Description of Plants

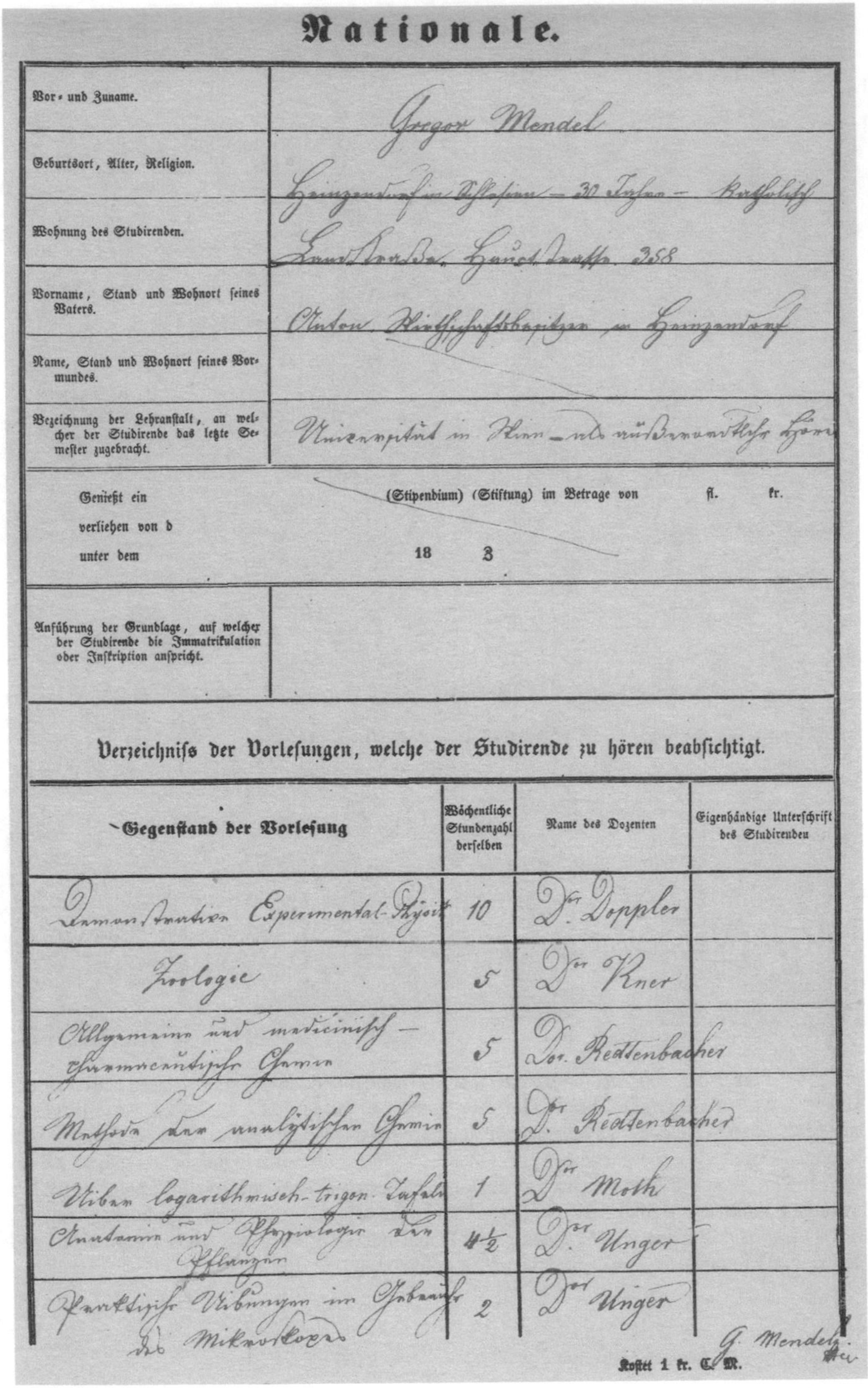

Nationale.

Vor- und Zuname.	Gregor Mendel
Geburtsort, Alter, Religion.	Heinzendorf in Schlesien – 30 Jahre – Katholisch
Wohnung des Studirenden.	Landstrasse, Hauptstrasse 358
Vorname, Stand und Wohnort seines Vaters.	Anton, Wirthschaftsbesitzer in Heinzendorf
Name, Stand und Wohnort seines Vormundes.	
Bezeichnung der Lehranstalt, an welcher der Studirende das letzte Semester zugebracht.	Universität in Wien – als außerordentlicher Hörer

Genießt ein (Stipendium) (Stiftung) im Betrage von fl. kr.
verliehen von d
unter dem 18 3

Anführung der Grundlage, auf welcher der Studirende die Immatrikulation oder Inskription anspricht.	

Verzeichniß der Vorlesungen, welche der Studirende zu hören beabsichtigt.

Gegenstand der Vorlesung	Wöchentliche Stundenzahl derselben	Name des Dozenten	Eigenhändige Unterschrift des Studirenden
Demonstrative Experimental-Physik	10	Dr. Doppler	
Zoologie	5	Dr. Kner	
Allgemeine und medicinisch-pharmaceutische Chemie	5	Dr. Redtenbacher	
Methoden der analytischen Chemie	5	Dr. Redtenbacher	
Ueber logarithmisch-trigon. Tafeln	1	Dr. Moth	
Anatomie und Physiologie der Pflanzen	4½	Dr. Unger	
Praktische Uebungen im Gebrauche des Mikroskops	2	Dr. Unger	

G. Mendel

Kostet 1 kr. C. M.

D29 In the Record of Studies (*Nationale*) Mendel filled in the name *Gregor Mendel, born at Heinzendorf in Silesia—30 years old—religion Catholic—place of living Landstrasser Hauptstrasse No. 358—father, Anton, farmstead owner in Heinzendorf*—the last semester finished *at the University of Wien/Vienna as an extraordinary student*

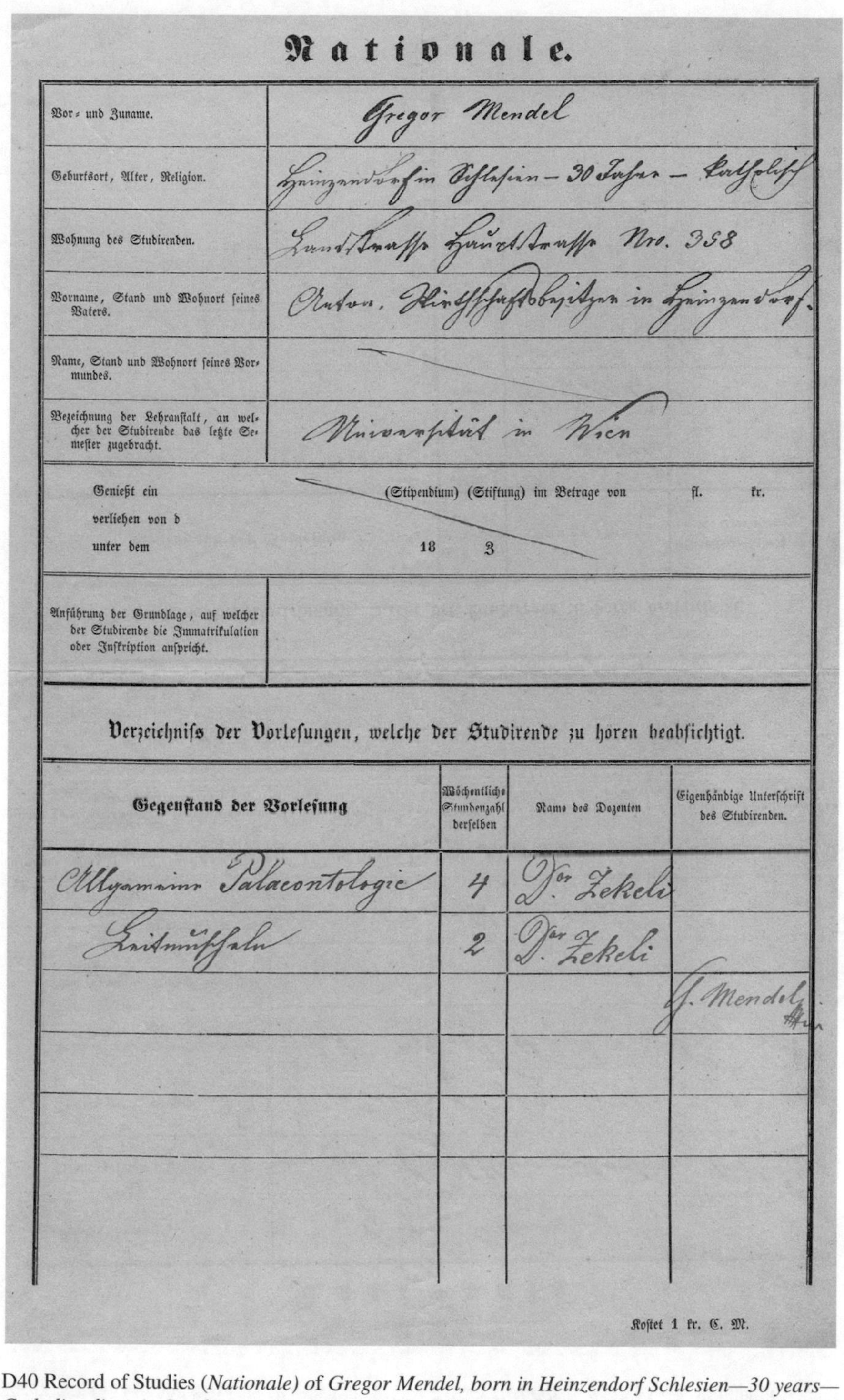

Nationale.

Vor- und Zuname.	Gregor Mendel
Geburtsort, Alter, Religion.	Heinzendorf in Schlesien – 30 Jahre – katholisch
Wohnung des Studirenden.	Landstrasse Hauptstrasse Nro. 358
Vorname, Stand und Wohnort seines Vaters.	Anton, Wirthschaftsbesitzer in Heinzendorf.
Name, Stand und Wohnort seines Vormundes.	
Bezeichnung der Lehranstalt, an welcher der Studirende das letzte Semester zugebracht.	Universität in Wien

Genießt ein (Stipendium) (Stiftung) im Betrage von fl. kr.
verliehen von d
unter dem 18 3

Anführung der Grundlage, auf welcher der Studirende die Immatrikulation oder Inskription anspricht.	

Verzeichniß der Vorlesungen, welche der Studirende zu hören beabsichtigt.

Gegenstand der Vorlesung	Wöchentliche Stundenzahl derselben	Name des Dozenten	Eigenhändige Unterschrift des Studirenden.
Allgemeine Palaeontologie	4	Dor Zekeli	
Leitmuscheln	2	Dor Zekeli	
			G. Mendel

Kostet 1 kr. C. M.

D40 Record of Studies (*Nationale) of Gregor Mendel, born in Heinzendorf Schlesien—30 years—Catholic—lives in Landstrasse Hauptstrasse No. 358, son of father Anton, farmstead holder in Heinzendorf*. In the winter semester of 1852/1853 Mendel intended to enrol in Zekeli's lectures on general palaeontology and guide shells that were a weak point in his 1850 examination in natural history

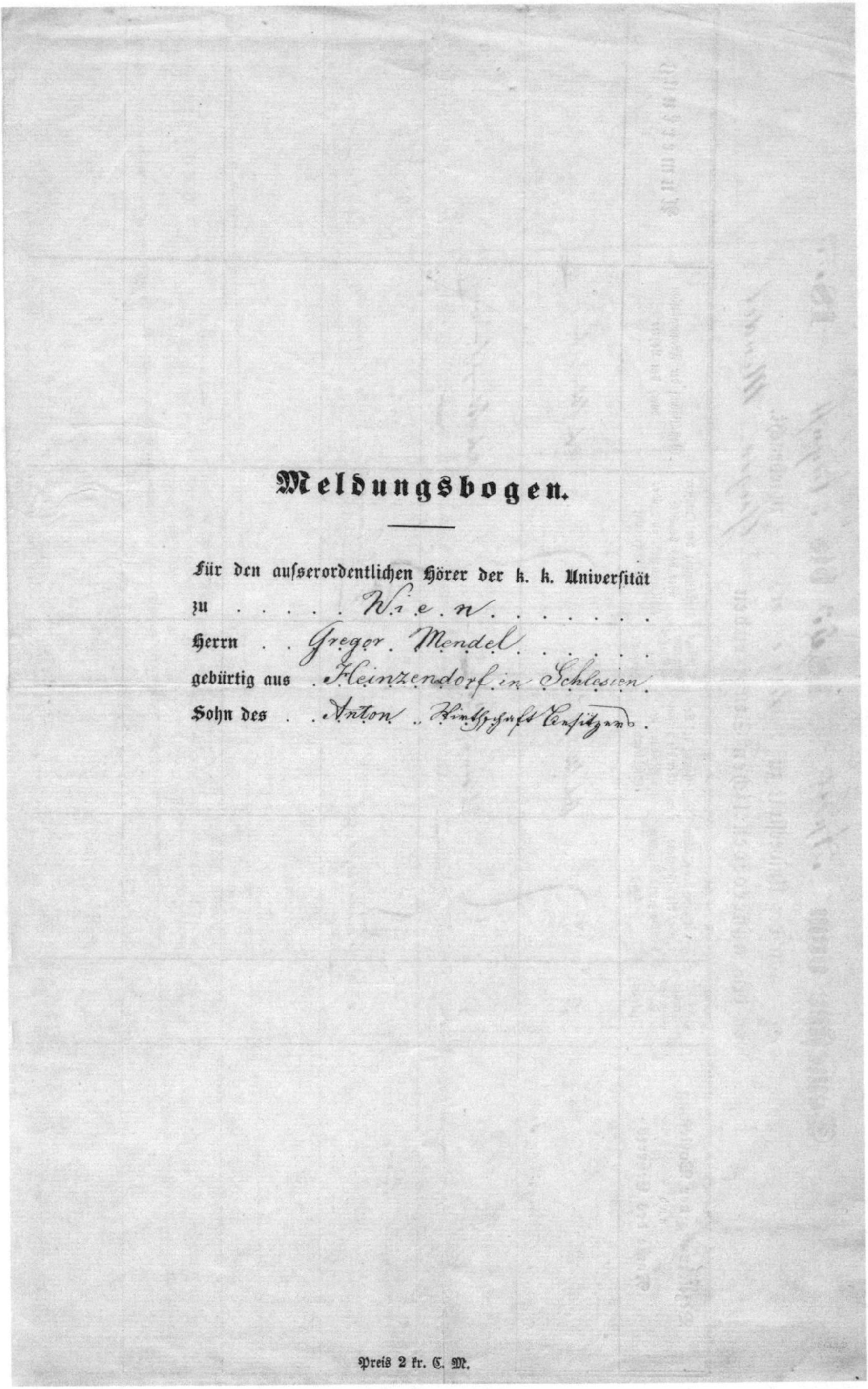

Meldungsbogen.

Für den ausserordentlichen Hörer der k. k. Universität

zu . . . . Wien . . . . . .

Herrn . . Gregor Mendel . . . . .

gebürtig aus . Heinzendorf in Schlesien .

Sohn des . . Anton . . Wirthschaftsbesitzers .

Preis 2 kr. C. M.

D86 Application form (*Meldungsbogen*) for the fourth semester as an extraordinary student at the University of Vienna. On its reverse side, Mendel listed the lectures in which he intended to enrol in the semester from April 1853 to August 1853

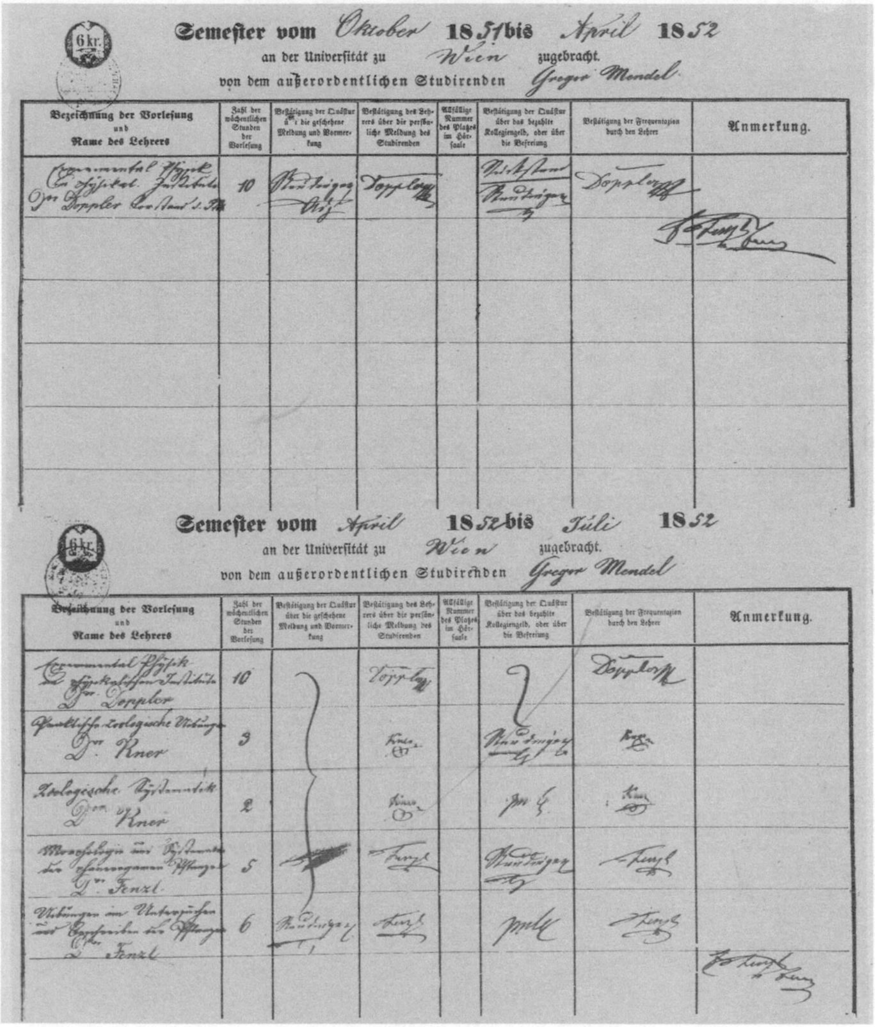

Semester vom Oktober 1851 bis April 1852
an der Universität zu Wien zugebracht.
von dem außerordentlichen Studirenden Gregor Mendel

Bezeichnung der Vorlesung und Name des Lehrers	Zahl der wöchentlichen Stunden der Vorlesung	Bestätigung der Quästur über die geschehene Meldung und Vormerkung	Bestätigung des Lehrers über die persönliche Meldung des Studirenden	Allfällige Nummer des Platzes im Hörsaale	Bestätigung der Quästur über das bezahlte Kollegiengeld, oder über die Befreiung	Bestätigung der Frequentation durch den Lehrer	Anmerkung.
Experimental Physik im physikal. Institute Dr. Doppler	10		Doppler			Doppler	Fenzl

Semester vom April 1852 bis Juli 1852
an der Universität zu Wien zugebracht.
von dem außerordentlichen Studirenden Gregor Mendel

Bezeichnung der Vorlesung und Name des Lehrers	Zahl der wöchentlichen Stunden der Vorlesung	Bestätigung der Quästur über die geschehene Meldung und Vormerkung	Bestätigung des Lehrers über die persönliche Meldung des Studirenden	Allfällige Nummer des Platzes im Hörsaale	Bestätigung der Quästur über das bezahlte Kollegiengeld, oder über die Befreiung	Bestätigung der Frequentation durch den Lehrer	Anmerkung.
Experimental Physik im physikalischen Institute Dr. Doppler	10		Doppler			Doppler	
Praktische zoologische Übungen Dr. Kner	3		Kner			Kner	
Zoologische Systematik Dr. Kner	2		Kner			Kner	
Morphologie und Systematik der phanerogamen Pflanzen Dr. Fenzl	5		Fenzl			Fenzl	
Übungen im Untersuchen und Beschreiben der Pflanzen Dr. Fenzl	6		Fenzl			Fenzl	

D86 On the reverse side of the application form (*Meldungsbogen*) are columns for the titles of the classes, the names of the teachers, the numbers of hours per week, confirmation by the bursar, confirmation by the teacher of the student's enrolment, personal number of the student's seat in the lecture hall, and confirmation from the bursar about fee payment or exemption

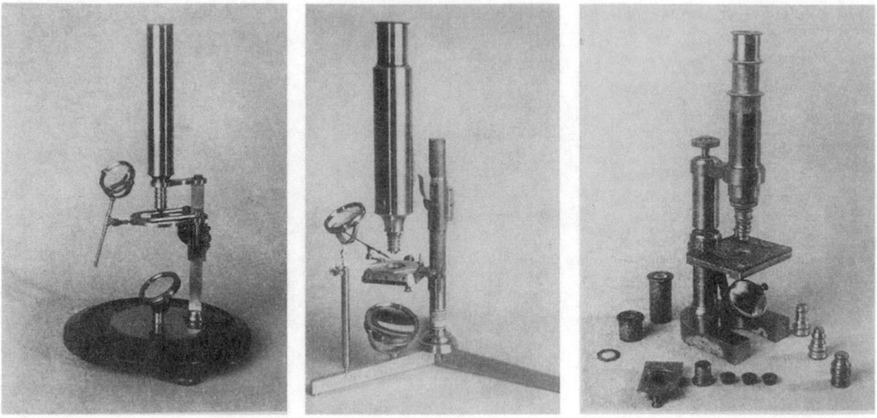

IM58 Mendel's microscopes (IM 1965) p. 35. From left: Simon Plössl, Vienna, type No. 1 produced in the years 1828–1831, Simon Plössl, Vienna, type No. 4 produced about 1863, and Carl Reichert, Vienna, produced in 1877

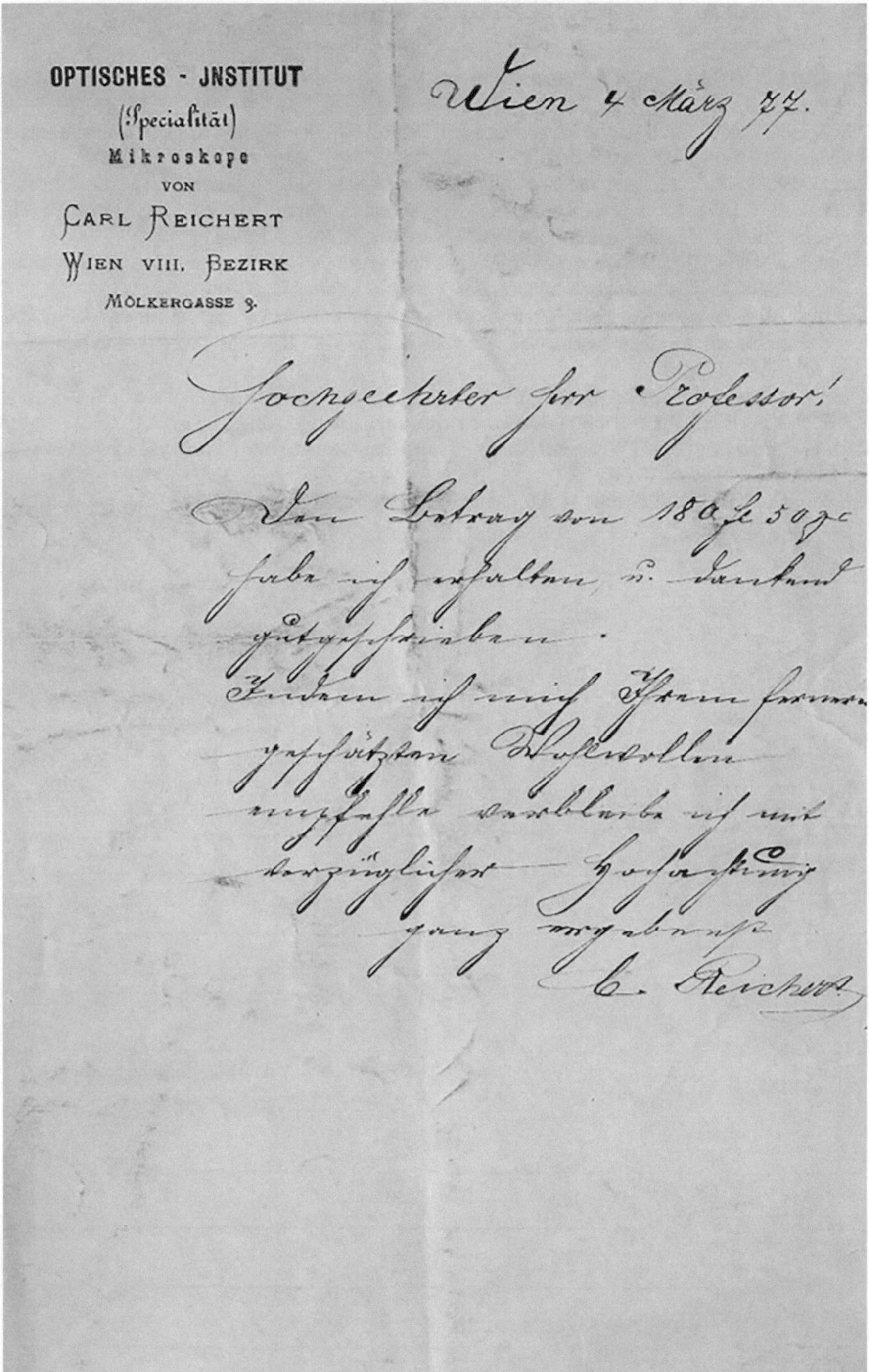

OPTISCHES - JNSTITUT
(Specialität)
Mikroskope
VON
CARL REICHERT
WIEN VIII. BEZIRK
MÖLKERGASSE 3.

Wien 4 März 77.

Hochgeehrter Herr Professor!

Den Betrag von 180 fl 50 xr habe ich erhalten, u. dankend gutgeschrieben.
Indem ich mich Ihrem ferneren geschätzten Wohlwollen empfehle verbleibe ich mit vorzüglicher Hochachtung

ganz ergebenst
C. Reichert

D357 The receipt of payment by the Optical Institute in Vienna. Dated in Wien/Vienna, 04 March 1877, signed by C. Reichert

References and Historical Printed Sources

Gicklhorn R (1969) Gregor Mendels Lehramtsprüfung und Studienzeit in Wien. Allgemeine Biologie der Medizinischen Fakultät der Universität Wien 7(4):145–159

Iconographia Mendeliana (1965) In: Jelínek J (ed) Prepared for publication by Orel V Marvanová L Sajner J. Moravian Museum Brno

Iltis H (1966) Life of Mendel. Translated by Eden and Cedar Paul. Hofner Publ. Co, New York

Mendel G (1853) Über Verwüstung am Gartenrettich durch Raupen Botys margaritalis. Verhandlungen des zoologisch-botanischen Vereines in Wien 3:116–118

Mendel G (1854) Über Bruchus pisi. Mendel's written communication reported by V. Kollar. Verhandlungen des zoologisch-botanischen Vereines in Wien 4:27–28

Mendel G (1867) Meteorologische Beobachtungen aus Mähren und Schlesien für das Jahr 1866. Verhandlungen des naturforschenden Vereines in Brünn für das Jahr 1866. Abhandlungen 5: 160–172

Milovidov T (1968) Gregor Mendel's Microscopic preparations. Folia Mendeliana 3:35–53

Nördlicher N (1852) Querschnitte von hundert Holzarten. Stuttgart and Tübingen

Orel V, Čunderlík I (1985) What was Mendel's intention in preparing microscopic slides? FM 20: 9–14

Procházka L (1985) Mikroskope von Johann Gregor Mendel. FM 20:15–27

Chapter 9
Balancing Between the Profane and the Sacral 1854–1855

As a consequence of the 1848 upheaval, the Augustinian Orders in the Austrian monarchy were tested on discipline. The Augustinians in Mendel's day belonged to an old order fundamentally influenced by the reforms of the Austrian monarch Joseph the Second. In 1783, their relatively new monastery building complex of St. Thomas, located inside the city walls, was appropriated to serve as Governor's office. The Augustinians were forced to relocate to the Brno suburbs (Old Brno) in a small, abandoned and dilapidated, Cistercian nunnery. Their religious life became limited, and their monastic discipline shattered. They had to focus predominantly on construction and restoration of their new home and on service to the state in education. Otherwise, they ran the risk of being abolished by the emperor. From 1808, on monarch's demand the Augustinians had supplied professors for the Philosophical and Theological Institutes, gymnasia and state schools. From the viewpoint of the church, the revolutionary year of 1848 caused much damage in the discipline in monasteries.

In the 1854–1855 apostolic visitation report, the Bishop of Brno concluded that the Old Brno Augustinians did not pursue lives of recluse according to the rules of St. Augustine. He determined that among 14 Augustinians six were teachers at state secondary schools or institutes of higher learning who achieved their qualifications at the expense of the Augustinian order. However, their intellectual lives were mostly invested outside the monastery. The bishop found that four Augustinians were living far from their community, and the practice of monastic religious virtue had suffered a loss of its original discipline.

After the revolutionary year of 1848, the church became exhaustively invested in defending Catholic faith against the threats of pantheism, liberalism, materialism and atheism. The Old Brno Augustinian friar, Matthaeus Klácel, a professor who had been dismissed from the Philosophical Institute in Brno in 1844, was known as a pantheist and free thinker. Bishop Schaaffgotsche had evidence that in 1848 Klácel was the spiritual father of the petition for freedoms for monk teachers. The Augustinian friar, Thomas Bratránek, one of Mendel's counter-candidates in the abbatial election in 1868, specialized in German *Naturphilosophie*. The church was

A. Matalová, E. Matalová, *Gregor Mendel - The Scientist*, Springer Biographies,
https://doi.org/10.1007/978-3-030-98923-1_9

attempting to restore old practices in the monasteries and, therefore, the pope decreed visitations by designated church officials to Austrian monasteries. Cardinal Schwarzenberg of Prague appointed Bishop Schaaffgotsche of Brno to carry out the visitation in the Old Brno Augustinian monastery. The bishop found the community unreformable, and proposed its abolition. At the same time, he proposed inviting another, well-disciplined order to take its place. Napp refuted the bishop's recommendation. He asked Cardinal Schwarzenberg to make a change in the statutes of the Augustinian order to allow its members more freedom to pursue the real character of their community.

The bishop's final report to Cardinal Schwarzenberg in Prague is dated on 07 September 1854. In it, the bishop accuses the Augustinians of being profane priests of the Austrian monarch rather than Roman Catholic monks. He insists that the Augustinian order is unreformable and should be dissolved. None of the Augustinians whom the bishop interrogated personally was willing to promise to change his way of life. Therefore, the abolition of the Old Brno Augustinians seemed to him as inevitable.

In the visitation report of 07 September 1854 by Bishop Schaaffgotsche of Brno to Cardinal Schwarzenberg of Prague (Czihak and Sladek, Wieseneder 1991/1992), Mendel is characterised as a profane priest: *I must mention Gregor Mendel who studies profane sciences in a worldly institute in Vienna at the expense of the monastery in order to work in a state institute as a professor; now he has been teaching physics and natural history in the so-called Realschule in Brno in a position of a substitute teacher.* The bishop was correct. Mendel studied physics and natural history at the University of Vienna at the expense of the monastery. He worked as a substitute teacher at the state *Realschule* in Brno, which was a modern technical high school intended for developers and industrialists.

Polarization between the church and the state, the holy orders and worldly priests, the sacred and the profane in the Austrian monarchy resulted in a revision of the life for Augustinian communities, decreed by Pope Pius IX and implemented by the bishops. The diplomatic Napp and his community of friars balanced their activities between monastic life and school and university activities, sometimes in remote places. Schaaffgotsche criticized not only Mendel for not being a real Augustinian. The Augustinian Professor Franz Xaver Wiesner had been active at the University in Olomouc. He was accused of not following the rules of monastic life in recluse. Mathematician Antonín František Alt was criticized for specializing in profane studies and teaching at the *Gymnasium* in Bratislava. Professor Phillipp Vincenz Gabriel was criticized for being director of the *Gymnasium* in Těšín in Silesia, all living far from the monastic community and neglecting the daily observance of the Augustinian principles. Professor Bratránek was accused of teaching worldly sciences at the Jagiellonian University in Krakow. In the official visitation report, Schaaffgotsche was of the opinion that the above-mentioned Augustinians would prefer to leave the order than to return to the monastery. Bishop Schaaffgotsche further informed Cardinal Schwarzenberg that Klácel was forced to resign from his position as a teacher at the Brno Philosophical Institute in 1844 for spreading pantheism in his lectures. He criticized him for taking active part on the barricades

O07 The Old Brno Augustinian Hermits St. Thomas. Photo by an unknown photographer, about 1863

in the 1848 revolution in Prague together with the mathematician Bernard Bolzano. On the last day of visitation, Abbot Napp wrote to Cardinal Schwarzenberg asking for a change in the statutes of the Augustinian Hermits to designate them as Regular Canons. On that occasion Napp noted that the Old Brno Augustinian monastery of St. Thomas had been exempt, i.e. not subordinate to the bishop. No other Augustinian community in the monarchy enjoyed such a privilege. The Apostolic Visitation had a zero effect on the life of the Augustinian community in Old Brno headed by Abbot Napp. No changes were realized.

O07: Standing from left: Benedict Fogler, Pavel Křížkovský, Tomáš Bratránek, Joseph Lindenthal, Gregor Mendel, Anselm Rambousek, Antonín Alt, Matouš Klácel. Seated from left: Baptist Vorthey, Cyrill Napp, Václav Šembera.

O36: Seated from left: Pavel Křížkovský, Baptist Vorthey, Cyrill Napp, Matouš Klácel. Standing from left: Benedict Fogler, Anselm Rambousek, Antonín Alt, Tomáš Bratránek, Joseph Lindenthal, Gregor Mendel, Václav Šembera.

O36 Members of the Old Brno Augustinian monastery (IM26). Photo by an unknown photographer, about 1863

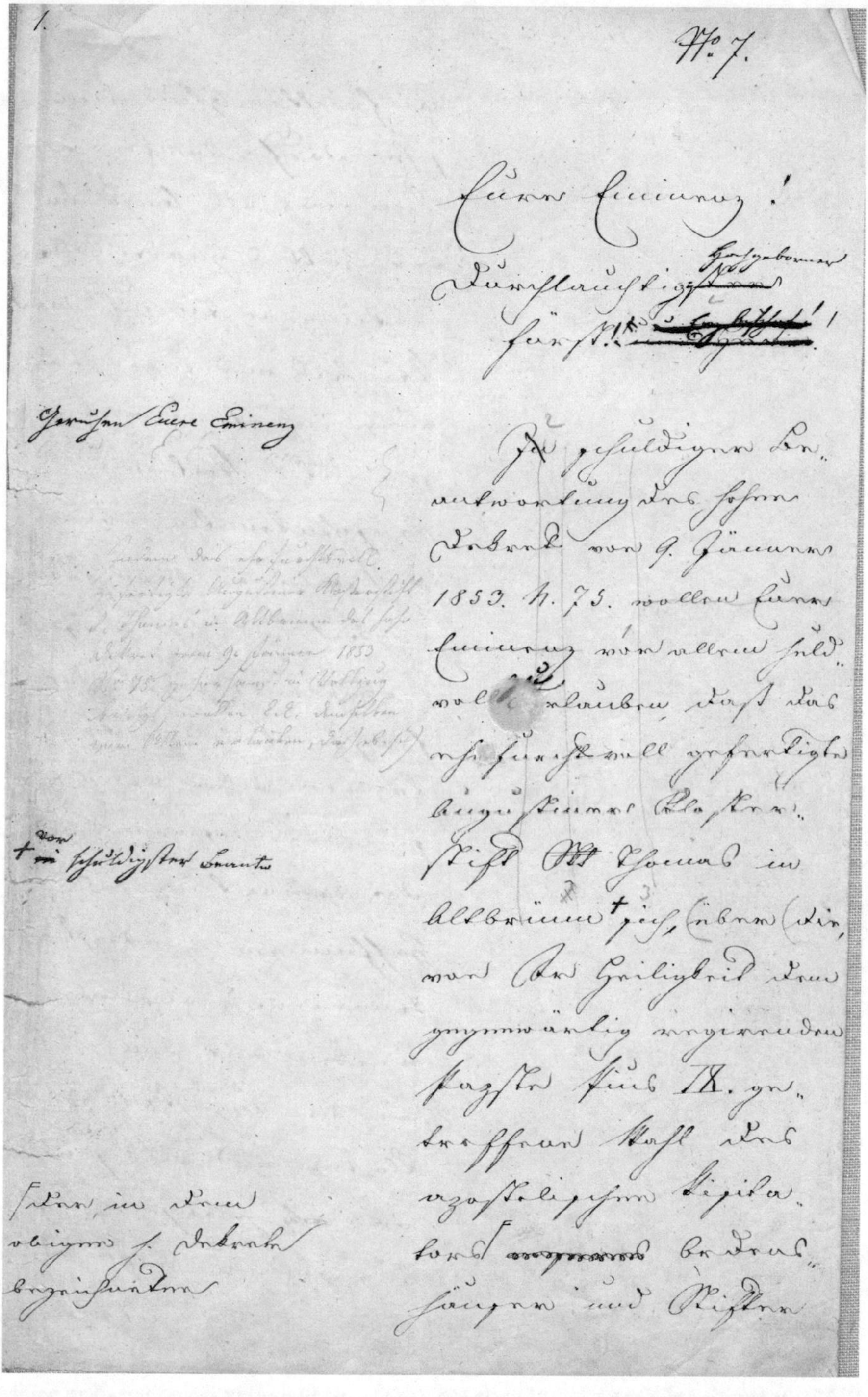

D972 The first page of Napp's draft of the defence of the Old Brno Augustinian monastery written for Bishop Schaaffgotsche of Brno

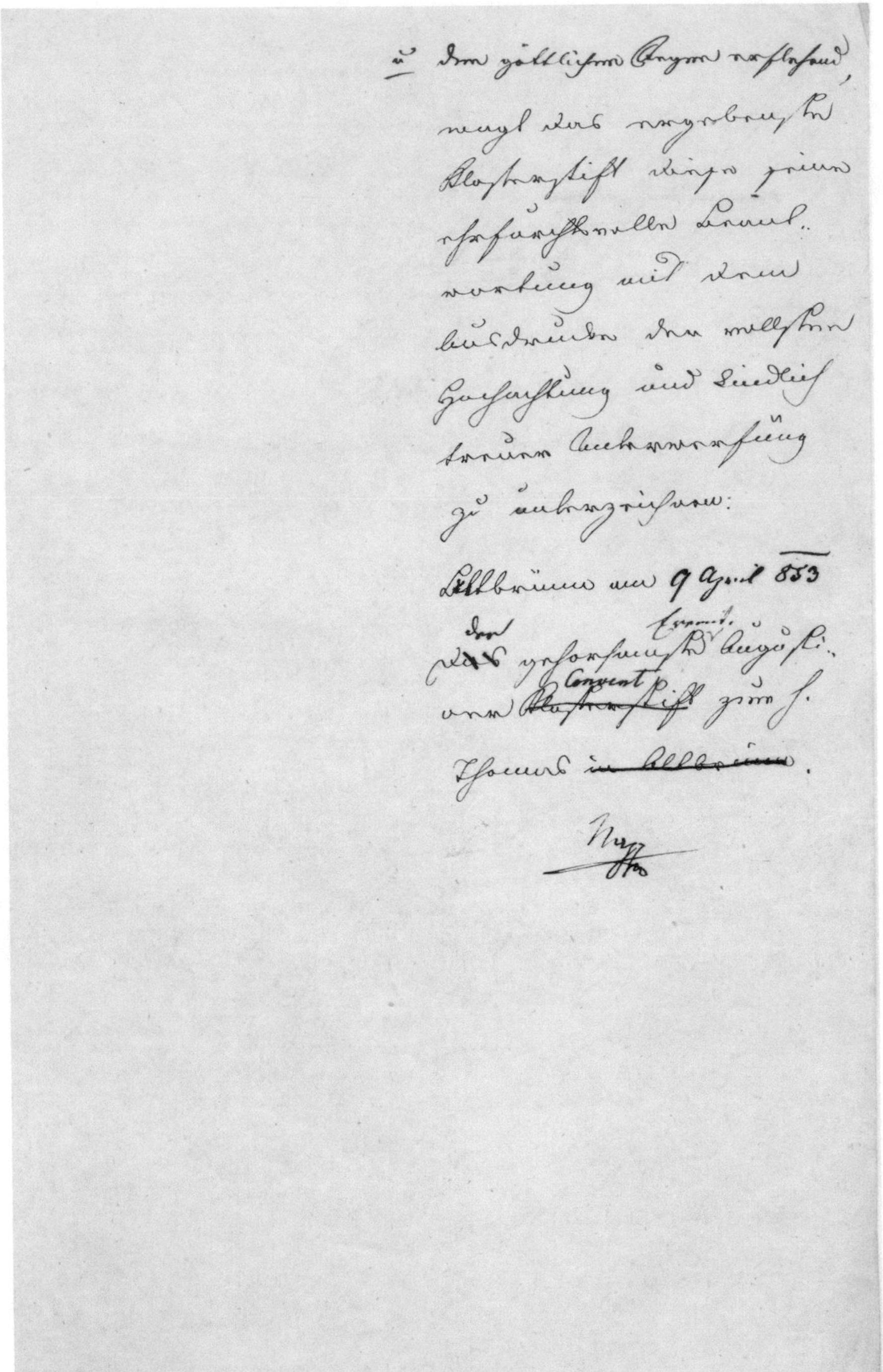

u. dem göttlichen Segen empfehlend,
möge das ergebenste
Klosterstift diese seine
ehrfurchtsvolle Beant-
wortung mit dem
Ausdrucke der vollsten
Hochachtung und kindlich
treuen Unterwerfung
zu unterbreiten:

Altbrünn am 9 April 853

der ~~das~~ gehorsamste August.
vom Convent ~~Klosterstift~~ zum h.
Thomas ~~in Altbrünn~~.

Napp

D972 The last page of Napp's draft of the defence of the Old Brno Augustinian monastery written for the Brno Bishop Schaaffgotsche. Dated in Altbrünn/Staré Brno, 09 April 1853, signed by Napp

References and Historical Printed Sources

Czihak G, Sladek P (1991/2) Die Persönlichkeit des Abtes Cyrill Franz Napp (1792-1862) und die innere Situation des Klosters zu Beginn der Versuche Gregor Mendels. FM 26(27):31–66

Chapter 10
Substitute Teacher at the *Realschule* 1854–1868

On 26 May 1854, Mendel began teaching as a substitute professor of physics and natural history in the lower grades of the *Realschule*. The letter written by the Director of the *Realschule*, Joseph Auspitz (1812–1889), informs his superior office regarding Mendel's employment as a substitute *professor in the vacancy left by Mr. Patek. He provides a testimonial of Mendel's personal and pedagogic qualities as a natural scientist.* The letter, dated 26 May 1854, was published in facsimile and transcription by Gicklhorn (1969, pp. 152-3, 157-8). The original is physically held in the archives of the University of Illinois in Urbana-Champaign.

The same year (1854), Mendel initiated the two-year trial with *Pisum* to test the varieties he was considering as parents in his crosses for constancy of type. In 1856, he made another attempt to complete the teacher's examination. According to Gicklhorn (1969), Mendel's application for the second attempt at taking the examination is entered in the Protocol Book Register of the Scientific *Gymnasium* Examination Commission under No. 209 on 03 July 1856, listing the date of the on-site examination in Vienna as 05 August 1856. The facsimile of the entry with the date of the examination is published by Gicklhorn (1969, pp. 154-5). No other primary source has been discovered pertaining to Mendel's second attempt at taking the examination. In the account book of the Augustinian monastery kept in the Brno State Archives there is a note on 09 April 1855 and 02 May 1856 that Mendel received 20 florins for each of his journeys to Vienna for his teachers´ closing examination (Richter 1943, p. 26). On 31 May 1856, a visit of Mendel's father and his brother was registered by the accountant (Richter 1943, p. 26). Their visit is given in relation to Mendel's ailment after the unsuccessful examination.

Křiženecký found an excerpt from the Czech letter by Klácel from Brno addressed to Bratránek in Krakow of 08 May 1856, copied by Anselm Matoušek, describing Mendel's nervous breakdown, preventing him from even starting the first part of the *Klausurprüfung* examination (written on-site) in Vienna. Křiženecký (1963a, b, p. 308) published Matoušek's passage from Klácel's Czech letter to Bratranek in German translation. Klein and Klein (2013, pp. 363-4) made the English translation of the passage from Klácel's letter pertaining to Mendel and

A. Matalová, E. Matalová, *Gregor Mendel - The Scientist*, Springer Biographies,
https://doi.org/10.1007/978-3-030-98923-1_10

his second attempt to take the examination: *P. Gregor was called to an examination in Vienna. He left and there was no chance that he would be back for the holidays, and since there was a shortage [of priests] for services, I had to stay home. P. Gregor was unlucky. Although he drew easy questions, he fell ill during the first Klausurprüfung and as a consequence was unable to write. He seems to have problems with his nerves generally since he endured several such insidious attacks already and they say that in his youth he suffered from epilepsy. The day passed and nothing was achieved. One has to feel sorry for him, since his homework* etc. *was graded as excellent. But formalities are formalities; in this case it was not possible to continue [with the exam]. Afraid that further attacks might follow, he returned home without accomplishing anything. I am very sorry for him, especially since he is anyhow discontented and so will eat himself even more.*

No primary sources have been discovered regarding Mendel's second attempt at taking the teacher certification examination at the University of Vienna. Mendel never attained the status of an ordinary *Gymnasium* professor.

In the Protocol (Register Book, Einreichungsprotokoll) of the Scientific *Gymnasium* Examination Commission of the University in Vienna from the years 1849–1859 Mendel's application is entered twice: on 17 April 1850 (Gicklhorn 1969, p.146) and 03 July 1856 Gicklhorn 1969, pp. 154–5). The entry of 03 July 1856 gives the date of the examination on 05 August.

A letter by Joseph Auspitz dated 23 June 1855 stamped by the Directorate of the Realschule in Brünn/Brno provides evidence of how Mendel was valued as an excellent teacher. The testimonial letter was written in connection with Mendel's repeated application to the Scientific *Gymnasium* Examination Commission in Vienna (Iltis 1924, pp. 59–60): *Von der gefertigten Direction wird hiemit bestättiget, dass der ehrwürdige Herr Gregor Mendel von der hiesigen k: k: Oberrealschule seit 26ten Mai 1854 als supplirender Lehrer für die Lehrfäche als Physik und Naturgeschichte in Verwendung steht. Während dieser Zeit war sein Benehmen in jeder Beziehung ein ausgezeichnetes. Von wahrer inniger Liebe zu Jugend erfüllt, weiss er durch einfache Mittel die Disziplin aufrechtzuerhalten; sein Vortrag ist deulich, logisch, seine Stimme genügend stark um dem Fassungsvermögen der Jugend vollkommen anpassend; Herr Mendel experimentirt sehr geschickt, und weiss mit geringem Mittel viel zu zeigen, zu demonstriren sowohl in der Physik als in der Naturgeschichte.—Die Direction der k: k: Oberrealschule—Brünn den 23 Juni 1855—Joseph Auspitz.*

The subjects of the *Realschule* were first taught in hired rooms in private buildings in Brno. The new building in which the *Realschule* would eventually be housed was initiated in 1858, with Auspitz as the *Realschule* director. The Viennese architect, Ludwig Foerster, designed a palace with a tower intended to house an astronomical observatory. In October 1859, the new school building was opened. On the occasion of the opening ceremony, an altar had been constructed with the dominating cross. A special table of freemasons´ luxury hammer and pinnacle was arranged. Joseph Auspitz was Director of the *Realschule* from the very beginning until 1869, when he advanced to the chief inspector of Moravian schools. In the position of inspector, Auspitz initiated the founding of further *Realschulen.* The

Brno *Realschule* became a model institution of education in the country. Mendel taught at the Brno *Realschule* for 14 years, both in its provisional conditions and its new building.

The governing administration and the director of the *Oberrealschule* acknowledged Mendel's excellent teaching several times for encouraging his pupils to pursue science: *Nach Erlass der hohen Statthalterei dto 13ten März 1858, Z. 6748, dient der Inhalt des Berichtes der gefertigten Direktion über den Zustand der Realschule im Schuljahre 1856/57 und insbesondere die auch durch anderweitige Wahrnehmungen bestätigte Versicherung der Direktion über den immer mehr zufrieden stellenden Stand des Unterrichtes und der Disziplin sowie über die fortschreitenden Bemühungen, die humanistischen Gegenstände mit den in einer erfreulichen Entwicklung begriffenen mathematischen, naturwissenschaftlichen und technischen Fächern immer mehr ins Gleichgewicht zu setzen, zur angenehmen Kenntnis.—Überdies wurde die gefertigte Direktion ermächtigt, Euer Hochwürden die volle Zufriedenheit mit Ihrem eifrigen und erfolgreichen Bemühungen im Namen der hohen k. k. Statthalterei bekanntzugeben.—Indem die gefertigte Direktion mit der Bekanntgebung dieser hohen Anerkennung einer angenehmen Pflicht nachkömmt, kann sie nicht umhin, auch in ihrem Namen Euer Hochwürden den Dank auszudrücken für den Eifer, mit dem Sie die Bemühungen der Direktion unterstützten, für die Liebe zur Wissenschaft und zur Jugend, die Sie bei jeder Gelegenheit an den Tag legten.—Die Direktion der k. k. Oberrealschule.—Brünn, den 18. März 1858.—Joseph A. Auspitz—Von der Direktion der k.k. Oberrealschule—An den hochwürdigen Herrn Gregor Mendl, Lehrer an der k. k. Oberrealschule in Brünn.*

Auspitz's letter to Mendel on 18 March 1858 is translated as follows by Abbott and Fairbanks (2016): *In the judgement of the governing Administration dated 13th March, 1858, line 6748, the content of the report by the authorized Directorate regarding the status of the Realschule in the school year 1856/57 and in particular the Directorate's assurance, confirmed by other observations, concerning the increasingly satisfactory status of instruction and, as well as the ongoing efforts to balance humanistic subjects with the mathematical, scientific and technical subjects currently undergoing a gratifying development, serves as welcome information.- Moreover, the authorized Directorate was given permission to disclose to your Reverend Father, in the name of the governing Administration, our complete satisfaction with your zealous and successful endeavours.- As the authorized Directorate is fulfilling its amiable duty in announcing this high recognition, it cannot omit expressing the gratitude to your Reverend Father, again in our name, for the zeal with which you support the efforts of the Directorate, for the love for science and for youth that you have exhibited at every opportunity.- The Directorate of the Imperial and Royal Oberrealschule—Brünn, March 18, 1858—Joseph A. Auspitz.—From the Directorate of the Imperial and Royal Oberrealschule—To the honorable Herr Gregor Mendl, teacher at the Imperial and Royal Oberrealschule in Brünn.*

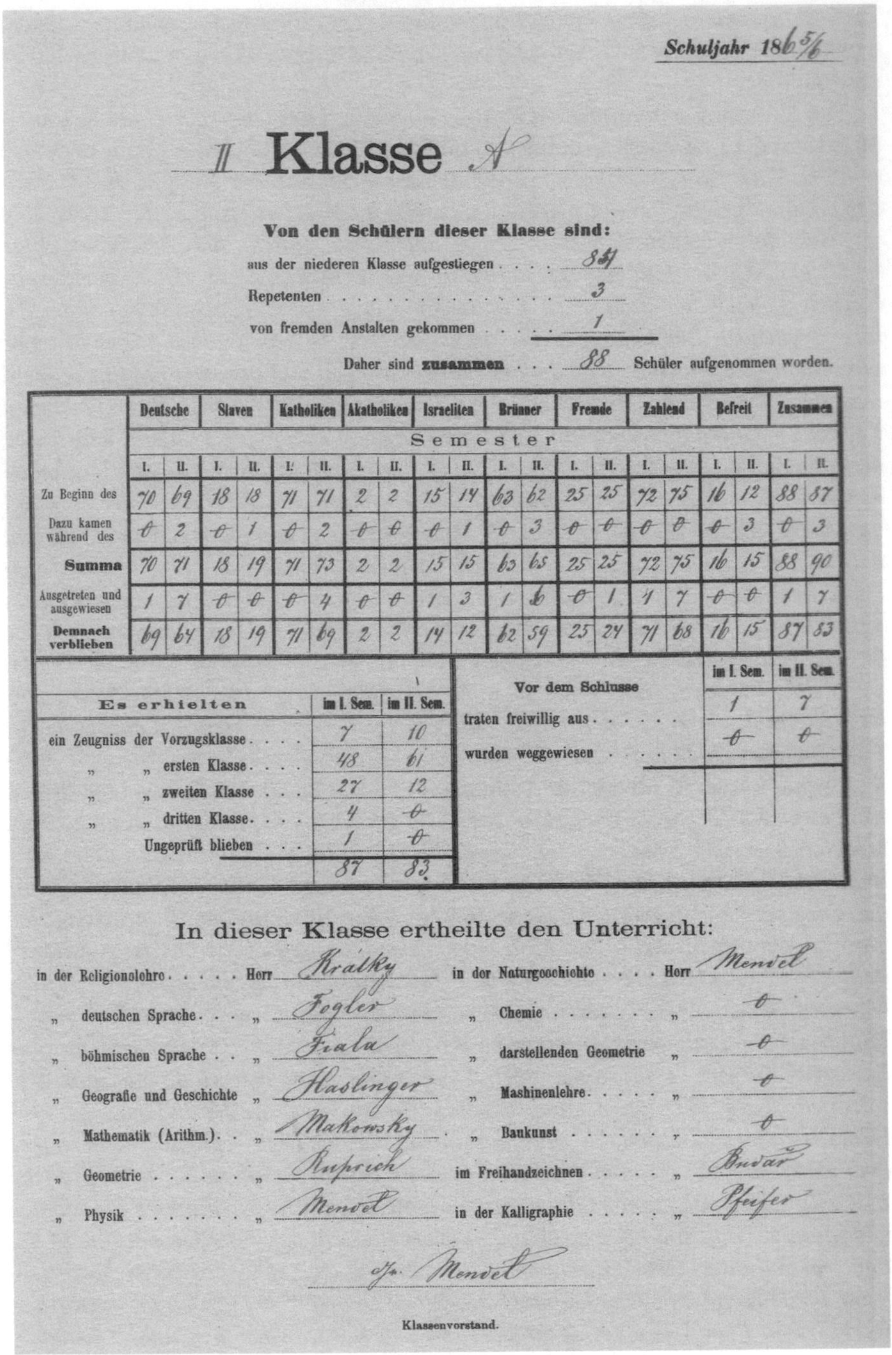

Schuljahr 1865/6

II Klasse A

Von den Schülern dieser Klasse sind:

aus der niederen Klasse aufgestiegen 84

Repetenten 3

von fremden Anstalten gekommen 1

Daher sind **zusammen** . . . 88 Schüler aufgenommen worden.

	Deutsche		Slaven		Katholiken		Akatholiken		Israeliten		Brünner		Fremde		Zahlend		Befreit		Zusammen	
	Semester																			
	I.	II.	I.	II.	I.	II.	I.	II.	I.	II.	I.	II.	I.	II.	I.	II.	I.	II.	I.	II.
Zu Beginn des	70	69	18	18	71	71	2	2	15	14	63	62	25	25	72	75	16	12	88	87
Dazu kamen während des	0	2	0	1	0	2	0	0	0	1	0	3	0	0	0	0	0	3	0	3
Summa	70	71	18	19	71	73	2	2	15	15	63	65	25	25	72	75	16	15	88	90
Ausgetreten und ausgewiesen	1	7	0	0	0	4	0	0	1	3	1	6	0	1	1	7	0	0	1	7
Demnach verblieben	69	64	18	19	71	69	2	2	14	12	62	59	25	24	71	68	16	15	87	83

Es erhielten	im I. Sem.	im II. Sem.
ein Zeugniss der Vorzugsklasse	7	10
„ „ ersten Klasse	48	61
„ „ zweiten Klasse	27	12
„ „ dritten Klasse	4	0
Ungeprüft blieben	1	0
	87	83

Vor dem Schlusse	im I. Sem.	im II. Sem.
traten freiwillig aus	1	7
wurden weggewiesen	0	0

In dieser Klasse ertheilte den Unterricht:

in der Religionslehre	Herr Krátky	in der Naturgeschichte	Herr Mendel
„ deutschen Sprache	„ Fogler	„ Chemie	„ 0
„ böhmischen Sprache	„ Fiala	„ darstellenden Geometrie	„ 0
„ Geografie und Geschichte	„ Haslinger	„ Mashinenlehre	„ 0
„ Mathematik (Arithm.)	„ Makowsky	„ Baukunst	„ 0
„ Geometrie	„ Ruprich	im Freihandzeichnen	„ Bradar
„ Physik	„ Mendel	in der Kalligraphie	„ Pfeifer

Gr. Mendel

Klassenvorstand.

IM47 Class register of the *Realchule*, second form A, school-year 1865/6, filled in by Mendel as the form master

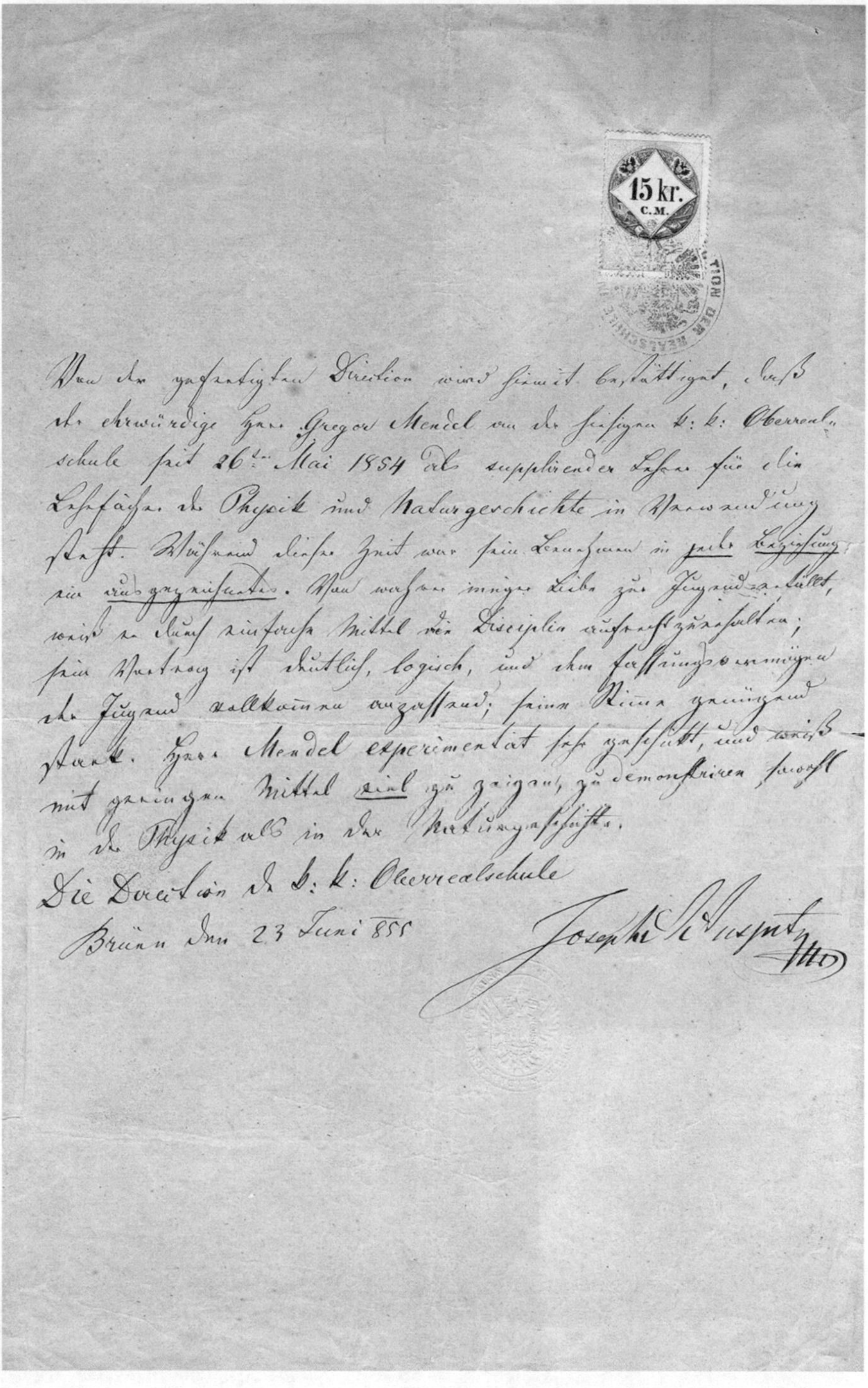

15 kr. C.M.

Von der gefertigten Direction wird hiemit bestättiget, daß der ehrwürdige Herr Gregor Mendel an der hiesigen k. k. Oberrealschule seit 26ten Mai 1854 als supplierender Lehrer für die Lehrfächer der Physik und Naturgeschichte in Verwendung steht. Während dieser Zeit war sein Benehmen in jeder Beziehung ein ausgezeichnetes. Mit wahrer inniger Liebe zur Jugend erfüllt, weiß er durch einfache Mittel die Disciplin aufrecht zu erhalten; sein Vortrag ist deutlich, logisch, und dem Fassungsvermögen der Jugend vollkommen angepasst; [illegible]. Herr Mendel experimentirt sehr geschickt, und weiß mit geringen Mittel viel zu zeigen, zu demonstriren sowohl in der Physik als in der Naturgeschichte.

Die Direction der k. k. Oberrealschule

Brünn den 23 Juni 855

Joseph A. Auspitz

D25 Auspitz's testimonial of Gregor Mendel written in support of his application for an attempt at accreditation as a *Gymnasium* teacher of physics for all grades and from natural history for lower grades only. Dated in Brünn, 23 June 1855, signed by Joseph A. Auspitz, stamped "Direction der Realschule in Brünn"

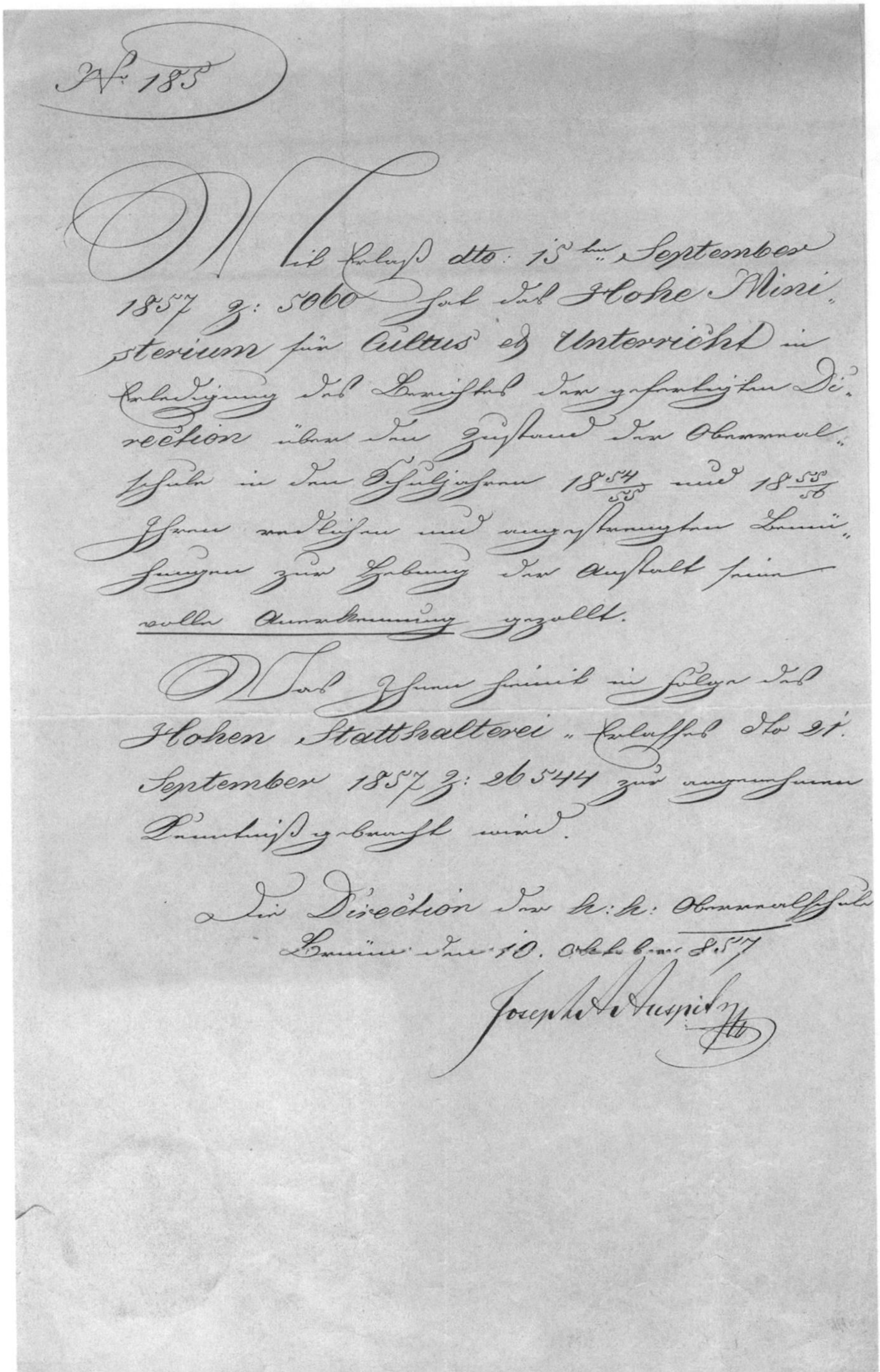

No. 185

Mit Erlaß dto: 15ten September 1857 Z: 5060 hat das Hohe Ministerium für Cultus u. Unterricht in Erledigung des Berichtes der gefertigten Direction über den Zustand der Oberrealschule in den Schuljahren 1854/55 und 1855/56 Ihren redlichen und angestrengten Bemühungen zur Hebung der Anstalt seine volle Anerkennung gezollt.

Was Ihnen hiemit in Folge des Hohen Statthalterei-Erlasses dto 21. September 1857 Z: 26544 zur angenehmen Kenntniß gebracht wird.

Die Direction der k: k: Oberrealschule
Brünn den 10. October 857
Joseph Auspitz

D78 A letter of acknowledgement by Auspitz to Mendel for his endeavour to raise the level of the *Realschule*, recognized in the report of the *Realschule* for the school years 1854/55 and 1855/56 by the Ministry of Worship and Education and the Governor's Office. Reference number 185. Dated in

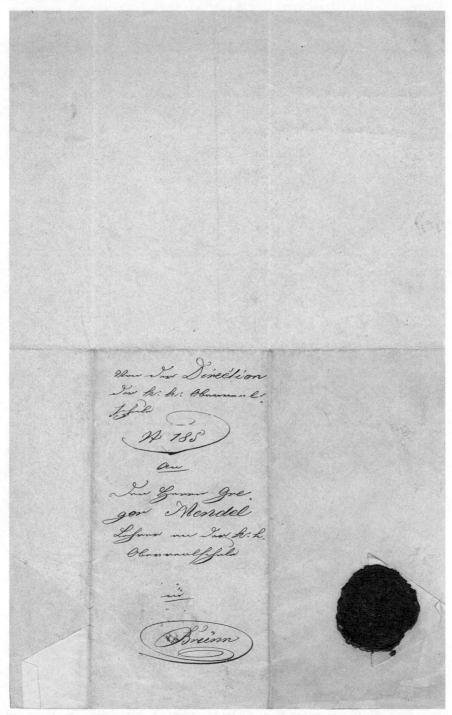

(continued) Brünn/Brno, 10 October 1857, signed by Joseph A. Auspitz. The envelope reads: *From the Directorate of the Imperial-Royal Oberrealschule No. 185 to Mr. Gregor Mendel—teacher at the Imperial-Royal Oberrealschule in Brünn/Brno. Sealed*

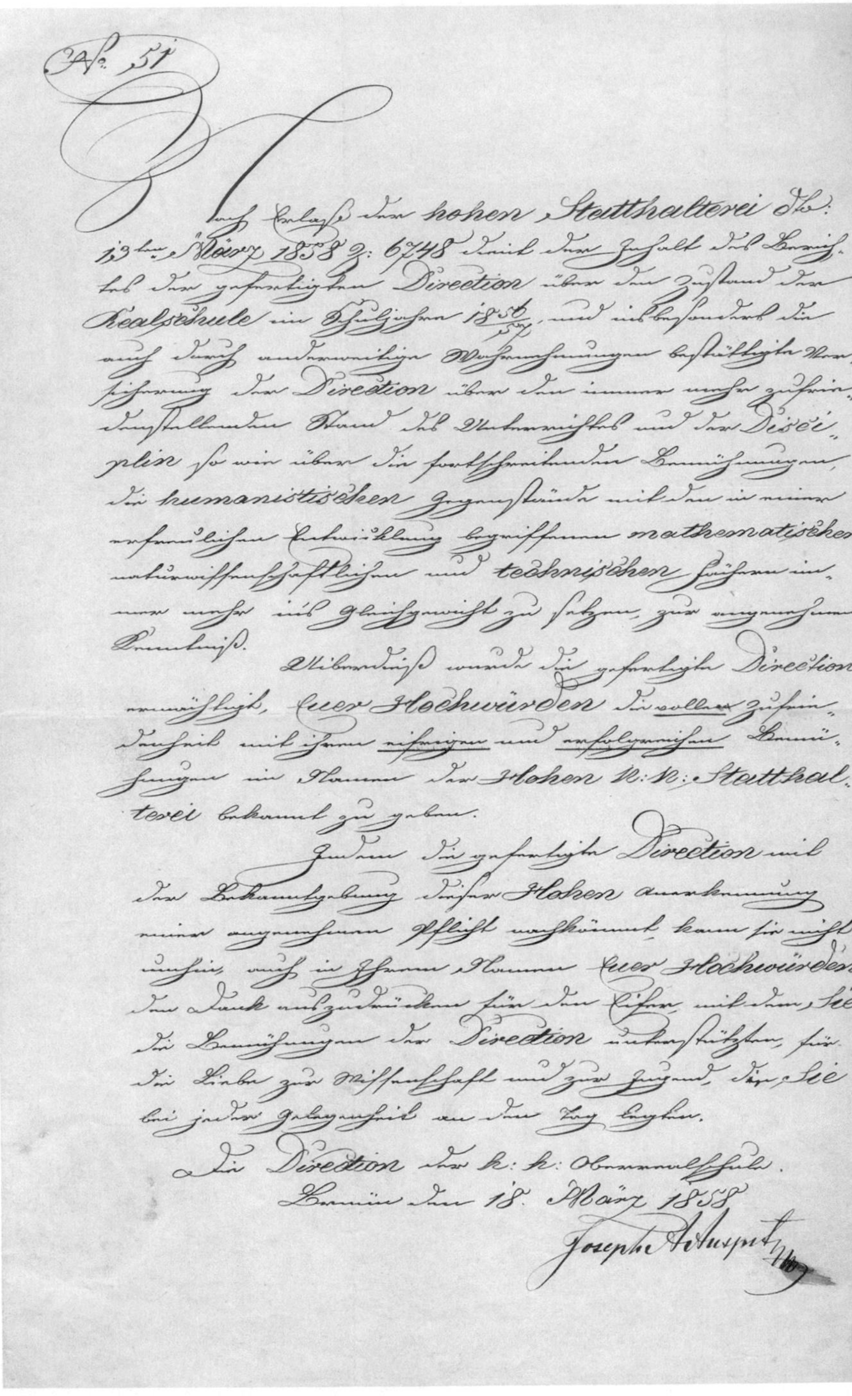

№ 51

Laut Erlaße der hohen Statthalterei dto: 13ten März 1858 Z: 6748 wird der Inhalt des Berichtes der gefertigten Direction über den Zustand der Realschule im Schuljahre 1856/57, und insbesondere die auch durch anderweitige Wahrnehmungen bestättigte Versicherung der Direction über den immer mehr zufriedenstellenden Stand des Unterrichtes und der Disciplin so wie über die fortschreitenden Bemühungen, die humanistischen Gegenstände mit den in einer erfreulichen Entwicklung begriffenen mathematischen naturwissenschaftlichen und technischen Fächern immer mehr ins Gleichgewicht zu setzen, zur angenehmen Kenntniß.

Überdieß wurde die gefertigte Direction ermächtiget, Euer Hochwürden die volle Zufriedenheit mit Ihren eifrigen und erfolgreichen Bemühungen im Namen der Hohen k: k: Statthalterei bekannt zu geben.

Indem die gefertigte Direction mit der Bekanntgebung dieser Hohen Anerkennung einer angenehmen Pflicht nachkömmt, kann sie nicht umhin, auch in ihrem Namen Euer Hochwürden den Dank auszudrücken für den Eifer mit dem Sie die Bemühungen der Direction unterstützten, für die Liebe zur Wissenschaft und zur Jugend, die Sie bei jeder Gelegenheit an den Tag legten.

Die Direction der k: k: Oberrealschule.

Brünn den 18. März 1858

Joseph A. Auspitz m/p

D76 A letter of acknowledgement by Auspitz to Mendel for his contribution to the balance of mathematical, natural scientific and technical subjects with the humanities. Reference number 51. Dated in Brünn/Brno, 18 March 1858, signed by Joseph A. Auspitz. The envelope is addressed *Von*

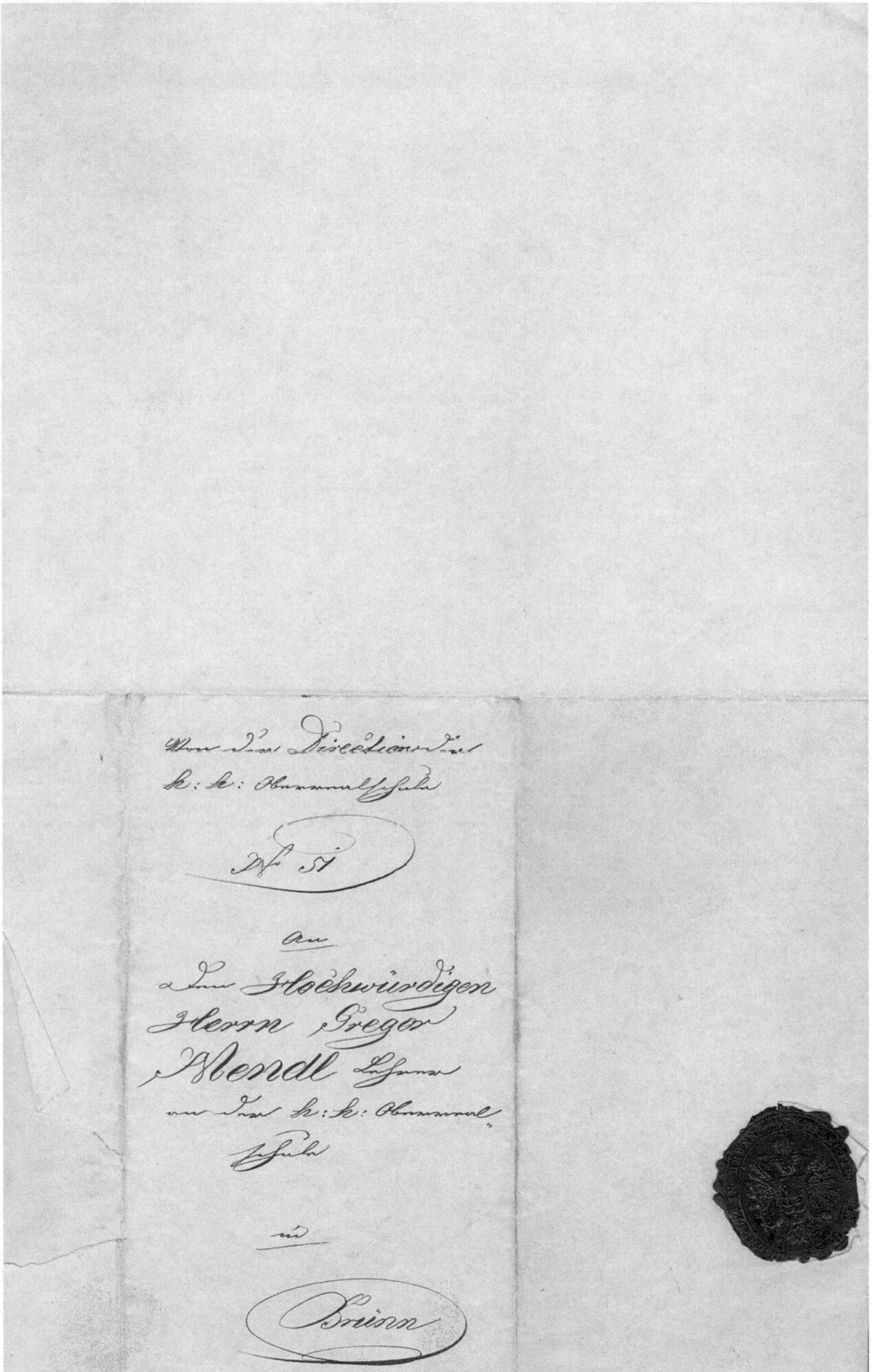

(continued) der Direction k: k: Oberrealschule an den Hochwurdigen Herrn Gregor Mendl Lehrer an der k: k: Oberrealschule in Brünn. Sealed

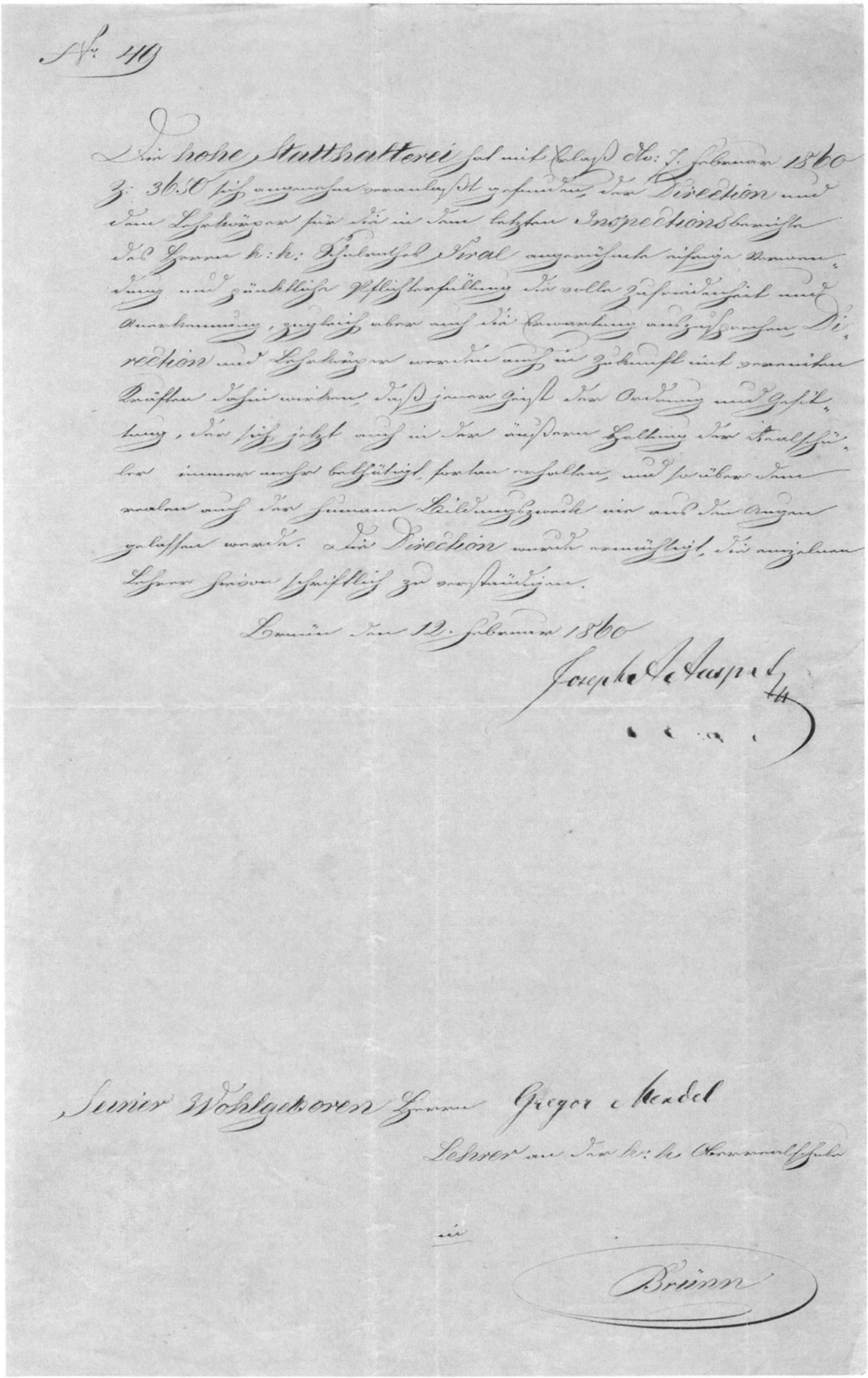

No. 49

Die hohe Statthalterei hat mit Erlaß vom 7. Februar 1860 Z: 3650 sich angenehm veranlaßt gefunden, der Direction und dem Lehrkörper für die in dem letzten Inspectionsberichte des Herrn k: k: Schulrathes Kral angerühmte eifrige Verwendung und pünktliche Pflichterfüllung die volle Zufriedenheit und Anerkennung, zugleich aber auch die Erwartung auszusprechen, Direction und Lehrkörper werden auch in Zukunft mit vereinten Kräften dahin wirken, daß jener Geist der Ordnung und Gesittung, der sich jetzt auch in der äußeren Haltung der Realschüler immer mehr bethätigt, forthin erhalten, und hierüber dennoch auch der höhere Bildungszweck nie aus den Augen gelassen werde. Die Direction macht es sich zur Pflicht, die einzelnen Lehrer hievon schriftlich zu verständigen.

Brünn den 12. Februar 1860

Joseph A. Auspitz

Seiner Wohlgeboren Herrn Gregor Mendel

Lehrer an der k: k: Oberrealschule

in

Brünn

D77 A letter of acknowledgement by Auspitz to Mendel for his pedagogic and teaching activities recognized in the Report on the state of the *Realschule* elaborated for the Governor's office. Reference number 49. Dated in Brünn/Brno, 12 February 1860, signed by Joseph A. Auspitz.

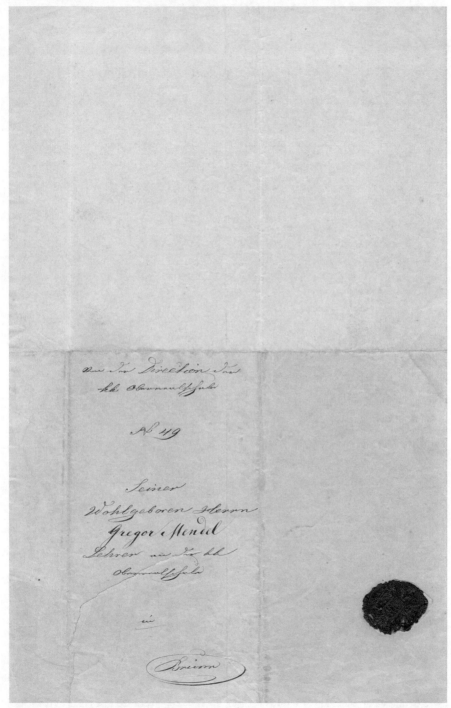

(continued) The envelope is addressed *Von der Direction der kk Oberrealschule. Seiner Wohlgeboren Herrn Gregor Mendel Lehrer an der kk Oberrealschule in Brünn. Sealed*

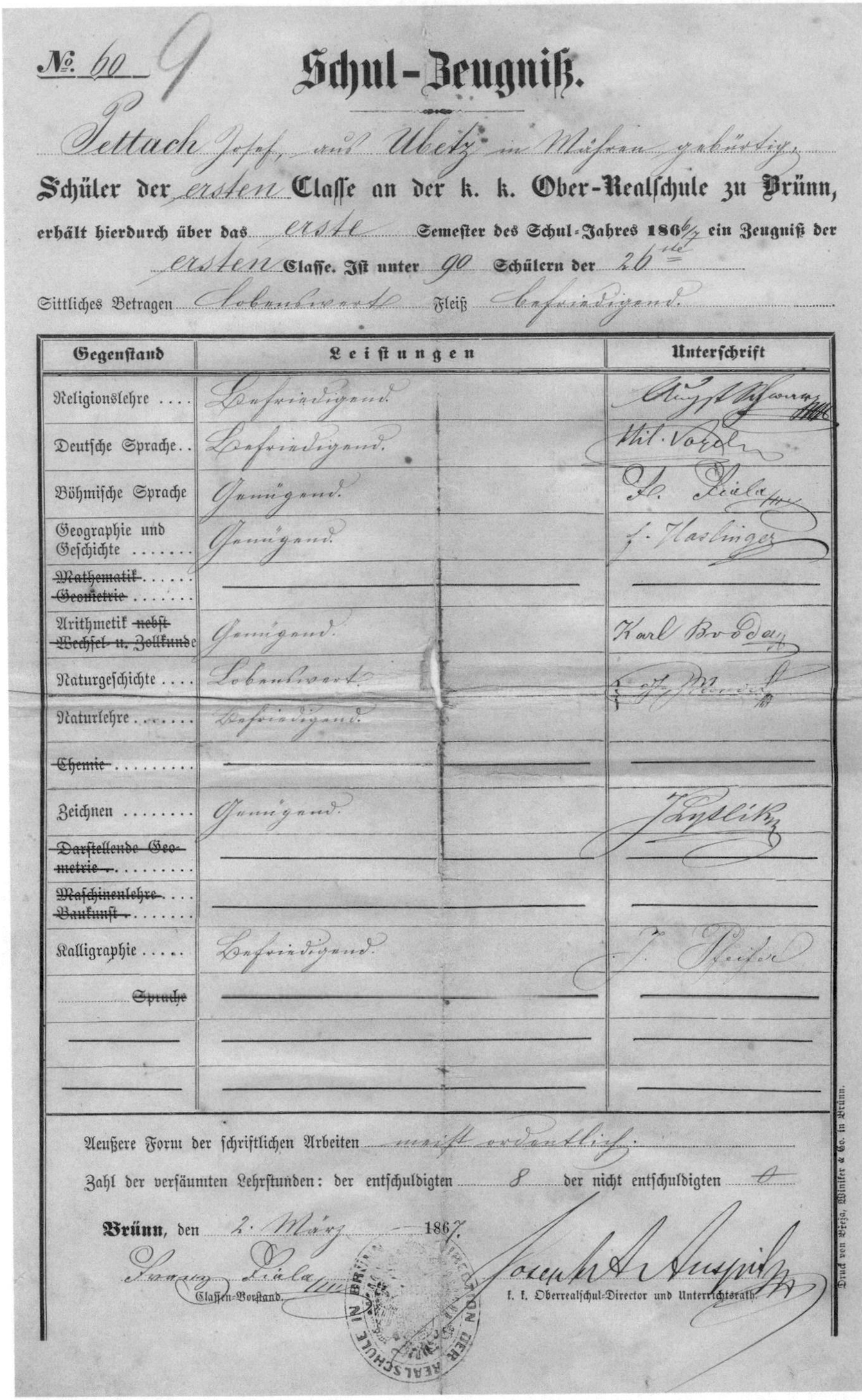

№ 60 9

Schul-Zeugniß.

Petlach Josef, aus Ubetz in Mähren gebürtig,

Schüler der ersten **Classe an der k. k. Ober-Realschule zu Brünn,**

erhält hierdurch über das erste **Semester des Schul-Jahres 1866/7 ein Zeugniß der** ersten **Classe. Ist unter** 90 **Schülern der** 26te

Sittliches Betragen lobenswert Fleiß befriedigend.

Gegenstand	Leistungen	Unterschrift
Religionslehre	befriedigend.	[illegible]
Deutsche Sprache	befriedigend.	[illegible]
Böhmische Sprache	genügend.	F. Fiala
Geographie und Geschichte	genügend.	J. Haslinger
~~Mathematik~~ ~~Geometrie~~	—	—
Arithmetik ~~nebst Wechsel- u. Zollkunde~~	genügend.	Karl [illegible]
Naturgeschichte	lobenswert.	[illegible]
Naturlehre	befriedigend.	
~~Chemie~~	—	—
Zeichnen	genügend.	[illegible]
~~Darstellende Geometrie~~	—	—
~~Maschinenlehre~~ ~~Baukunst~~	—	—
Kalligraphie	befriedigend.	[illegible]
~~Sprache~~	—	—
—	—	—
—	—	—

Aeußere Form der schriftlichen Arbeiten meist ordentlich.

Zahl der versäumten Lehrstunden: der entschuldigten 8 der nicht entschuldigten 0

Brünn, den 2. März 1867.

Franz Fiala
Classen-Vorstand.

Joseph A. Auspitz
k. k. Oberrealschul-Director und Unterrichtsrath.

Druck von Preza, Winiker & Co. in Brünn.

D418 Certificate of Education of Josef Petlach for the first semester of the first grade of the school year 1866/7 with Mendel's signature as professor of physics and natural history. Dated in Brünn/Brno, 02 March 1867, signed by Joseph A. Auspitz as director and Franz Fiala as class master; stamp of the Directorate of the Realschule in Brünn/Brno/Direction der Realschule in Brünn/

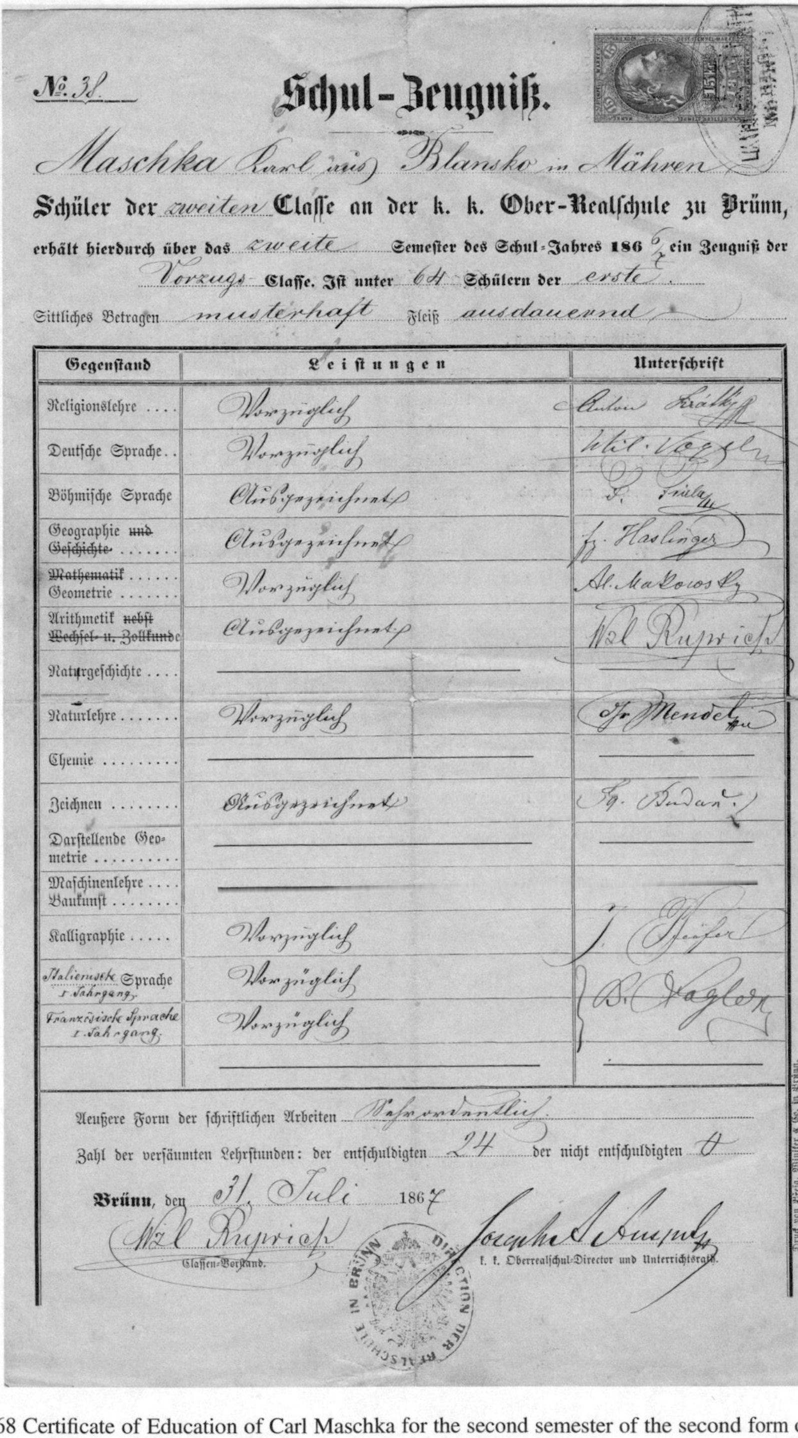

№ 38

Schul-Zeugniß.

Maschka Karl aus Blansko in Mähren

Schüler der zweiten Classe an der k. k. Ober-Realschule zu Brünn,

erhält hierdurch über das zweite Semester des Schul-Jahres 186 6/7 ein Zeugniß der Vorzugs-Classe. Ist unter 64 Schülern der erste.

Sittliches Betragen musterhaft Fleiß ausdauernd

Gegenstand	Leistungen	Unterschrift
Religionslehre	Vorzüglich	Anton Krátký
Deutsche Sprache	Vorzüglich	[illegible]
Böhmische Sprache	Ausgezeichnet	[illegible]
Geographie ~~und Geschichte~~	Ausgezeichnet	Fr. Haslinger
~~Mathematik~~ Geometrie	Vorzüglich	Al. Makowsky
Arithmetik ~~nebst Wechsel- u. Zollkunde~~	Ausgezeichnet	Wzl Ruprich
Naturgeschichte		
Naturlehre	Vorzüglich	Gr. Mendel
Chemie		
Zeichnen	Ausgezeichnet	[illegible]
Darstellende Geometrie		
Maschinenlehre Baukunst		
Kalligraphie	Vorzüglich	[illegible]
Italienische Sprache I. Jahrgang	Vorzüglich	[illegible]
Französische Sprache I. Jahrgang	Vorzüglich	

Aeußere Form der schriftlichen Arbeiten Sehr ordentlich

Zahl der versäumten Lehrstunden: der entschuldigten 24 der nicht entschuldigten 0

Brünn, den 31. Juli 1867

Wzl Ruprich
Classen-Vorstand.

Joseph A. Auspitz
k. k. Oberrealschul-Director und Unterrichtsrath.

DIRECTION DER REALSCHULE IN BRÜNN

Druck von [illegible], Winiker & Co. in Brünn.

D468 Certificate of Education of Carl Maschka for the second semester of the second form of the *Realschule* with Mendel's signature as teacher of physics. Dated in Brünn/Brno, 31 July 1867, signed by Joseph A. Auspitz as director and Wzl Ruprich as class teacher; stamp of the Directorate of the *Realschule* in Brünn/Brno/*Direction der Realschule in Brünn*

IM48 The body of teachers of the Realschule in 1864. Seated from left: Franz Matzek, mathematics, geometry; Josef Rotter, German, geography, calligraphy; Dr. Richard Rotter, geography, history; Anton Krátký, catholic religion; Joseph Auspitz, arithmetic, accounting, customs; Dr. Alexander Zawadski, natural history, physics; Benedikt Fogler, German, Italian, French; Franz Berr, chemistry, physics; Gregor Mendel, natural history, physics; Anton Mayssl, painting, modelling; Standing from left: Franz Fiala, German, Czech; Friedrich Haslinger, geography, history; Hilarius Vogel, German, geography, history; Budař unspecified; Alexander Makowsky, mathematics, natural history; Jakob Bratkovič, mathematics, physics; Josef Hoffmann, mechanics, geometry, drawing; Pfeifer, unspecified; Johann Pytlik, geometry, painting; Wenzel Ruprich, arithmetic, geometry, architecture; Josef Roller, painting, modelling

O13 Mendel's portrait enlarged from a group photograph of travellers to the London Industrial Exhibition in 1862. The *Realschule* of Brno exhibited there crystallographic wall tables prepared by their students (Richter 1943). The group photograph was taken in front of the Grand Hotel in Paris on 06 August 1862 by Pierre Petit. The group photograph is published in IM 1965, picture No. 49, pp. 30–31

References and Historical Printed Sources

Abbott S, Fairbanks DJ (2016) Experiments on plant hybrids by Gregor Mendel. Genetics 204:407–422

Gicklhorn R (1969) Gregor Mendels Lehramtsprüfung und Studienzeit in Wien. Allgemeine Biologie der Medizinischen Fakultät der Universität Wien 7(4):145–159

Iltis H (1924) Gregor Johann Mendel. Leben, Werk und Wirkung. Julius Springer, Berlin. Life of Mendel. English translation by Eden and Cedar Paul. First edition. Norton, New York 1932; Second edition Allen and Unwin London, 1966

Klein J, Klein N (2013) Solitude of a Humble Genius – Gregor Johann Mendel: Volume 1 Formative Years. In: Klein P (ed) Springer, Berlin. ISBN 978-3-642-35254-6 (eBook). https://doi.org/10.1007/978-3-642-35254-6. Library of Congress Control Number 2013948128

Křiženecký J (1963a) Genetische Abteilung Gregor Mendel im Mährischen Museum in Brünn. Naturwissenschaftliche Rundschau 16:477–480

Křiženecký J (1963b) Mendels zweite erfolglose Lehramtsprüfung im Jahre 1856. Sudhoffs Archiv für Geschichte der Medizin und der Naturwissenschaften 47:305–310. Franz Steiner Verlag. Wiesbaden

Richter O (1943) Johann Gregor Mendel wie er wirklich war. Neue Beiträge zur Biographie des berühmten Biologen aus Brünns Archiven mit 31 Abbildungen im Texte. Verhandlungen des Naturforschenden Vereines in Brünn, Abteilung für Naturforschung der Deutschen Gesellschaft für Wissenschaft und Volkstumforschung in Mähren 74 (für das Jahr 1942):1-262. Druck von Josef Klär, Brno

Chapter 11
Founding Member of the *Naturforschender Verein* in Brno 1861

Under the new social conditions that arose after the 1848 revolution, Zawadski initiated a transformation of the Natural Science Section of the Agricultural Society into an independent research society dedicated to "pure" science. Such concept of pure scientific research deviated from the statutes of the Agricultural Society, in which research served practical needs.

Some Natural Science Section members of the Agriculture Society preferred to maintain its original status, i.e. writing expert opinions on practical actual problems for the Agricultural Society central board.

Zawadski worked intensely to free the Natural Science Section from its subordinate position, dedicated solely to addressing practical questions in agriculture. Negotiations between the Natural Science Section, represented by A. Zawadski, and the Agricultural Society, represented by C. Napp, were controversial. The conversion of the Natural Science Section into an independent scientific society, proposed by Zawadski, was at first rejected by Napp. He would eventually consider the loss of the Natural Science Section as his personal failure. Much later, however, Napp accepted the compromise offered by Zawadski to establish a new independent society of researchers without affecting the traditional structure of the Agricultural Society. For tactic reasons, the conversion of the Natural Science Section of the Agricultural Society into a self-contained *Naturforschender Verein of researchers* was realized in two steps. First, the Natural Science Section and the *Naturforschender Verein* existed simultaneously as two parallel organizations. This was possible because some section members preferred to continue working in the original section and some simultaneously in both the *Naturforschender Verein* and the Natural Science Section. Mendel was a member of both organizations. The second step was to quietly close the activity of the Natural Science Section.

At the end of 1861, Zawadski organized a meeting in the museum, the seat of the Agricultural Society. He proposed the establishment of a new independent society devoted to scientific research on the material basis of nature. The overwhelming support of that proposal among Brno scientists resulted in the establishment of the *Naturforschender Verein*. The term r*esearch (forschender)* symbolized its

A. Matalová, E. Matalová, *Gregor Mendel - The Scientist*, Springer Biographies,
https://doi.org/10.1007/978-3-030-98923-1_11

progressive nature of experimentation. After 2 years of parallel existence of the two bodies, the activity of the Natural Science Section of the Agricultural Society was closed. Napp never officially announced the end of the Natural Science Section, maintaining his recognition of its existence until his death in 1867.

The *Naturforschender Verein* had its seat in the *Realschule*. Director Auspitz was elected as a member of its committee. He expected that the new methods of research discussed in the Society would be implemented in the teaching at his school, paving the way for technical literacy in Moravia.

The declared objective of the *Naturforschender Verein* was to discover the laws of nature and to explain its creative forces on a material basis.

Mendel participated in the foundational meeting. Honorary members of the *Naturforschender Verein* were physiologists Unger, Virchow and Purkyně. Unger had been Mendel's teacher at the University of Vienna in anatomy and physiology of plants and taught him the science of microscopy. Purkyně had been thinking about creating his own theory of the cell nucleus (Körnchentheorie). Mendel's brethren, Bratránek and Klácel, stressed the significance of works by Goethe and Humboldt for the society's library. It is probable that the verses from Goethe's play *Lilla*, copied in Mendel's handwriting, date from those days full of enthusiasm for opening a new era of science within the frame of the German *Naturphilosophie*.

In the *Naturforschender Verein*, Mendel lectured on his research in plant hybridization on 08 February and 08 March 1865. In its journal *Verhandlungen des naturforschenden Vereines in Brünn*, Mendel published his *opus magnum* on hybridization on the garden pea, reports on meteorological observations and his paper on *Hieracium* hybrids. It is evident that Mendel probably would not have published his paper Experiments on Plant Hybrids, at least in a journal based in Brno, if there were no *Naturforschender Verein* with Niessl as its secretary. The first English translation of Mendel's article was titled *Experiments in Plant Hybridisation*, published in the *Journal of the Royal Horticultural Society* in 1901. The translation was made by Charles T. Druery and commissioned by William Bateson. In 1902, Bateson revised Druery's translation in his booklet *Mendel's Principles of Heredity: A Defence*. This English version of Mendel's major work was repeatedly published to be accessible for all generations of geneticists. In 1966, genetics was officially renewed in Czechoslovakia by means of a governmental decree (Matalová and Sekerák 2004).

Eva R. Sherwood published her translation of Mendel's *Experiments on Plant Hybrids* in a book with Curt Stern *The Origin of Genetics: A Mendel Source Book* in 1966 to mark the hundredth anniversary of the publication of Mendel's major work. A new translation of Mendel's *Experiments on Plant Hybrids* was prepared by Abbott and Fairbanks 2016, the sesquicentennial year of the Mendel's publication, for *Genetics* (Vol. 204, pp. 407–422). Gregor Mendel's *Experiments on Plant Hybrids* was also prepared that same sesquicentennial year in English translation by Staffan Müller-Wille and Kersten Hall for the *British Society for the History of Science Translation Series*.

Transcription of the draft of Mendel's letter to Kerner von Marilaun: *Hochgeehrter Herr! Die anerkannten Verdienste, welche Eur Wohlgeboren um die*

Bestimmung und Einreihung wild wachsender Pflanzenbastarde erworben haben, machen es mir zur angenehmen Pflicht, die Beschreibung einiger Versuche über künstliche Befruchtung an Pflanzen zur gütigen Kenntnissnahme vorzulegen.—Mit dem Ausdrucke der grössten Hochachtung für Eur Wohlgeboren zeichnet sich—Gregor Mendel—Stifts-Capitular und Lehrer an der Oberrealschule.—Brünn am 1. Jänner 867.

Mendel's note analyzed by Heimans in 1969 is one of the two surviving fragments. It is generally called "*Notizblatt*". It is written on the reverse side of the paper containing Abbot Mendel's arguments against the application of the law of 07 May 1874, increasing taxation on rich monasteries. It probably dates from 1876 or 1877 and provides evidence of Mendel's thinking about his experiments with plant hybrids some 10 years after its publication.

In the absence of Mendel's notes from his research, Correns (1900) points to important details from Mendel's Pisum experiments in Mendel's letter to Carl Wilhelm von Nägeli, dated 18 April 1867 (in Stern and Sherwood 1966, pp. 60–61): ... *the experiments which are discussed were conducted from 1856 to 1863. I knew that the results I obtained were not easily compatible with our contemporary scientific knowledge, and that under the circumstances publication of one such isolated experiment was doubly dangerous; dangerous for the experimenter and for the cause he represented. Thus I made every effort to verify, with other plants, the results obtained with Pisum. A number of hybridizations undertaken in 1863 and 1864 convinced me of the difficulty of finding plants suitable for an extended series of experiments, and that under unfavourable circumstances years might elapse without my obtaining the desired information. I attempted to inspire some control experiments, and for that reason discussed the Pisum experiments at the meeting of the local society of naturalists. I encountered, as was to be expected, divided opinion; however, as far as I know, no one undertook to repeat the experiments. When, last year, I was asked to publish my lecture in the proceedings of the society, I agreed to do so, after having re-examined my records for the various years of experimentation, and not having been able to find a source of error. The paper which was submitted to you is the unchanged reprint of the draft of the lecture mentioned; thus the brevity of the exposition, as is essential for a public lecture.—I am not surprised to hear your honor speak of my experiments with mistrustful caution; I would not do otherwise in a similar case.*

The following year, Mendel informed Nägeli in a letter dated 04 May 1868 about his election as abbot of the Augustinian monastery:

Recently there has been a completely unexpected turn in my affairs. On March 30 my unimportant self was elected life-long head, by the chapter of the monastery to which I belong. From the very modest position of teacher of experimental physics I thus find myself moved into a sphere in which much appears strange to me, and it will take some time and effort before I feel at home in it. This shall not prevent me from continuing the hybridization experiments of which I have become so fond; I even hope to be able to devote more time and attention to them, once I have become familiar with my new position.

(From Stern and Sherwood 1966, p. 79)

Statuten

des

naturforschenden Vereines

in Brünn.

1861.

Druck von Georg Gastl in Brünn.

The Statutes of the *Naturforschender Verein in Brünn* were published in 1861 (Druck bei Georg Gastl in Brünn). Paragraph 23 states that, upon dissolving the Society, its resources should serve the natural scientific interests of the country, and its collection and library should be transferred to what was then the Franz Museum (now the Moravian Museum). Dated in Brünn/Brno, 08 June 1861. Approved by Minister Lasser in Wien/Vienna on 23 October 1861

In 1867 Auspitz was appointed the chief school inspector of the country. Zawadski died in 1868. Mendel was elected abbot and left the *Realschule* in the

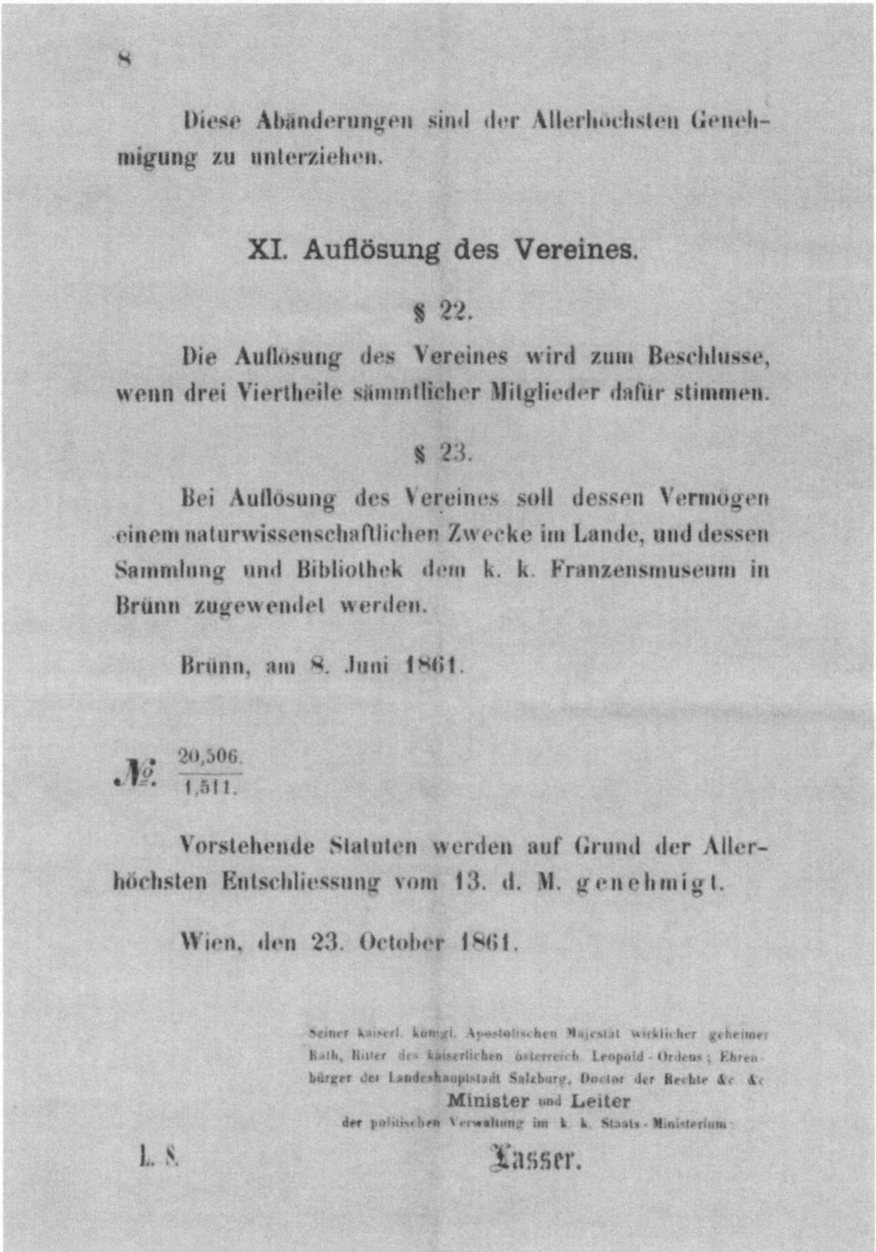

8

Diese Abänderungen sind der Allerhöchsten Genehmigung zu unterziehen.

XI. Auflösung des Vereines.

§ 22.

Die Auflösung des Vereines wird zum Beschlusse, wenn drei Viertheile sämmtlicher Mitglieder dafür stimmen.

§ 23.

Bei Auflösung des Vereines soll dessen Vermögen einem naturwissenschaftlichen Zwecke im Lande, und dessen Sammlung und Bibliothek dem k. k. Franzensmuseum in Brünn zugewendet werden.

Brünn, am 8. Juni 1861.

№ 20,506. / 1,511.

Vorstehende Statuten werden auf Grund der Allerhöchsten Entschliessung vom 13. d. M. genehmigt.

Wien, den 23. October 1861.

Seiner kaiserl. königl. Apostolischen Majestät wirklicher geheimer Rath, Ritter des kaiserlichen österreich. Leopold-Ordens; Ehrenbürger der Landeshauptstadt Salzburg, Doctor der Rechte &c. &c.

Minister und Leiter

der politischen Verwaltung im k. k. Staats-Ministerium:

L. S. Lasser.

(continued)

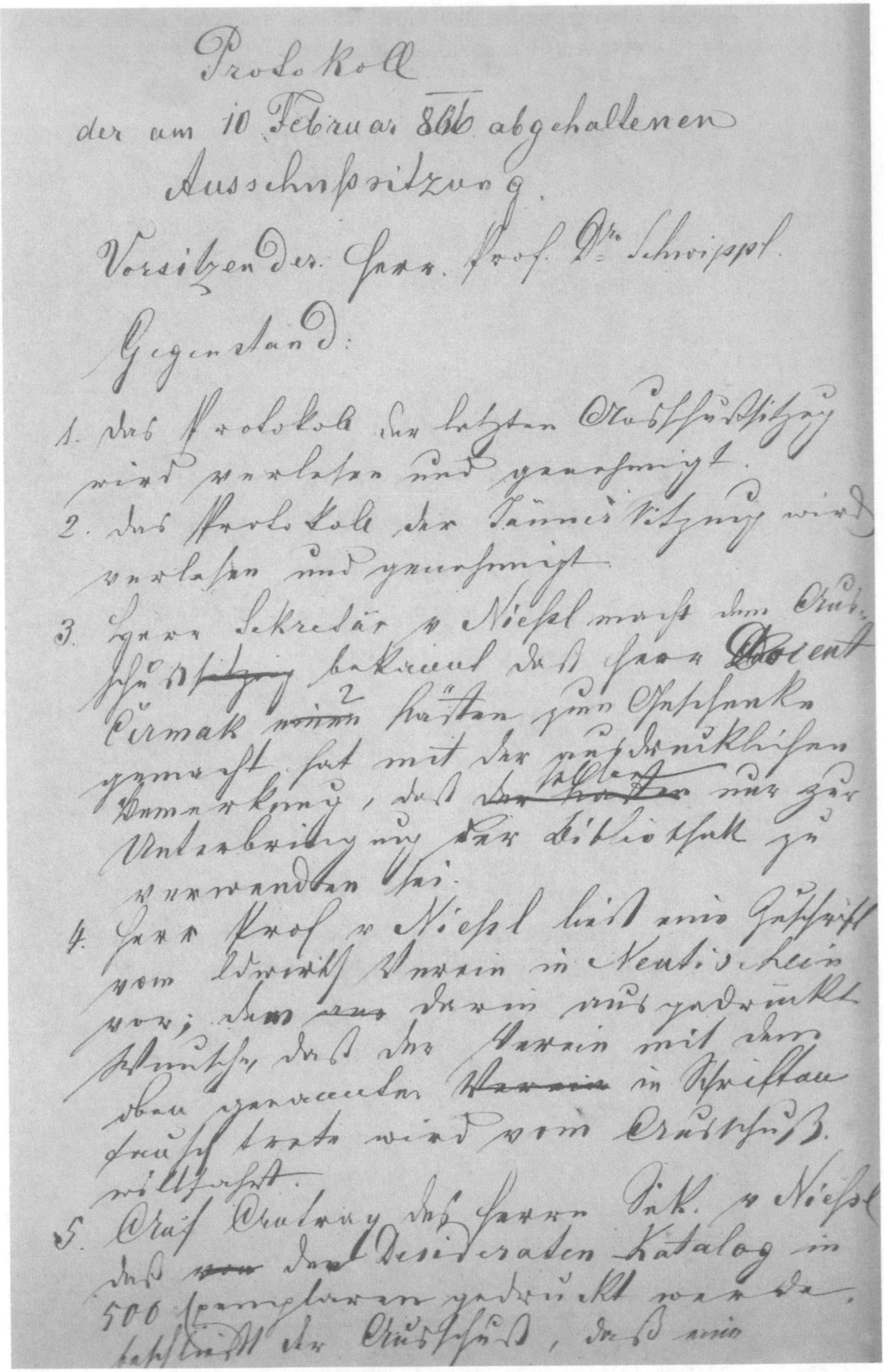

Protokoll

der am 10 Februar 866 abgehaltenen

Ausschußsitzung.

Vorsitzender Herr Prof. Dr. Schwippel.

Gegenstand:

1. Das Protokoll der letzten Ausschußsitzung wird verlesen und genehmigt.
2. Das Protokoll der Jahressitzung wird verlesen und genehmigt.
3. [illegible]
4. [illegible]
5. [illegible]

Mann 1992 The minutes of the *Naturforschender Verein*. Four pages. *Item No. 6, Prof. Dr. Niessl presents the work on hybrids by Prof. Mendel. The committee approves the contribution for print in the journal of the association.* Dated in Brno, 10 February 1866, signed by Dr. K. Schwippel. The minutes were in possession of the *Naturforschender Verein* in Brno and attached to the original manuscript of Mendel's Experiments on Plant Hybrids purchased from the Brno Augustinians

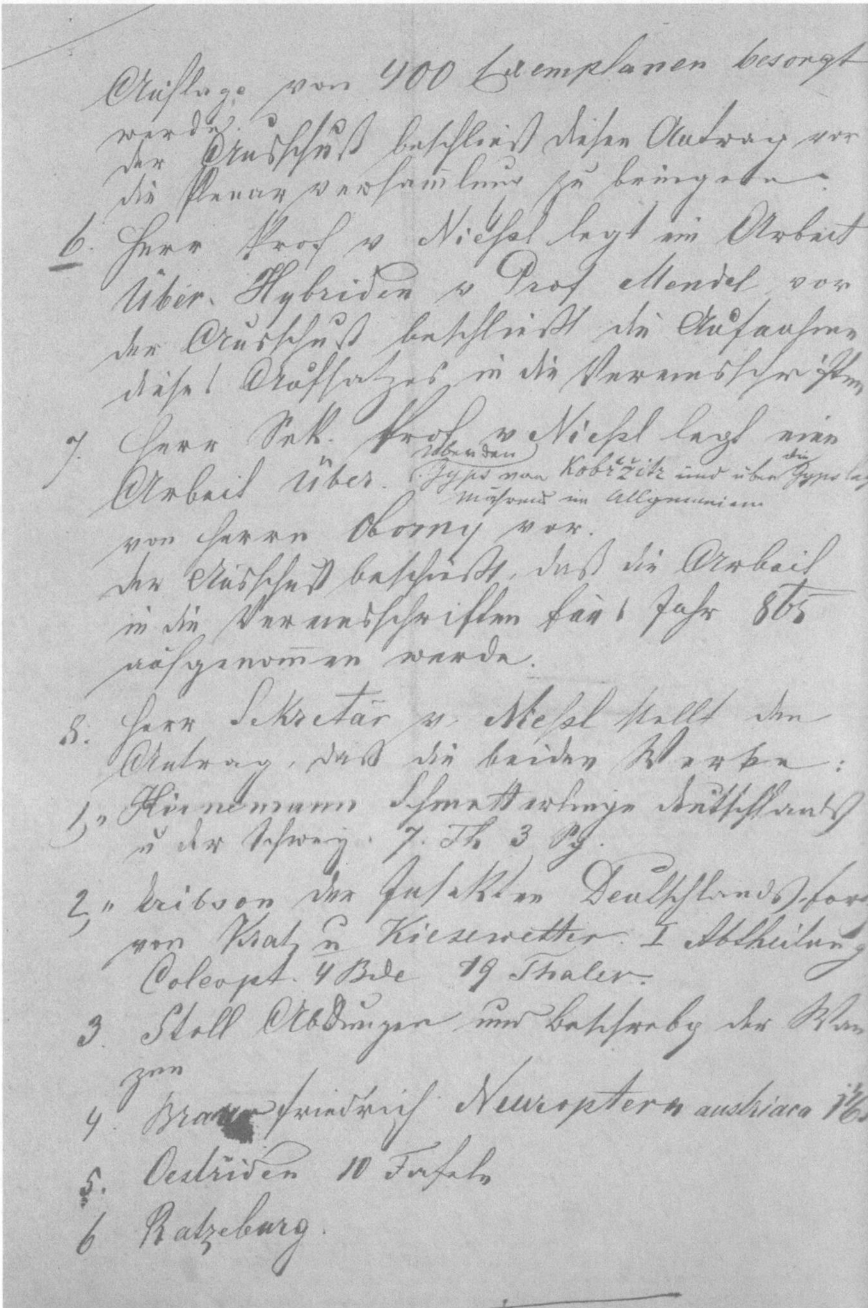

Auflage von 400 Exemplaren besorgt werde. Der Ausschuss beschliesst diesen Antrag vor die Plenarversammlung zu bringen.

6. Herr Prof. v. Niessl legt eine Arbeit über Hybriden v. Prof. Mendel vor. Der Ausschuss beschliesst die Aufnahme dieses Aufsatzes in die Verhandlungen.

7. Herr Prof. v. Niessl legt eine Arbeit über abnorme Typen von Kobřzik und über die Typen während im Allgemeinen von Herrn Oborny vor. Der Ausschuss beschliesst, dass die Arbeit in die Verhandlungen fürs Jahr 865 aufgenommen werde.

8. Herr Sekretär v. Niessl stellt den Antrag, dass die beiden Werke:
1. Heinemann Schmetterlinge Deutschlands u. der Schweiz 7 Th. 3 Sg.
2. Erichson der Insekten Deutschlands fortg. von Kraatz u. Kiesenwetter I Abtheilung Coleopt. 4 Bde 19 Thaler.
3. Stoll Abbildungen von ... der Wanzen
4. Brauer Friedrich Neuroptera austriaca 1857
5. Oestriden 10 Tafeln
6. Ratzeburg.

(continued)

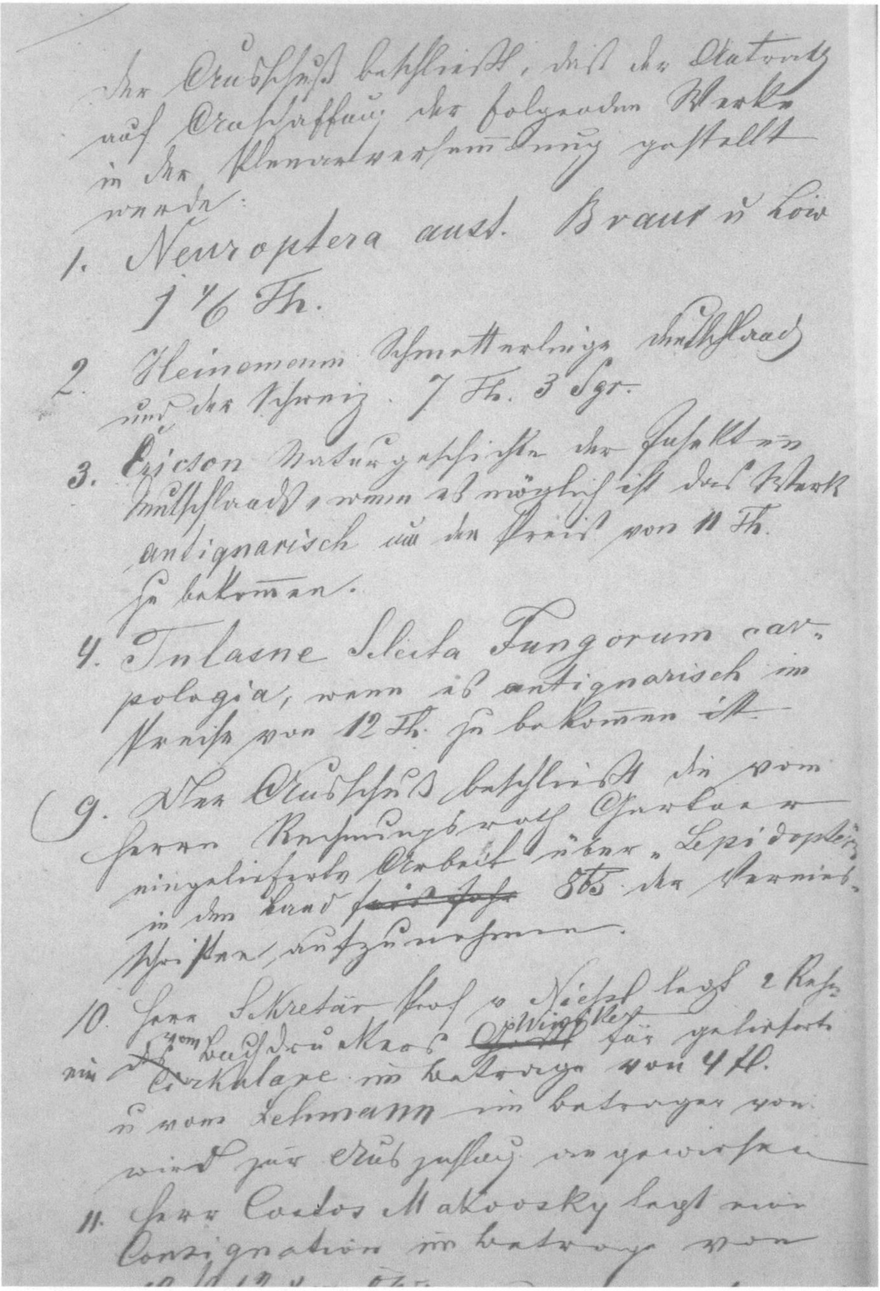
der Ausschuß beschließt, daß der [illegible] nach Anschaffung der folgenden Werke in der Monatsversammlung gestellt werde:

1. Neuroptera austr. Brauer u. Löw 1 1/6 Th.

2. Heinemann Schmetterlinge Deutschlands und der Schweiz 7 Th. 3 Sgr.

3. Ericson Naturgeschichte der Insekten Deutschlands, wenn es möglich ist das Werk antiquarisch um den Preis von 11 Th. zu bekommen.

4. Tulasne Selecta Fungorum carpologia, wenn es antiquarisch im Preise von 12 Th. zu bekommen ist.

9. Der Ausschuß beschließt die vom Herrn [illegible] Gartner eingelieferte Arbeit über „Lepidopteren" in den Band [illegible] der Vereinsschriften aufzunehmen.

10. Herr Sekretär Prof. v. Niessl legt 2 Rechnungen vom Buchbinder [illegible] für gelieferte Circulare im Betrage von 4 fl. und vom Lehmann im Betrage von [illegible] wird zur Auszahlung angewiesen.

11. Herr Custos Makowsky legt eine Consignation im Betrage von [illegible]

(continued)

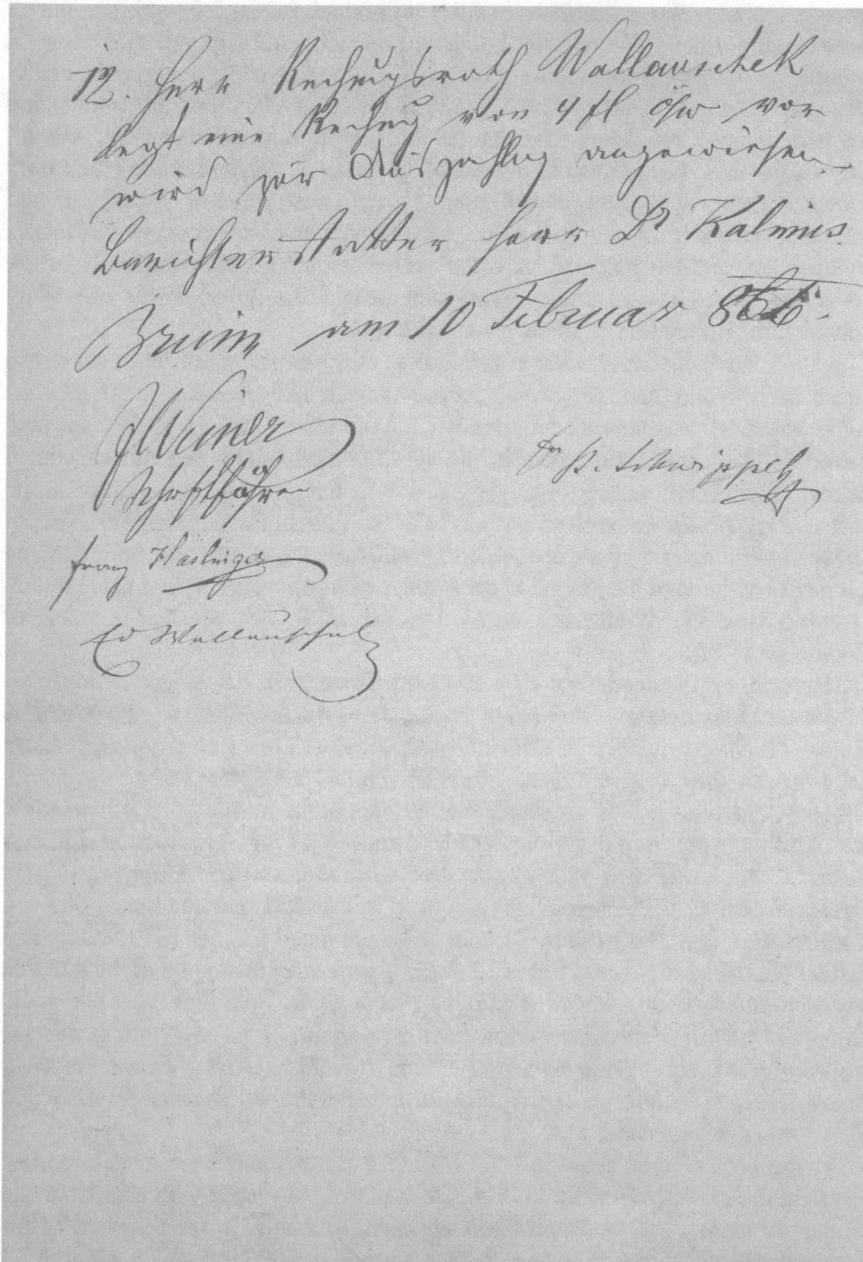

(continued)

spring of 1868. The *Naturforschender Verein* had fulfilled its programme of implementing a new type of demonstrational teaching and experimental research. Its initiative in making collections of botanical, zoological, entomological, mineralogical and geological materials for educational purposes at schools, and introducing new experimental and demonstrational methods in physics and chemistry, substantially modernized the process of instruction and turned the *Realschule* into a truly modern school. The *Naturforschender Verein* was oriented locally toward instructing teachers on how to conduct experiments and to encourage students to explore their creative potential. A better understanding of the complexity of the world, the interrelations among living beings and the development of society facilitated the orientation of students in their careers.

In 1869 the *Naturforschender Verein* had to move to new premises in the Urban House. Mendel was made responsible for the transfer. The second floor in the Urban House was no favourable location neither for the library nor for the lecture hall. Therefore, the lectures were held in the new building of the Technical Institute. Mendel gave his lecture on tornadoes in 1869 in that place. It was printed in the journal of the *Natuforschender Verein* in 1870. Its title, in English translation is: *On Hieracium-Hybrids Obtained By Artificial Fertilisation*. His last publication (printed in the 1871 volume of the journal of the *Naturforschender Verein*) was his report on a tornado titled Die Windhose vom 13. October 1870 (The whirlwind/tornado of 13 October 1870).

Transcription of Mendel's copied text from Goethe's Lilla: *Feiger Gedanken—Bangliches Schwanken,—Weibisches Zagen,—Ängstliches Klagen,—Wendet kein Elend,—Macht dich nicht frei.—Allen Gewalten—Zum Trotz sich erhalten, Nimmer sich beugen,—Kräftig sich zeigen,—Rufet die Arme—der Götter herbei.*

Mendel gave his lecture on Hieracium bastards in the meeting on 09 June 1869. The reprints were introduced under the heading in the brackets as follows: (*Sonderabdruck aus dem VIII. Bande der Verhandlungen des naturforschenden Vereines in Brünn.*). The reprint is reproduced in IM 1965 picture number 60.

An excerpt from the letter by Mendel to Nägeli dated 03 July 1870 reveals that Mendel published the results of his *Hieracium* experiments in spite of the fact that they were not to be in agreement with his *Pisum* work, from Stern and Sherwood 1966, p. 90: *On this occasion I cannot resist remarking how striking it is that the hybrids of Hieracium show a behaviour exactly opposite to those of Pisum. Evidently we are here dealing only with individual phenomena, that are the manifestation of a higher, more fundamental, law.*

Having been elected abbot in 1868 Mendel became the second most important person in the church hierarchy in Brno. He was fully burdened with charitable and political functions. Abbot Mendel was expected to follow in the footsteps of his respected multifaceted predecessor, Abbot Napp, who had been much admired for his skills in diplomacy, culture and politics. In the newly transformed political conditions, after the origin of Austria–Hungary, public life became decisively influenced by the Liberal Party, whose members supported progressive trends in society, industrialization, trade and banking. Auspitz, Mendel's headmaster at the *Realschule*, was the leader of Brno's liberals. He published a liberal daily *Tagesbote* to which Mendel subscribed. Mendel had taught natural history and physics for 14 years at Auspitz´ *Realschule* and had enjoyed its liberal atmosphere. In the 1871 elections, Abbot Mendel voted for the Liberal Party, which became the winner. Leaders of the victorious Liberal Party attempted to impose limitations on the church's influence on society. Changes in both the private lives of citizens and in the public education system were intended to encourage individual creativity and capabilities. In this context, Mendel's signature on the petition to the constitutional assembly of 1848 shows that both Klácel and Auspitz influenced him considerably in his way of thinking. It was unusual that a Roman Catholic abbot sided with the liberals against the conservatives. Shortly after his engagement in politics, Mendel was awarded the Comthur Cross of the Emperor Franz Joseph for his outstanding political work and meritorious accomplishments as a teacher at the *Realschule*.

In 1874, the liberal government issued a law increasing the contribution of wealthy ecclesiastic institutions to "the religious fund" (Vybral 1971) to make the church less dependent on the state. For 10 years Mendel consistently refused the application of this law on his monastery. He wished to contribute voluntarily but with a much lower sum of money than the state demanded. From his American exile, Klácel criticized the "freethinker" Mendel for opposing the "liberal" law and accused him of worshipping the golden calf. But Mendel needed funds for the reconstruction and maintenance of the archaic monastery buildings, which had been heavily damaged by a tornado in 1870. Mendel believed that the law to increase taxation on rich monasteries was unjust, especially given the fact that his monastery had been actively and fundamentally assisting the state in its progress in science and education, at its own expense for many decades. Mendel hired attorneys and persistently resisted paying the tax. He did not succeed in preventing the liberal government from sequestering income and placing a lien on property of the Augustinian monastery. In 1877, the monetary sequestration of the monastery's income for payment of the tax began in Neuhwiesdliz (now Nové Hvězdlice) on 30 April.

Despite the fact that sequestrations could have been higher than the officially established sum of money, Mendel continued refusing to respect the law. He was becoming further entrenched in social isolation. Even his Augustinian brethren did not understand his stubbornness in his so-called "struggle for the right" that caused him to age prematurely. In the times of his struggle against the increased taxation, Mendel remarked on a paper scribbled with his arguments against the so-called religious fund a proverbial rhyme that reads: *Wer durch die Welt will rücken, der muss sich auch hübsch bücken* (If in this world you wish to move up, you have to nicely bow down).

A fundamental change had transpired from the time of Mendel the novice to the time of Mendel the abbot. Brno had developed into a busy industrial and commercial city. It attracted entrepreneurs from various parts of Europe of diverse religious denominations who successfully implemented the English model of social structure. The approach towards science was systematic, introducing new and positive analytical methods into the educational institutes. Mendel contributed fundamentally to this progressive endeavour by teaching physics and natural history. Agriculture, as a subject, was merged with natural history, composed of zoology, botany, mineralogy,

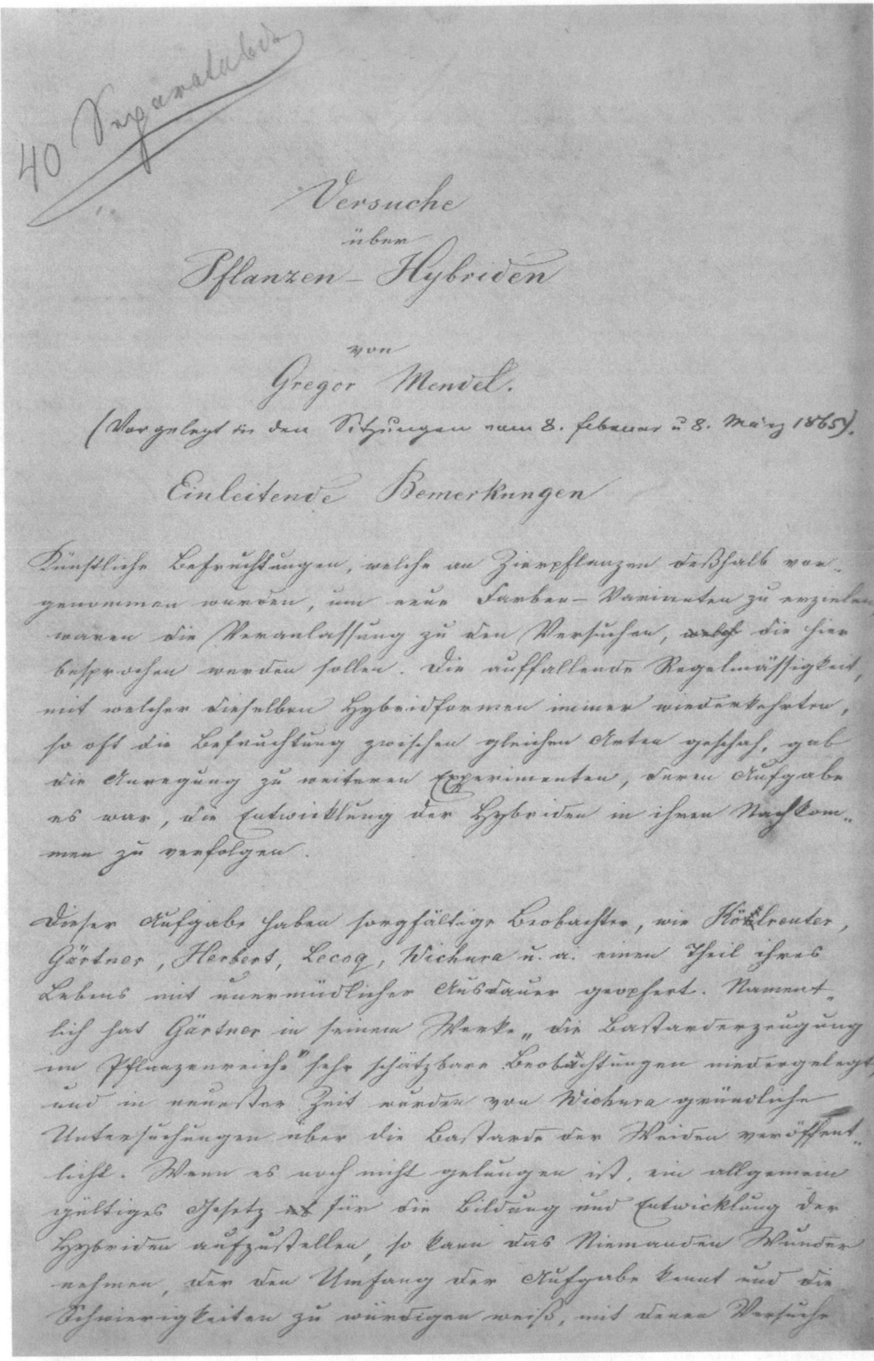

40 Separatabd.

Versuche
über
Pflanzen-Hybriden
von
Gregor Mendel.
(Vorgelegt in den Sitzungen vom 8. Februar u. 8. März 1865).

Einleitende Bemerkungen

Künstliche Befruchtungen, welche an Zierpflanzen deshalb vorgenommen wurden, um neue Farben-Varianten zu erzielen, waren die Veranlassung zu den Versuchen, welche hier besprochen werden sollen. Die auffallende Regelmässigkeit, mit welcher dieselben Hybridformen immer wiederkehrten, so oft die Befruchtung zwischen gleichen Arten geschah, gab die Anregung zu weiteren Experimenten, deren Aufgabe es war, die Entwicklung der Hybriden in ihren Nachkommen zu verfolgen.

Dieser Aufgabe haben sorgfältige Beobachter, wie Kölreuter, Gärtner, Herbert, Lecoq, Wichura u. a. einen Theil ihres Lebens mit unermüdlicher Ausdauer geopfert. Namentlich hat Gärtner in seinem Werke „die Bastarderzeugung im Pflanzenreiche" sehr schätzbare Beobachtungen niedergelegt, und in neuester Zeit wurden von Wichura gründliche Untersuchungen über die Bastarde der Weiden veröffentlicht. Wenn es noch nicht gelungen ist, ein allgemein gültiges Gesetz für die Bildung und Entwicklung der Hybriden aufzustellen, so kann das Niemanden Wunder nehmen, der den Umfang der Aufgabe kennt und die Schwierigkeiten zu würdigen weiss, mit denen Versuche

Iltis 1924. The first page of the manuscript of Mendel's lectures on *Experiments on Plant Hybrids*. In the left-hand upper corner there is Niessl's note 40 reprints *(40 Sonderabd)*

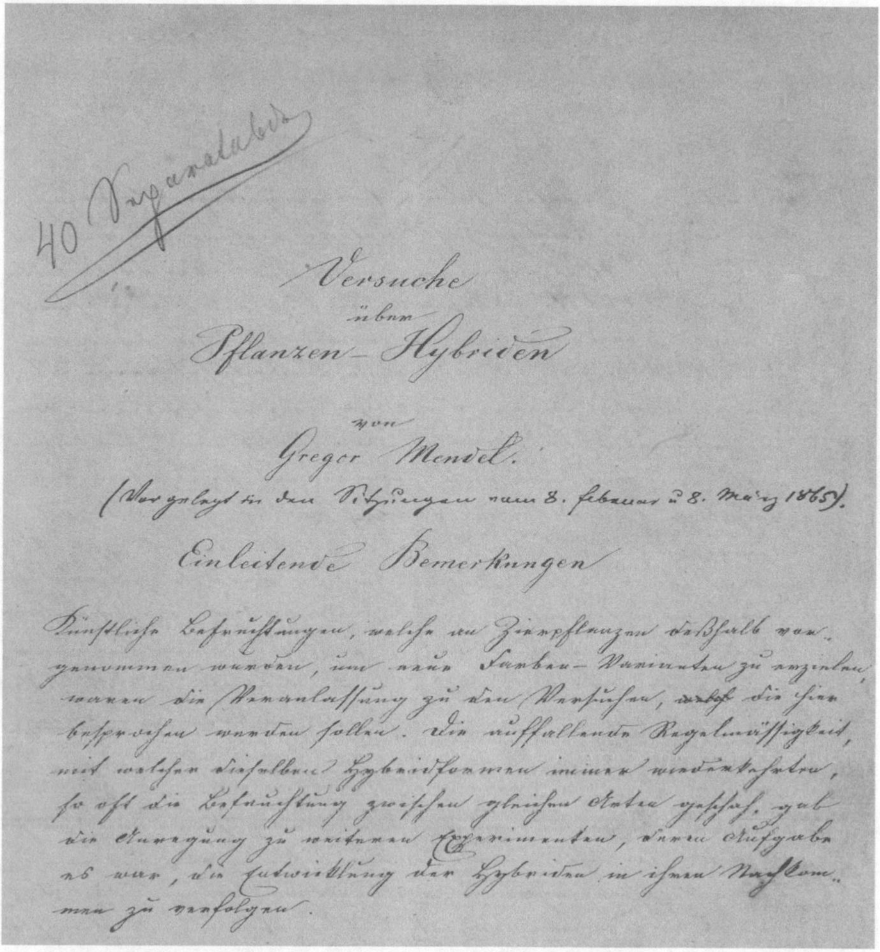

40 Separatabdr.

Versuche
über
Pflanzen-Hybriden
von
Gregor Mendel.
(Vorgelegt in den Sitzungen vom 8. Februar u. 8. März 1865).

Einleitende Bemerkungen

Künstliche Befruchtungen, welche an Zierpflanzen deshalb vorgenommen wurden, um neue Farben-Varianten zu erzielen, waren die Veranlassung zu den Versuchen, welche hier besprochen werden sollen. Die auffallende Regelmässigkeit, mit welcher dieselben Hybridformen immer wiederkehrten, so oft die Befruchtung zwischen gleichen Arten geschah, gab die Anregung zu weiteren Experimenten, deren Aufgabe es war, die Entwicklung der Hybriden in ihren Nachkommen zu verfolgen.

The first page of the manuscript of Mendel's lectures on *Experiments on Plant Hybrids* was published by Walther Mann in Darmstadt in 1992. Note the paperclip mark in the left-hand upper corner. The mark is a remnant of the secret hiding of Mendel's manuscript, originally kept in the safe of the *Naturforschender Verein in Brünn* in the Escompt Bank in Brno, before the Nazis took control of the city

Versuche über Pflanzen-Hybriden.

Von

Gregor Mendel.

(Vorgelegt in den Sitzungen vom 8. Februar und 8. März 1865.)

Einleitende Bemerkungen.

Künstliche Befruchtungen, welche an Zierpflanzen desshalb vorgenommen wurden, um neue Farben-Varianten zu erzielen, waren die Veranlassung zu den Versuchen, die her besprochen werden sollen. Die auffallende Regelmässigkeit, mit welcher dieselben Hybridformen immer wiederkehrten, so oft die Befruchtung zwischen gleichen Arten geschah, gab die Anregung zu weiteren Experimenten, deren Aufgabe es war, die Entwicklung der Hybriden in ihren Nachkommen zu verfolgen.

Dieser Aufgabe haben sorgfältige Beobachter, wie Kölreuter, Gärtner, Herbert, Lecocq, Wichura u. a. einen Theil ihres Lebens mit unermüdlicher Ausdauer geopfert. Namentlich hat Gärtner in seinem Werke „die Bastarderzeugung im Pflanzenreiche" sehr schätzbare Beobachtungen niedergelegt, und in neuester Zeit wurden von Wichura gründliche Untersuchungen über die Bastarde der Weiden veröffentlicht. Wenn es noch nicht gelungen ist, ein allgemein giltiges Gesetz für die Bildung und Entwicklung der Hybriden aufzustellen, so kann das Niemanden Wunder nehmen, der den Umfang der Aufgabe kennt und die Schwierigkeiten zu würdigen weiss, mit denen Versuche dieser Art zu kämpfen haben. Eine endgiltige Entscheidung kann erst dann erfolgen, bis Detail-Versuche aus den verschiedensten Pflanzen-Familien vorliegen. Wer die Ar-

1*

The title page of the printed *Experiments on Plant Hybrids* by Gregor Mendel in the *Verhandlungen des naturforschenden Vereines in Brünn for 1865*, published by the *Naturforschender Verein* (*Im Verlage des Vereines*) Brünn/Brno, 1866

Verhandlungen

des

naturforschenden Vereines

in Brünn.

IV. Band

1865.

Brünn, 1866.

Im Verlage des Vereines.

(continued)

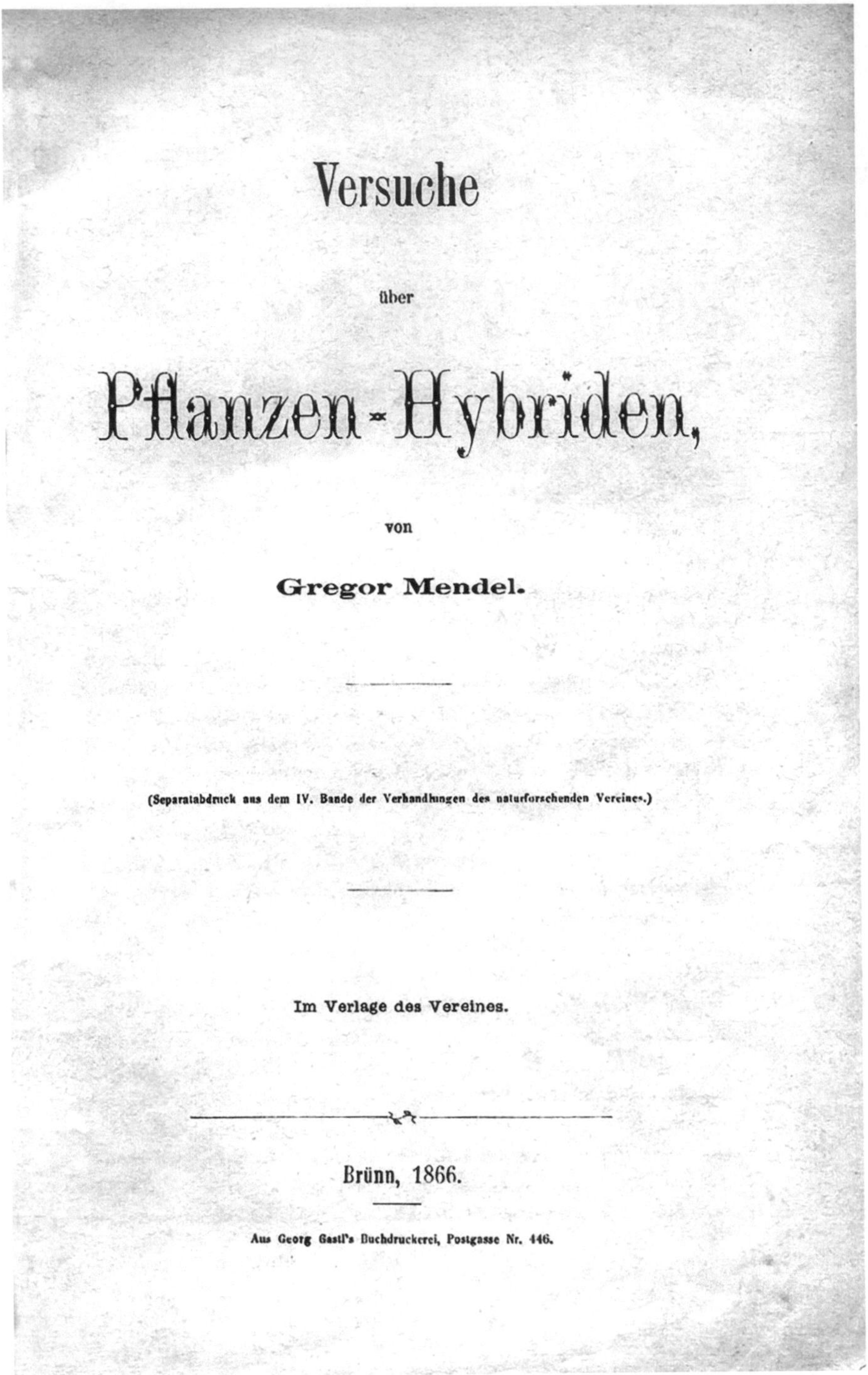

Versuche

über

Pflanzen-Hybriden,

von

Gregor Mendel.

(Separatabdruck aus dem IV. Bande der Verhandlungen des naturforschenden Vereines.)

Im Verlage des Vereines.

Brünn, 1866.

Aus Georg Gastl's Buchdruckerei, Postgasse Nr. 446.

The title page of the reprint of Mendel's paper *Experiments on Plant Hybrids,* 1866. Printed by Georg Gastl Printing House in Brno, Postgasse (Post Street) No. 446

Für Brünn:
Vierteljährig / Monatlich sammt Montagsblatt 2 fl. 40 kr. / – „ 80 „
Zustellung ins Haus pr. Monat . . 10 „
Einzelne Blätter . . . . 3 „
In Verlust gerathene Pränumerationsscheine werden unter keiner Bedingung ersetzt.

NEUIGKEITEN.

Für Auswärtige:
Mit täglicher portofreier Zusendung durch die k. k. Post:
Vierteljährig / Monatlich sammt Montagsblatt 3 fl. 48 kr. / 1 „ 16 „
In Iglau bei Herrn L. v. Löwenthal / Znaim bei Herren Fournier u. Haberler monatlich 1 fl.

Nr. 68. | Inserate kosten bei einmaliger Einschaltung 5 Kr. pr. Petitspaltzeile. | Brünn, Freitag den 10. März 1865. | Jede folgende Insertion 3 Kr. per Spaltzeile. Eingesendet pr. Zeile 15 Kr. | Jahrg. XV.

Sitzung des naturforschenden Vereins.

Z. **Brünn**, 9. März. Nach Eröffnung der Sitzung durch den Vizepräsidenten Herrn Karl Theimer und Mittheilung der seit der letzten Versammlung eingegangenen Geschenke und Sendungen hielt Herr Professor G. Mendl seinen zweiten Vortrag über Pflanzenhybriden. Anknüpfend an die bezüglichen Mittheilungen in der letzten Vereinsversammlung am 8. v. M. sprach er über Zellenbildung, Befruchtung und Samenbildung überhaupt und bei den Hybriden insbesondere unter Hinweisung auf seine bei Pisum (Erbse) mit eben so viel Umsicht, als Erfolg angestellten Versuche, welche er auch im nächsten Sommer fortzusetzen erklärte.

Zum Schluße theilte er mit, daß er auch mit vielen anderen, namentlich angezeigten, stammverwandten Pflanzen künstliche Befruchtungen zur Erzielung von Bastarden in den letzten Jahren vorgenommen habe, und sich durch die erlangten günstigen Resultate aufgemuntert fühle, derlei Bastardirungen nicht nur weiter zu versuchen, sondern auch hierüber eingehende Berichte zu erstatten.

Diesem mit vielfacher Anerkennung belohnten Vortrage fügte Herr Professor v. Nießl bei, daß auch von ihm bei Pilzen, Mosen und Algen mit Hilfe des Mikroskopes Hybridisationen beobachtet worden seien, und daß weitere dießfällige Beobachtungen nicht nur bisherige Hypothesen begründen, sondern auch weitere interessante Aufklärungen bringen werden.

Hierauf zeigte Herr Professor Makowsky einige ihm aus Kunstadt zugekommene Mineralien und zwar lose Thon-Eisenstein-Granaten in einer Größe, wie sie bisher in Mähren nicht vorgefunden wurden. Die bezüglichen Exemplare wiegen 21 Wiener Lothe und die Achsenlänge derselben beträgt $2^1/_2$ Wiener Zolle.

Besonders interessant waren die aus derselben Gegend stammenden Sandsteinkonglomerate von ungewöhnlicher Größe, abgerundet, ganz ausgehöhlt und mit einer kreisrunden Oeffnung versehen.

Ueber Antrag des Ausschusses wurden endlich die Volksschulen zu Littau und Bistritz, Igl. Kreis, welche sich bittlich an den Verein wandten, mit Geschenken von Naturalien zur Förderung des naturwissenschaftlichen Unterrichtes bedacht.

Zum Beweise der regen Theilnahme, deren sich dieser strebsame Verein erfreut, sei noch erwähnt, daß sich neuerlich zehn Herrn zu Mitgliedern wählen ließen.

Meeting of the Natural History Society.

Brünn, March 9. After opening of the session by the Vice-President Karl Theimer and the report about gifts and communications received since the last meeting, Professor G. Mendel held his second lecture on plant hybrids. In connection with the respective reports at the last meeting of the 8th of the past month, he spoke on cell division, fertilization and seed formation in general, and in the case of hybrids especially in regard to his experiments with Pisum (peas), conducted with as much circumspection as success.

He concluded that during the past years he undertook artificial fertilization with many other indicated species related plants to obtain hybrids, and that he feels gratified by the positive results, not only to go on with such experiments, but also report about these in detailed presentations.

To this much appreciated acclaimed lecture Professor v. Niessl added that hybridization was observed with the aid of a microscope in fungi, ferns and algae, and that more similar observations are not only lay foundation to the present hypotheses, but that they will produce more interesting explanations.

Following that Professor Makowsky demonstrated minerals from Kunstadt, that is loose clay iron garnets of a size not previously found in Moravia. The specimen weighed 21 Vienna pounds (Lothe) and the axis measured 2 1/2 Vienna inches (Zolle).

Of special interest, from the same locality, were sandstone conglomerates of unusual size, round, entirely hollow with a round aperture.

At the proposal of the executive committee, the pleadings of the grammar schools of Littau and Bistritz, Iglau County, to the society in regard to gifts of specimens, to promote the teaching of natural sciences, was finally fulfilled.

As prove of the lively activities of this industrious society we must mention that ten gentlemen accepted election to membership.

[Mendel adds to his lecture some anatomical discussion of plant growth, including seed formation, which is not contained in his published paper! Was it due to Mendel recognizing that his audience did not understand his mathematical elaboration? - Gustav v. Niessl was a mathematician and possibly the author of the two newspaper reports. - Alexander Makowsky was professor of natural history, geology].

167

Neuigkeiten reports on Mendel's lectures Experiments on Plant Hybrids delivered on 08 February and 08 March 1865, on serialization in the *Naturforschender Verein* meeting in the Oberrealschule in Janská Street. The reporter's signature is Z. (Zawadski) was the chairperson of the *Naturforschender Verein*. The reports are published in Brno 10 February 1865 and 10 March 1865 signed Z. (Zawadski)

NEUIGKEITEN.

Für Brünn:

Für Auswärtige:

. 40. | Brünn, Freitag den 10. Februar 1865. | Jahrg. XV.

Sitzung des naturforschenden Vereins.

Z. **Brünn**, 9. Febr. In der gestern abgehaltenen, abermals sehr zahlreich besuchten Monatssitzung führte der neu gewählte Vicepräsident Hr. Theimer den Vorsitz.

Nach Bekanntgabe der Einläufe hielt Herr Prof. G. Mendel einen längeren, besonders für Botaniker interessanten Vortrag über Pflanzenhybriden, welche durch künstliche Befruchtungen stammverwandten Arten und zwar durch Uebertragung des männlichen Blüthenstaubes auf die Samenpflanze hervorgebracht werden.

Er hob dabei hervor, daß die Fruchtbarkeit der Pflanzenhybriden, oder Bastarde zwar erwiesen sei, aber nicht konstant bleibe, und daß dieselben stets geneigt wären, zur Stammart rückzukehren, welche Rückkehr eben auch durch wiederholte künstliche Befruchtungen mit dem Blüthenstaube der Stammpflanzen beschleunigt werden kann.

Der Vortragende betonte hierauf seine durch mehrere Jahre mit Erfolg gemachten Versuche die er namentlich mit mehreren Erbsengattungen (Pisum sativum, P. sacharatum und P. quadratum) anstellte und zeigte die Proben aus den bezüglichen Generationen vor, wornach gemeinsame Merkmale gegenseitig übergangen waren, differirende Merkmale aber ganz neue Charaktere hervorbrachten. Die Differenzmerkmale der Erbsenhybriden zeigten sich in der Gestalt, dann Färbung des reifen Samens und der Samenschale, in der Farbe der Blüthen, in der Form der reifen und in der Farbe der unreifen Samenhülsen, in der Stellung der Blüthen und im Unterschiede der Achsenlänge. Beachtenswerth waren die ziffermäßigen Zusammenstellungen mit Rücksicht auf die eingetretenen Differenz-Merkmale der Hybriden und deren Verhältniß gegenüber der Stammarten.

Daß der Vorwurf des Vortrages ein glücklicher und die Durchführung desselben eine ganz befriedigende war, bewies die rege Theilnahme des Auditoriums.

Ueber Antrag des Vereins-Ausschusses wurde ferner beschlossen, die Pfarrhauptschule in Weißkirchen auf ihre Bitte mit einer Kollektion von Pflanzen und Käfern zu beschenken, dann mit einer Wiener und Leipziger Pflanzentauschanstalt behufs Komplettirung des Vereinsherbariums in Verbindung zu treten.

Der Verein selbst erhielt schließlich durch die Wahl von fünf neuen Mitgliedern einen weiteren Zuwachs.

Meeting of the Natural History Society.

Brünn, February 9.Yesterday's monthly meeting was again very well attended; Mr. Theimer, the newly elected Vice-President, chaired it.

After presenting of the mail Prof. G. Mendel presented a lengthy lecture about plant hybrids, of special interest to botanists, which are produced by artificial insemination of related species, namely by transmitting the male flower pollen on the seed plant.

He emphasized at the same time that the fertility or the plant hybrids, or bastards was documented, but that it does not remain constant, and that these always tended to reverse to the original type, that this reversal is made faster by repeated artificial insemination with the flower pollen of the original type.

The lecturer emphasized thereafter his successful experiments of a number of years duration which he carried out with a number of pea species (Pisum sativum, P. sacharatum and P. quadratum) and demonstrated samples from he respective generations, following which the common features mutually were transferred, differentiating features, however, produced entirely new characters. The differentiating features of the pea hybrids showed themselves in the shape, then the color of the adult seeds and seed shell, in the color of he flowers, in the form of the mature and in the color of the immature seed shells, in the position of the flowers and in the differences of the axis length. Noticeable was the numeral relationships in regard to the developed differential characters of the hybrids and their relationship to the types of origin.

That the subject of the lecture was fortunately chosen and the presentation entirely satisfactory was shown by the lively participation of the audience.

Upon the proposal of the society committee it was decided to present as gift a collection of plants and beetles to the main school in Weisskirchen following their request and then relate to the Vienna and Leipzig exchange institutions to complete the herbarium of the society.

The membership of the society increased by the election of five new members.

[Karl Theimer, the chair, was a botanist doing research regarding hybrids.]

166

(continued)

Iltis 1924. Committee members of the *Naturforschender Verein in Brünn*. Seated from the left: (1) Karl Theimer, pharmacy, botany; (2) Joseph Auspitz, Headmaster of *Realschule*, mathematics, bookkeeping; (3) Dr. Alexander Zawadski, natural history, physics; (4) Johann Nave, botany; (5) Eduard Wallauschek, entomology. Standing from the right: (6) Julius Muller, entomology; (7) Franz Tschermak, chemistry, mineralogy; (8) Dr. Karl Schwippel, natural history, physics; (9) Gustav von Niessl, mathematics; (11) Ignaz Weiner, physics; (12) Dr. Jakob Kalmus, botany

geology and palaeontology. Natural history, as a topic, took on a more prominent role, balancing the natural sciences with the humanities in the curricula of modern schools, compared to its previously subordinate status. Support of free thinking and free exchange of ideas represented progress, and would challenge poetic romanticism and the traditions of medieval Catholicism. Flourishing liberalism tended to eradicate all forms of phantasm and mysticism, supporting the development of the free and creative human mind founded on reason and the productive strength of science. Roger Bacon's famous proclamation *Wissenschaft ist Macht* (Science is Might) was the mantra of the day. Abbot Mendel undoubtedly understood that most members of the clergy considered liberalism, pantheism and materialism to be inimical to the church, whose representative he was.

4

Brünn, am 26. Mai 1879

Bestätigung.

Ueber Dreissig *fl. öst. W. als Jahresbeitrag für das Jahr* 1879

Naturforschender Verein.

Jos. Kafka m.p.

Rechnungsführer.

Für P. T. Herrn Prälaten Gregor Mendel Hochwürden in Brünn

Druck v. W. Burkart in Brünn

D767 Receipt of payment for Mendel's annual dues to the *Naturforschender Verein*. Dated 26 May 1879, signed by Jos. Kafka

NATURFORSCHENDER VEREIN IN BRÜNN

6

Brünn, am 13. April 1882

Bestätigung.

Ueber dreißig fl. öst. W. als ~~Eintrittsgebühr~~ ~~und~~ Jahresbeitrag für das Jahr 1882.

Naturforschender Verein.

Woharek

Rechnungsführer.

Für P. T. Herrn Gregor Mendl, Hochwürden, Prälat des Stiftes St. Thomas in Brünn

Druck v. W. Burkart in Brünn

D768 Receipt of payment of Mendel's annual dues to the *Naturforschender Verein*. Dated 13 April 1882, signed by Woharek. In the upper left hand corner stamp of the Naturforschender Verein in Brünn

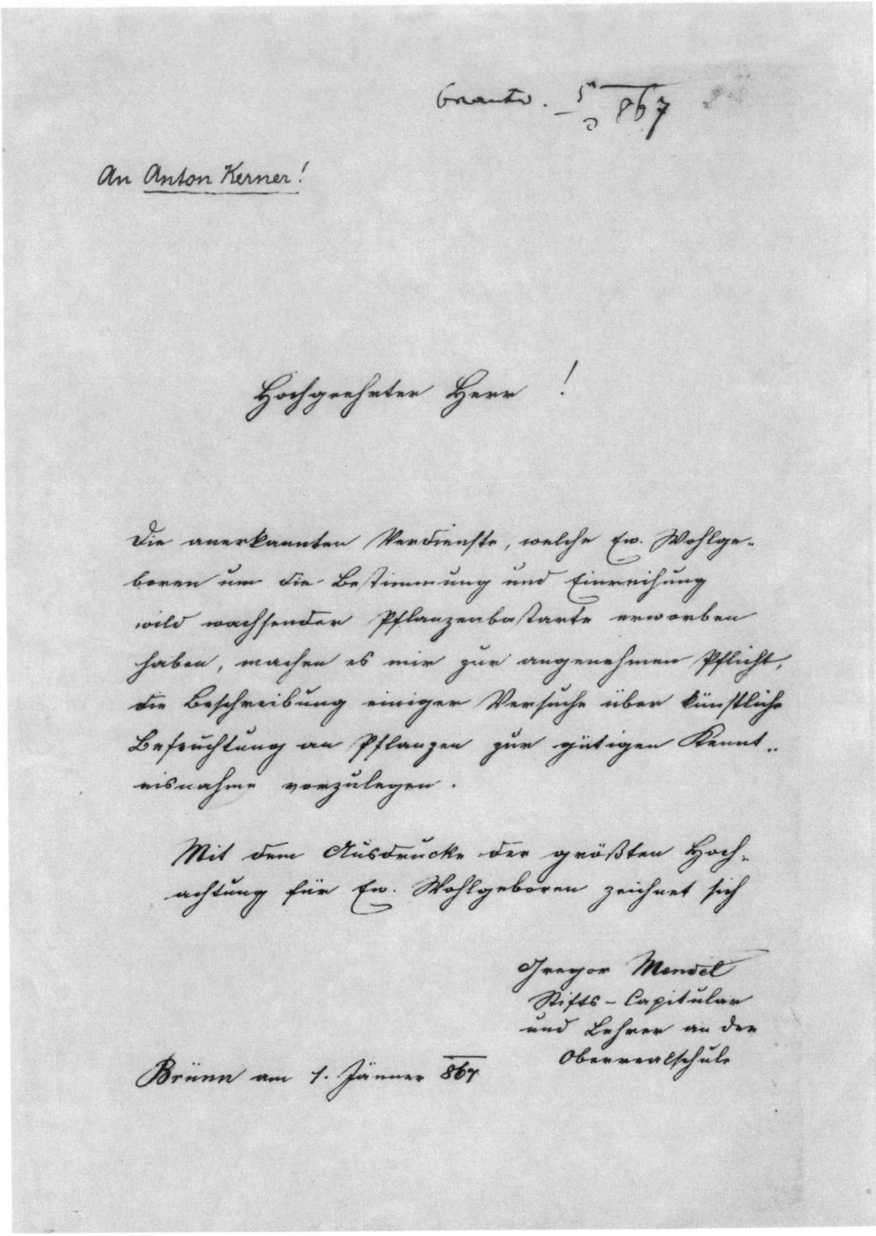

Beantw. – 5/3 867

An Anton Kerner!

Hochgeehrter Herr!

Die anerkannten Verdienste, welche Ew. Wohlgeboren um die Bestimmung und Einreihung wild wachsender Pflanzenbastarte erworben haben, machen es mir zur angenehmen Pflicht, die Beschreibung einiger Versuche über künstliche Befruchtung an Pflanzen zur gütigen Kenntnisnahme vorzulegen.

Mit dem Ausdrucke der größten Hochachtung für Ew. Wohlgeboren zeichnet sich

Gregor Mendel
Stifts-Capitular
und Lehrer an der
Oberrealschule

Brünn am 1. Jänner 867

A draft of a letter in Mendel's hand to Anton Kerner to whom he sent a reprint of his Pisum paper immediately after publishing. Dated in Brno, 01 January 1867, signed by Gregor Mendel

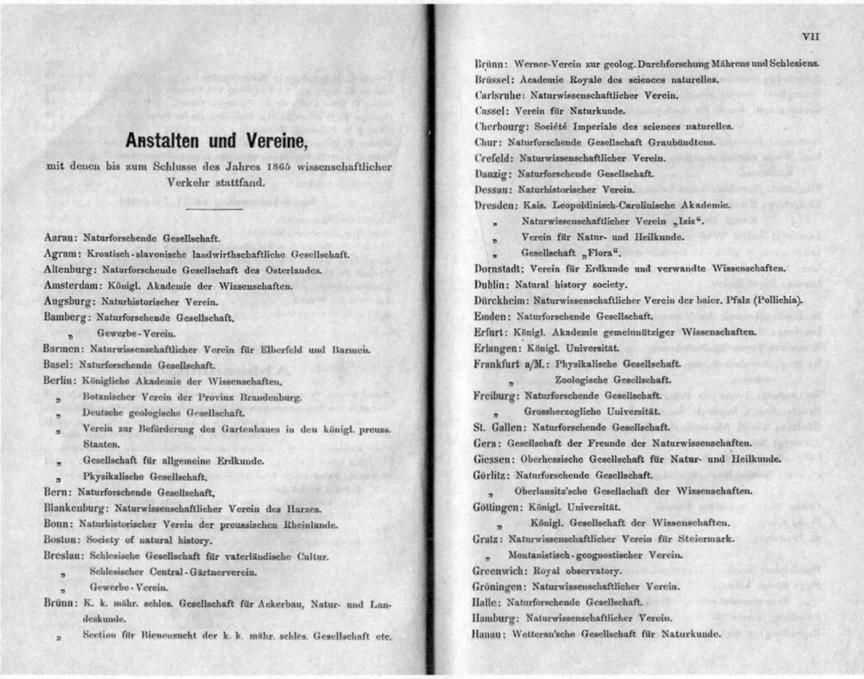

Anstalten und Vereine,

mit denen bis zum Schlusse des Jahres 1865 wissenschaftlicher Verkehr stattfand.

Aarau: Naturforschende Gesellschaft.
Agram: Kroatisch-slavonische landwirthschaftliche Gesellschaft.
Altenburg: Naturforschende Gesellschaft des Osterlandes.
Amsterdam: Königl. Akademie der Wissenschaften.
Augsburg: Naturhistorischer Verein.
Bamberg: Naturforschende Gesellschaft.
„ Gewerbe-Verein.
Barmen: Naturwissenschaftlicher Verein für Elberfeld und Barmen.
Basel: Naturforschende Gesellschaft.
Berlin: Königliche Akademie der Wissenschaften.
„ Botanischer Verein der Provinz Brandenburg.
„ Deutsche geologische Gesellschaft.
„ Verein zur Beförderung des Gartenbaues in den königl. preuss. Staaten.
„ Gesellschaft für allgemeine Erdkunde.
„ Pkysikalische Gesellschaft.
Bern: Naturforschende Gesellschaft.
Blankenburg: Naturwissenschaftlicher Verein des Harzes.
Bonn: Naturhistorischer Verein der preussischen Rheinlande.
Boston: Society of natural history.
Breslau: Schlesische Gesellschaft für vaterländische Cultur.
„ Schlesischer Central-Gärtnerverein.
„ Gewerbe-Verein.
Brünn: K. k. mähr. schles. Gesellschaft für Ackerbau, Natur- und Landeskunde.
„ Section für Bienenzucht der k. k. mähr. schles. Gesellschaft etc.

VII

Brünn: Werner-Verein zur geolog. Durchforschung Mährens und Schlesiens.
Brüssel: Academie Royale des sciences naturelles.
Carlsruhe: Naturwissenschaftlicher Verein.
Cassel: Verein für Naturkunde.
Cherbourg: Société Imperiale des sciences naturelles.
Chur: Naturforschende Gesellschaft Graubündtens.
Crefeld: Naturwissenschaftlicher Verein.
Danzig: Naturforschende Gesellschaft.
Dessau: Naturhistorischer Verein.
Dresden: Kais. Leopoldinisch-Carolinische Akademie.
„ Naturwissenschaftlicher Verein „Isis".
„ Verein für Natur- und Heilkunde.
„ Gesellschaft „Flora".
Dornstadt: Verein für Erdkunde und verwandte Wissenschaften.
Dublin: Natural history society.
Dürckheim: Naturwissenschaftlicher Verein der baier. Pfalz (Pollichia).
Emden: Naturforschende Gesellschaft.
Erfurt: Königl. Akademie gemeinnütziger Wissenschaften.
Erlangen: Königl. Universität.
Frankfurt a/M.: Physikalische Gesellschaft.
„ Zoologische Gesellschaft.
Freiburg: Naturforschende Gesellschaft.
„ Grossherzogliche Universität.
St. Gallen: Naturforschende Gesellschaft.
Gera: Gesellschaft der Freunde der Naturwissenschaften.
Giessen: Oberhessische Gesellschaft für Natur- und Heilkunde.
Görlitz: Naturforschende Gesellschaft.
„ Oberlausitz'sche Gesellschaft der Wissenschaften.
Göttingen: Königl. Universität.
„ Königl. Gesellschaft der Wissenschaften.
Gratz: Naturwissenschaftlicher Verein für Steiermark.
„ Montanistisch-geognostischer Verein.
Greenwich: Royal observatory.
Gröningen: Naturwissenschaftlicher Verein.
Halle: Naturforschende Gesellschaft.
Hamburg: Naturwissenschaftlicher Verein.
Hanau: Wetterau'sche Gesellschaft für Naturkunde.

The list of institutes and societies in contact with the *Naturforschender Verein* in 1865. Published in the 1866 volume of *Verhandlungen des naturforschenden Vereines in Brünn/Brno*. Four pages

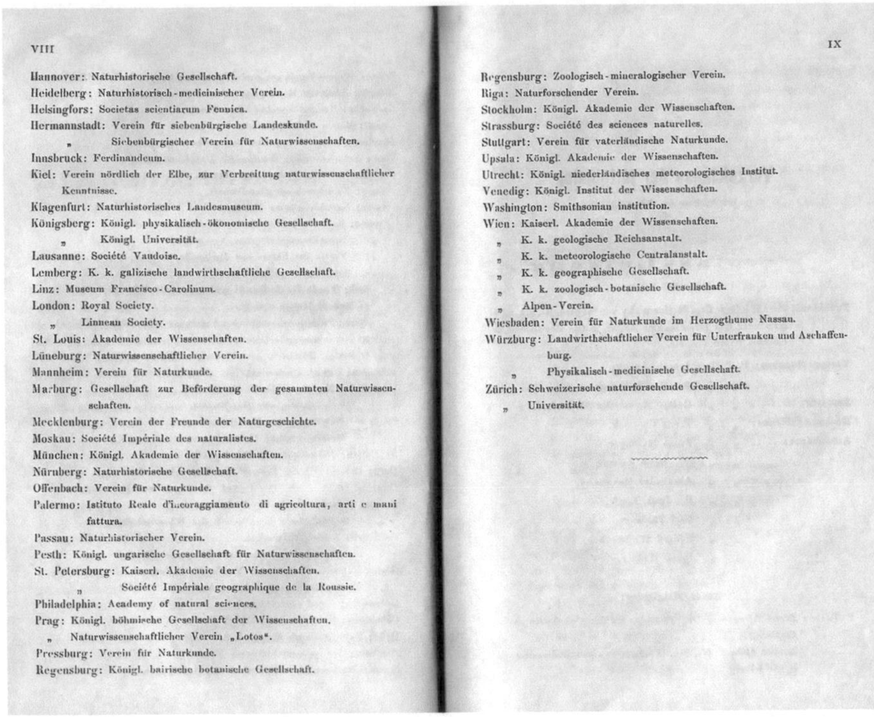

VIII

Hannover: Naturhistorische Gesellschaft.
Heidelberg: Naturhistorisch-medicinischer Verein.
Helsingfors: Societas scientiarum Fennica.
Hermannstadt: Verein für siebenbürgische Landeskunde.
" Siebenbürgischer Verein für Naturwissenschaften.
Innsbruck: Ferdinandeum.
Kiel: Verein nördlich der Elbe, zur Verbreitung naturwissenschaftlicher Kenntnisse.
Klagenfurt: Naturhistorisches Landesmuseum.
Königsberg: Königl. physikalisch-ökonomische Gesellschaft.
" Königl. Universität.
Lausanne: Société Vaudoise.
Lemberg: K. k. galizische landwirthschaftliche Gesellschaft.
Linz: Museum Francisco-Carolinum.
London: Royal Society.
" Linnean Society.
St. Louis: Akademie der Wissenschaften.
Lüneburg: Naturwissenschaftlicher Verein.
Mannheim: Verein für Naturkunde.
Marburg: Gesellschaft zur Beförderung der gesammten Naturwissenschaften.
Mecklenburg: Verein der Freunde der Naturgeschichte.
Moskau: Société Impériale des naturalistes.
München: Königl. Akademie der Wissenschaften.
Nürnberg: Naturhistorische Gesellschaft.
Offenbach: Verein für Naturkunde.
Palermo: Istituto Reale d'incoraggiamento di agricoltura, arti e manifattura.
Passau: Naturhistorischer Verein.
Pesth: Königl. ungarische Gesellschaft für Naturwissenschaften.
St. Petersburg: Kaiserl. Akademie der Wissenschaften.
" Société Impériale geographique de la Roussie.
Philadelphia: Academy of natural sciences.
Prag: Königl. böhmische Gesellschaft der Wissenschaften.
" Naturwissenschaftlicher Verein „Lotos".
Pressburg: Verein für Naturkunde.
Regensburg: Königl. bairische botanische Gesellschaft.

IX

Regensburg: Zoologisch-mineralogischer Verein.
Riga: Naturforschender Verein.
Stockholm: Königl. Akademie der Wissenschaften.
Strassburg: Société des sciences naturelles.
Stuttgart: Verein für vaterländische Naturkunde.
Upsala: Königl. Akademie der Wissenschaften.
Utrecht: Königl. niederländisches meteorologisches Institut.
Venedig: Königl. Institut der Wissenschaften.
Washington: Smithsonian institution.
Wien: Kaiserl. Akademie der Wissenschaften.
" K. k. geologische Reichsanstalt.
" K. k. meteorologische Centralanstalt.
" K. k. geographische Gesellschaft.
" K. k. zoologisch-botanische Gesellschaft.
" Alpen-Verein.
Wiesbaden: Verein für Naturkunde im Herzogthume Nassau.
Würzburg: Landwirthschaftlicher Verein für Unterfranken und Aschaffenburg.
" Physikalisch-medicinische Gesellschaft.
Zürich: Schweizerische naturforschende Gesellschaft.
" Universität.

(continued)

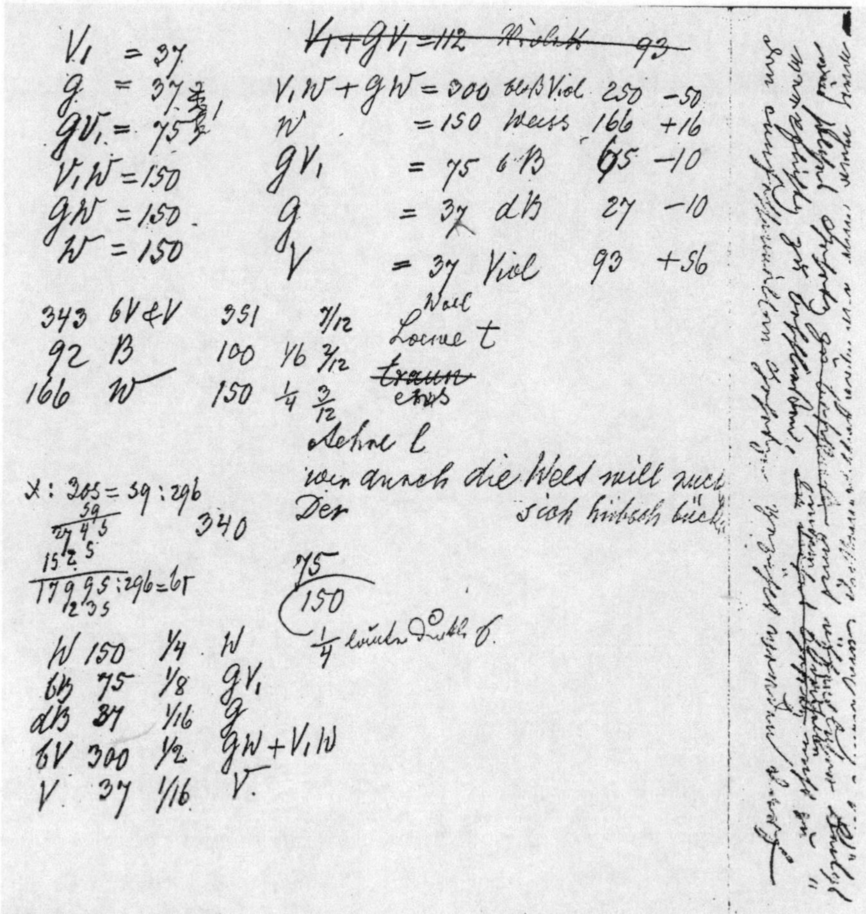

D125 Mendel's analysis of one of his hybridization experiments (Heimans 1969)

D229 Mendel's second note called *Notizblatt Two*, referring to Hieracium, Salix and to Wichura. Mendel is most probably comparing his results in hawkweed with Wichura's experiments with willow (Heimans 1970)

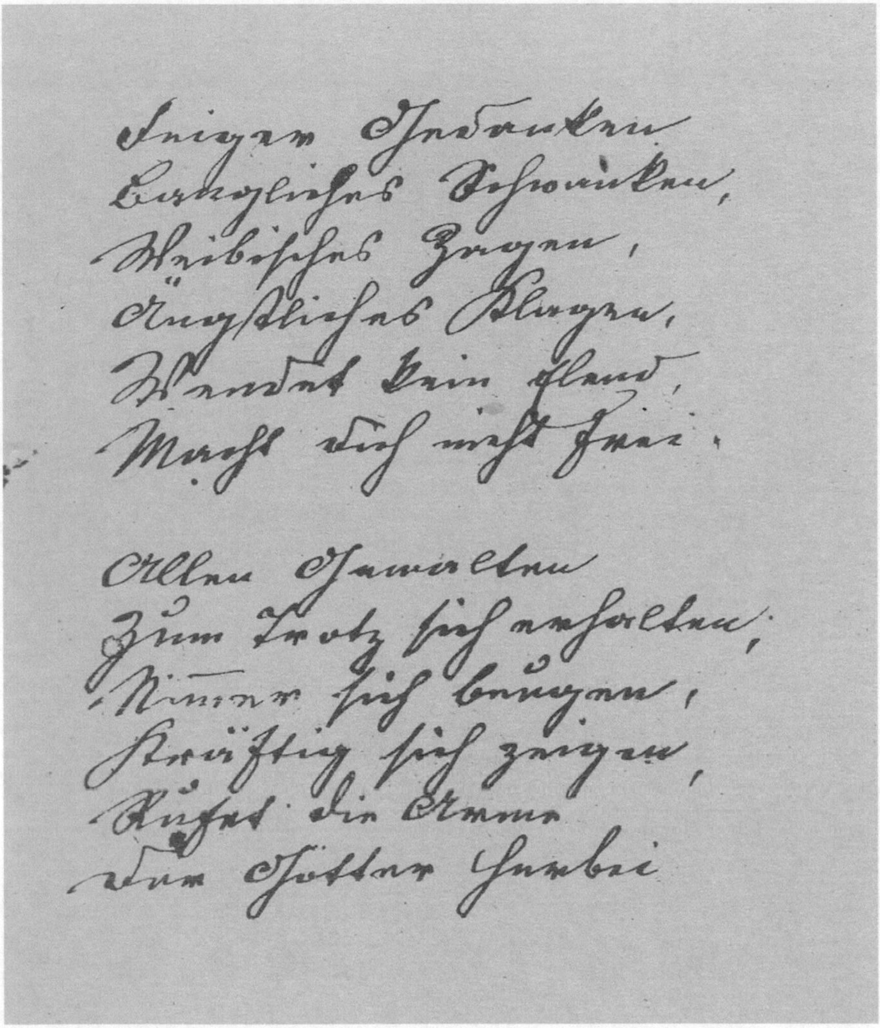
Feigen Gedanken
Bängliches Schwanken,
Weibisches Zagen,
Ängstliches Klagen,
Wendet kein Elend,
Macht dich nicht frei.

Allen Gewalten
Zum Trotz sich erhalten,
Nimmer sich beugen,
Kräftig sich zeigen,
Rufet die Arme
Der Götter herbei

D72 Mendel's handwritten text copied from Goethe's play *Lilla*

Ueber

einige aus künstlicher Befruchtung gewonnenen

Hieracium-Bastarde

von

G. Mendel.

(Mitgetheilt in der Sitzung vom 9. Juni 1869.)

Wiewohl ich schon mehrfache Befruchtungsversuche zwischen verschiedenen Arten aus dem Genus Hieracium vorgenommen habe, ist es mir bis jetzt doch nur gelungen, folgende 6 Bastarde und diese bloss in einem bis drei Exemplaren zu erhalten:

H. Auricula + H. aurantiacum *)
H. Auricula + H. Pilosella,
H. Auricula + H. pratense,
H. echioides **) + H. aurantiacum,
H. praealtum + H. flagellare Rchb.,
H. praealtum + H. aurantiacum.

Die Schwierigkeit, Bastarde in einer grösseren Anzahl zu gewinnen, liegt in dem Umstande, dass es bei der Kleinheit der Blüthen und dem eigenthümlichen Baue derselben nur selten gelingt, die Antheren aus der zu befruchtenden Blüthe zu entfernen, ohne dass der eigene Pollen auf die Narbe gelangt, oder der Griffel verletzt wird und abstirbt. Bekanntlich sind die Antheren in ein Röhrchen verwachsen,

*) Durch diese Bezeichnung wird angedeutet, dass der Bastard aus der Befruchtung des H. Auricula mit dem Pollen des H. aurantiacum erhalten wurde.

**) Diese Versuchspflanze ist nicht genau das typische H. echiodes. Sie scheint der Uebergangsreihe zu H. praealtum anzugehören, steht jedoch dem H. echioides näher, wesshalb sie auch den Formenkreis des letzteren eingestellt wurde.

Mendel's printed paper *On Hieracium Hybrids Obtained By Artificial Fertilisation. Verhandlungen des naturforschenden Vereines in Brünn. Abhandlungen.* Vol. 8 (for 1869), 26–31, 1870

O27 Gregor Mendel as abbot and prelate (elected 1868). Reproduction by Carl Pietzner Brünn/Brno, Franzensglacis 9

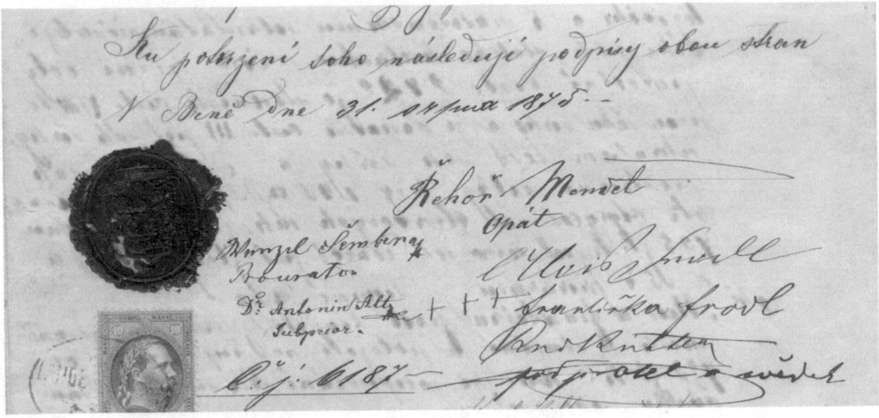
Ku potvrzení toho následují podpisy obou stran

Řehoř Mendel
Opát

Dr Antonín Alt
Subprior

D502 A contract dated in Brno, 21 August 1875, was written in Czech, and signed by Mendel in Czech *Řehoř Mendel—opát*

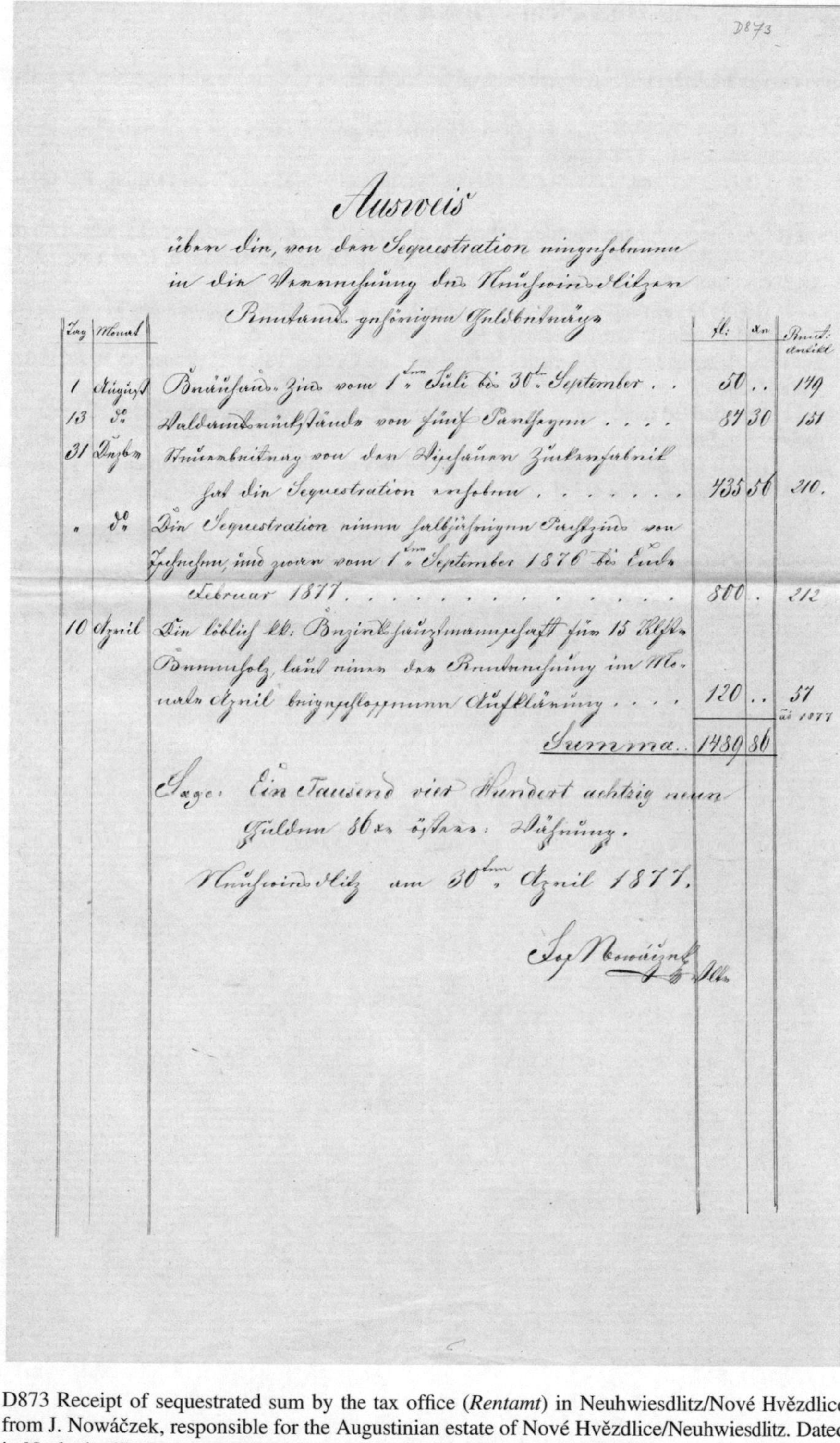

D873

Ausweis

über die, von der Sequestration eingehobene in die Verrechnung des Neuwiesdlitzer Rentamt gehörige Geldbeträge

Tag	Monat		fl.	xr	Rent. Artikel
1	August	Bräuhaus-Zins vom 1ten Juli bis 30t. September . .	50	..	149
13	do	Waldamtsrückstände von fünf Partheyen	87	30	151
31	Dezbr	Neuerbeitrag von der Wischauer Zuckerfabrik hat die Sequestration erhoben	435	56	210.
"	do	Die Sequestration einen halbjährigen Pachtzins von Zeschna, und zwar vom 1ten September 1876 bis Ende Februar 1877.	800	.	212
10	April	Die löbliche k.k. Bezirkshauptmannschaft für 15 Klftr. Brennholz, laut einer der Rentrechnung im Monate April beigeschlossenen Aufklärung . . .	120	..	57 ad 1877
		Summa..	1489	86	

Sage: Ein Tausend vier Hundert achtzig neun Gulden 86 xr österr: Währung.

Neuwiesdlitz am 30ten April 1877.

Jos Nowáczek

D873 Receipt of sequestrated sum by the tax office (*Rentamt*) in Neuhwiesdlitz/Nové Hvězdlice from J. Nowáčzek, responsible for the Augustinian estate of Nové Hvězdlice/Neuhwiesdlitz. Dated in Neuhwiesdlitz/Nové Hvězdlice, 30 April 1877, signed by Jos. Nowáčzek

References and Historical Printed Sources

Abbott S, Fairbanks DJ (2016) Experiments on plant hybrids by Gregor Mendel. Genetics 204:407–422

Heimans J (1969) Ein Notizblatt aus dem Nachlass Gregor Mendels mit Analysen eines seiner Kreuzungsversuche. FM 4:5–36

Heimans J (1970) A recently discovered note on hybridization in Mendel's handwriting. FM 6:91–103

Iltis H (1924) Gregor Johann Mendel. Leben, Werk und Wirkung. Julius Springer, Berlin. Life of Mendel. English translation by Eden and Cedar Paul. First edition. Norton, New York 1932; Second edition Allen and Unwin London, 1966

Mann W (1992) Erinnerungen an Johann Gregor Mendel mit einem Faksimile des Manuskriptes von Mendels Arbeit "Versuche über Pflanzenhybriden", Darmstadt

Matalová A, Sekerák J (2004) Genetics Behind the Iron Curtain. Moravian Museum Brno. ISBN 9788070282465

Stern C, Sherwood ER (eds) (1966) The origin of genetics. A Mendel Source Book. W. H. Freeman and co, San Francisco

Vybral V (1971) Der Streit Mendels mit der Staatsverwaltung über die Beitragspflicht des Klosters zum Religionsfonde. FM 6:231–237

Chapter 12
Ceiling Paintings 1875

The ceiling paintings in the Great Chapter Hall and Library Hall in the Old Brno Augustinian monastery represent Mendel's scientific activities. In the Library Hall ceiling paintings, Mendel's emblem and abbot Napp's emblem symbolize the golden era of natural science in the Augustinian monastery.

Mendel had his monogram GM and the date 1875 painted on the ceiling in the Great Chapter Hall. He used the opportunity that arose when restoration work became necessary in the prelate quarters whose roofs were damaged by the tornado in 1870.

The artist of the ceiling paintings has not been identified. The paintings were completed in accordance with Mendel's proposal. He is the spiritual father of the paintings. The Great Chapter Hall and the Library Hall are situated on the first floor of the frontal wing of the monastery, facing Pekařská Street. The statue of St. Augustine in the facade of the first floor is located in the middle of the four windows leading to the Great Chapter Hall. The paintings in the Great Chapter Hall no longer exist. They were removed by the Augustinians while redecorating the Great Chapter Hall. Fortunately, Anselm Matoušek, the then curator of the Mendel Museum in the monastery, had photographs made of them. Thanks to him images of these paintings were published by Richter (1943, pp. 121-3). The ceiling paintings in the Library Hall were painted by the same artist as those in the Great Chapter Hall. The original paintings in the Library Hall ceiling have been preserved in their original shape. They have been photographed in colour by Filip Fojtík in 2020 as actualization of Mendel's relics and deposited in the Mendelianum in the Moravian Museum.

A. Matalová, E. Matalová, *Gregor Mendel - The Scientist*, Springer Biographies,
https://doi.org/10.1007/978-3-030-98923-1_12

IM22 The Great Chapter Hall in the Augustinian monastery with the original ceiling paintings commissioned by Mendel. Photo of unknown date and origin

Four paintings in the ceiling corners of the Great Chapter Hall, representing agriculture, apiculture and pomology, portrayed Mendel's three scientific specializations in the Agricultural Society and meteorology in the *Naturforschender Verein*. In the centre of the ceiling of the Great Chapter Hall close to the chandelier, Mendel commissioned portraits of St. Augustine and St. Monica. They represent the Order of St. Augustine. The paintings in the Library Hall depict Mendel's significant role in organization of exhibits of flowers and fruits organized by the Agricultural Society each year. They were the top events in the social and cultural life in Brno and Moravia

MMM 1218 The Augustinian Library Hall is situated on the first floor (the five arched windows) above the refectory

References and Historical Printed Sources

Richter O (1943) Johann Gregor Mendel wie er wirklich war. Neue Beiträge zur Biographie des berühmten Biologen aus Brünns Archiven mit 31 Abbildungen im Texte. Verhandlungen des Naturforschenden Vereines in Brünn, Abteilung für Naturforschung der Deutschen Gesellschaft für Wissenschaft und Volkstumforschung in Mähren 74 (für das Jahr 1942):1-262. Druck von Josef Klär, Brno

Chapter 13
Supporter of Agriculture

After his return from Znojmo to the Old Brno monastery in 1850, Mendel began to work in the Agricultural Society. He joined its Natural Science Section on 23 July 1851. His superior, Abbot Napp, was a significant member of the Agricultural Society which established the *first* museum in Moravia that merged with the Society in 1817. Mendel co-signed the annual reports of the museum, reviewed books to be purchased for the museum library and influenced its acquisition policy. The museum library opened three reading rooms for the public in 1828.

The Agricultural Society had many relatively independent sections or associations that were created or abolished depending on current needs. After 1848, the Agricultural Society was restructured. The Sheep Breeders´ Association, controlled by Napp, was important during the boom of the textile industry, but it became outdated and was not included in the new structure of the Society. The sections of the Agricultural Society that were active in Mendel's day were specialized in pomiculture, viniculture and gardening, forestry, apiculture, agriculture, meteorology, history and statistics. Mendel, and the physician/naturalist Olexik, who had his private greenhouses and a large garden, organized with Mendel the Society's exhibits of flowers, vegetables and fruits. Formerly, the exhibits had been a domain of Napp and Diebl. The organizers had the privilege of announcing prizes awarded to the best-displayed flowers, fruits and vegetables. Many of the finest products were achieved through hybridization.

Specialized schools of gardening, forestry, vine- and fruit tree growing were founded. Experimental fields and plots were available to researchers. Mendel conducted his experiments in the gardens of the Old Brno monastery surrounded by tall brick walls.

Mendel was a fully qualified member of the Agricultural Society. He completed examinations given by F. Diebl in Brno in agriculture and was also awarded a certificate in pomiculture and viniculture in 1846. The symbol of agriculture in one corner of the ceiling in the Great Chapter Hall is symbolized by the kneeling figure of St. Isidor with a pilgrim's stick. He is the patron of peasants. An angel with a plough drawn by two bulls is reminiscent of the plough in Mendel's abbatial emblem. It

A. Matalová, E. Matalová, *Gregor Mendel - The Scientist*, Springer Biographies,
https://doi.org/10.1007/978-3-030-98923-1_13

symbolizes his peasant origin. It may also be considered as a remembrance of Mendel saying farewell to his home country. The improvement of agriculture stood at the forefront of the endeavour of the Agricultural Society. Special attention was paid to production of best seed of useful crop varieties. Different qualities of crop seeds were on display in the museum. After 1872 Mendel became the vice-chairman of the Agricultural Society and made decisions about agricultural subsidies. He was also responsible for distributing high-quality seeds to responsible farmers.

During his student years in Vienna, he published a report on *Rafanus sativus* (Mendel 1853), the larva of the mother-of-pearl moth, which he observed in Brno in a garden plot of devastated radishes. Mendel presented his findings in the *Imperial-Royal Zoological-Botanical Society* in the Court Museum in Vienna headed by director V. Kollar (1797–1860), an entomologist and Mendel's countryman from Silesia. Mendel also conducted research on the pea weevil *Bruchus pisi* (Mendel 1854), a pest that consumes pea seeds.

Mendel reported his findings to Kollar in a letter, which Kollar, in turn, presented in a meeting of the *Imperial-Royal Zoological Botanical Society* in Vienna in 1854. Mendel's report stressed the possibility that peas, a nutritionally valuable seed crop, might disappear from the market. The inspectors did not allow peas affected by this pest to be sold. In his contribution, Mendel treated an entomological topic and added practical aspects, as was typical for experts of the Agricultural Society. Mendel followed in Napp's footsteps in the sense that Napp had studied *Melilotus vulgaris, Phalaema noctua segetum* and *Justus terrestis* in cooperation with Diebl (1839). Those pests caused damage to seed production in the monastery's farms. In 1874, the Agricultural Society made a proposal to the central organs in Vienna to create an institution for plant pest prevention (Orel 1996).

Mendel was a recognized European expert in seed production. C. W. Eichling, a representative of one of the large French plant specialty establishments, visited Mendel in the Old Brno Monastery on recommendation of Ernst Benary from Erfurt in the summer of 1878 (Eichling 1942) to find out information about pea seed varieties.

Mendel, as a member of the Central Board of the Agricultural Society in Brno, chaired the plenary sessions and made important decisions in the name of the president of the Agricultural Society. Mendel was a respected authority in the Agricultural Society, proposing candidates for medals awarded by the Agricultural Society for special merits. In 1882, he was offered the position of Director of the Agricultural Society, but he declined for health reasons.

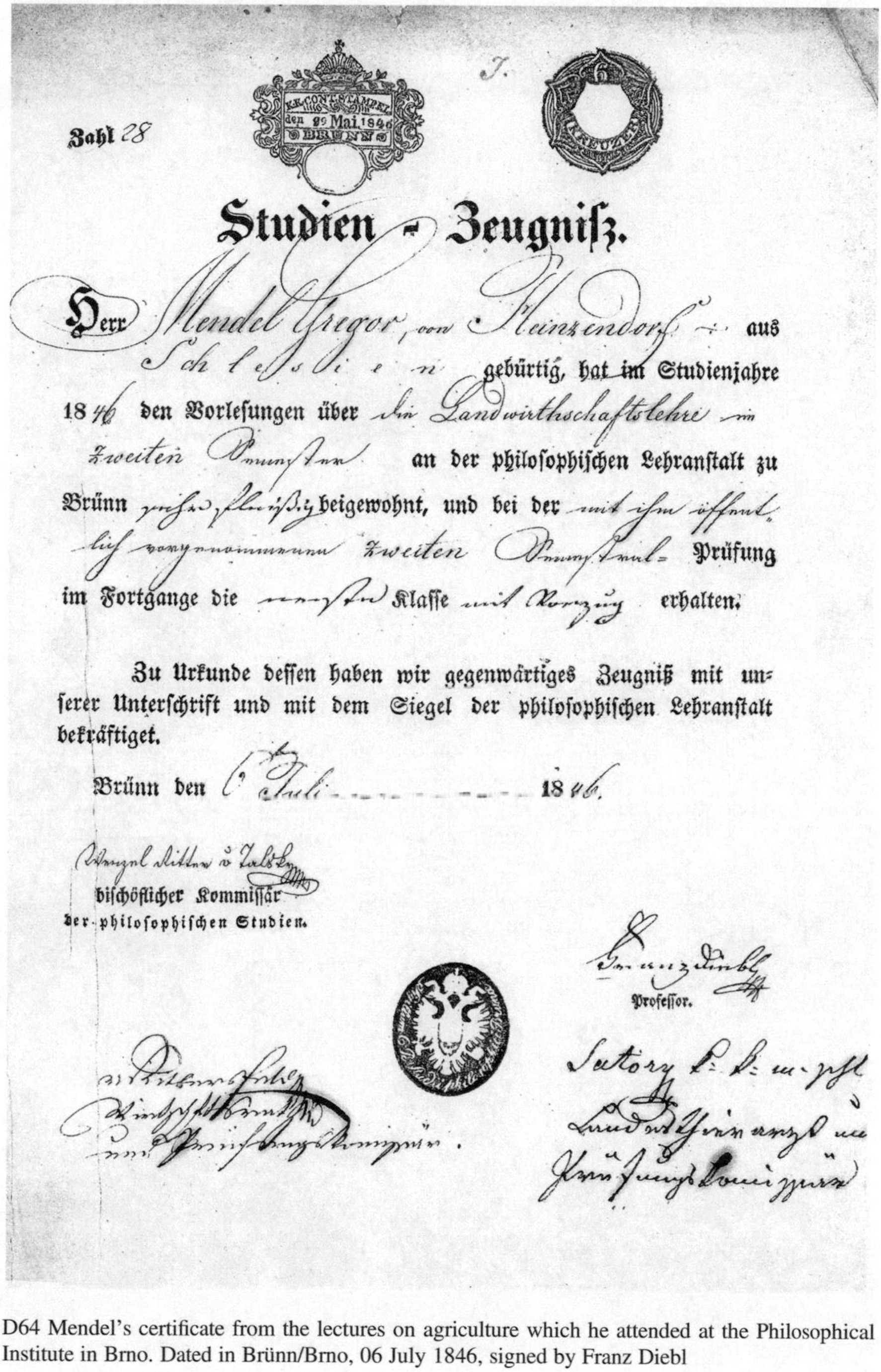

Zahl 28

Studien-Zeugniß.

Herr Mendel Gregor, von Heinzendorf aus Schlesien gebürtig, hat im Studienjahre 1846 den Vorlesungen über die Landwirthschaftslehre im zweiten Semester an der philosophischen Lehranstalt zu Brünn sehr fleißig beigewohnt, und bei der mit ihm öffentlich vorgenommenen zweiten Semestral-Prüfung im Fortgange die erste Klasse mit Vorzug erhalten.

Zu Urkunde dessen haben wir gegenwärtiges Zeugniß mit unserer Unterschrift und mit dem Siegel der philosophischen Lehranstalt bekräftiget.

Brünn den 6 Juli 1846.

bischöflicher Kommissär der philosophischen Studien.

Franz Diebl
Professor.

D64 Mendel's certificate from the lectures on agriculture which he attended at the Philosophical Institute in Brno. Dated in Brünn/Brno, 06 July 1846, signed by Franz Diebl

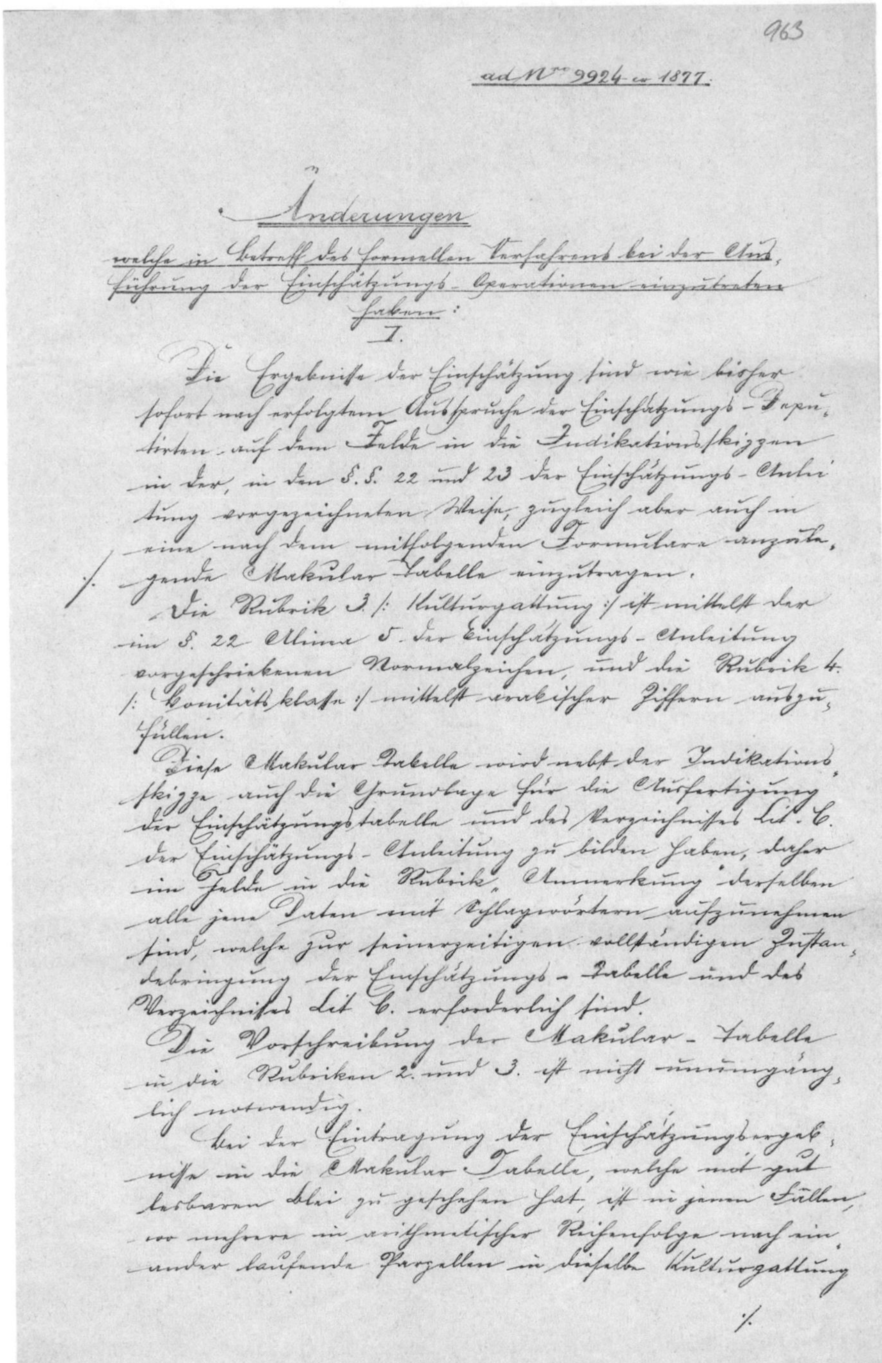

963

ad Nro 9924 ex 1877.

Änderungen

welche in Betreff des formellen Verfahrens bei der Ausführung der Einschätzungs-Operationen einzutreten haben:

I.

Die Ergebnisse der Einschätzung sind wie bisher sofort nach erfolgtem Ausspruche der Einschätzungs-Deputirten auf dem Felde in die Indikationsskizzen in der, in den §. §. 22 und 23 der Einschätzungs-Anleitung vorgezeichneten Weise, zugleich aber auch in eine nach dem mitfolgenden Formulare anzulegende Makular-Tabelle einzutragen. /.

Die Rubrik 3. /: Kulturgattung :/ ist mittelst der im §. 22 Alinea 5. der Einschätzungs-Anleitung vorgeschriebenen Normalzeichen, und die Rubrik 4. /: Bonitätsklasse :/ mittelst arabischer Ziffern auszufüllen.

Diese Makular-Tabelle wird nebst der Indikationsskizze auf die Grundlage für die Anfertigung der Einschätzungstabelle und des Verzeichnisses Lit. C. der Einschätzungs-Anleitung zu bilden haben, daher im Felde in die Rubrik „Anmerkung" derselben alle jene Daten mit Schlagwörtern aufzunehmen sind, welche zur seinerzeitigen vollständigen Zustandebringung der Einschätzungs-Tabelle und des Verzeichnisses Lit C. erforderlich sind.

Die Vorschreibung der Makular-Tabelle in die Rubriken 2. und 3. ist nicht unumgänglich notwendig.

Bei der Eintragung der Einschätzungsergebnisse in die Makular-Tabelle, welche mit gut lesbarem Blei zu geschehen hat, ist in jenen Fällen, wo mehrere in arithmetischer Reihenfolge nach einander laufende Parzellen in dieselbe Kulturgattung

/.

D963 The first page of the methods of the Commission of the Provincial Diet responsible for fixed agricultural taxation. Mendel was nominated as a member of the Moravian commission by the Minister of Finance on 29 January 1870

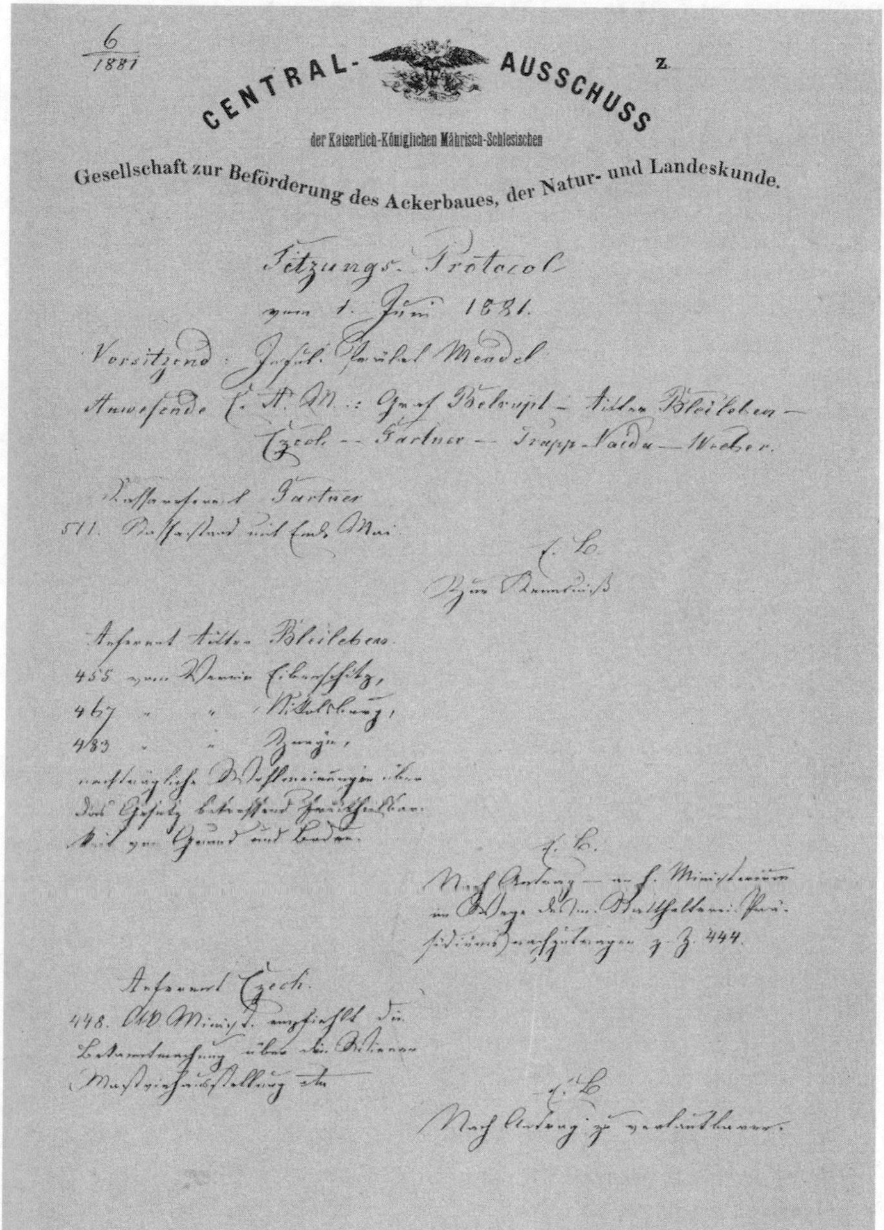

6/1881

CENTRAL-AUSSCHUSS

z.

der Kaiserlich-Königlichen Mährisch-Schlesischen

Gesellschaft zur Beförderung des Ackerbaues, der Natur- und Landeskunde.

Sitzungs-Protocol

vom 1. Juni 1881.

Vorsitzend: Inful. Prälat Mendel

Anwesende C. A. M.: Graf Belcredi – Ritter Bleileben – Czech – Gartner – Trapp – Vaida – Weber.

[illegible]

The first page of the minutes of the session of the Central Board of the Agricultural Society written in Mendel's hand. "The meeting was presided by Infulated Prelate Mendel" (Inful. Prälat Mendel). Dated 01 June 1881

References and Historical Printed Sources

Eichling GW (1942) I talked with Mendel. J Hered 33:243–256

Mendel G (1853) Über Verwüstung am Gartenrettich durch Raupen Botys margaritalis. Verhandlungen des zoologisch-botanischen Vereines in Wien 3:116–118

Mendel G (1854) Über Bruchus pisi. Mendel's written communication reported by V. Kollar. Verhandlungen des zoologisch-botanischen Vereines in Wien 4:27–28

Orel V (1996) Gregor Mendel the first geneticist. Translated by Stephen Finn. Oxford University Press. ISBN: 0 19 854774 9

Chapter 14
Expert in Meteorology

The symbol of meteorology in the Great Chapter Hall paintings shows a globe, compass, pinnacle, books and maps. The pyramids, palms and boat represent geographical discoveries, the telescope points to astronomy, the thermometer to meteorology.

Mendel was invited to participate in meteorological observations by P. Olexik who was appointed the Brno's meteorologist by the Viennese Institute for Meteorology and Earth Magnetism in 1857. Meteorological activities started in Brno in 1849.

Mendel was invited by Olexik to elaborate graphic tabular overviews of the data from his meteorological observations. In this connection, Mendel's name was first published in the report of 1862. In addition to its publication in the *Verhandlungen des naturforschenden Vereines*, it was also appended to Makowsky's publication on the Flora of the Brno District (Makowsky 1863), published in 500 copies. Abbot Mendel purchased founding membership in the Viennese Institute for Meteorology and Earth Magnetism. On the official obituary notice of Mendel's death issued by the Augustinian monastery in January 1884, his founding membership in that institute is specifically stated, along with the most significant positions he held during his life. After Olexik's passing, Mendel founded a meteorological observatory in the Old Brno Augustinian Monastery. Some meteorological instruments were transferred from Olexik's meteorological station on Pekařská street No. 100 to the nearby Augustinian monastery. When Mendel helped Olexik with drawing tables and graphs for publication, he used to add data from his own meteorological measurements. The network of meteorological stations created by Mendel expanded precision of meteorological observations in the country and popularized meteorology through weather forecasting.

From 1865–1880, Mendel measured the groundwater level in the monastery well. He reported his measurements to Liznar who published them (1902). In October 1870, Mendel observed a tornado from the windows of his prelate quarters and published the first remarkable analysis of this rare phenomenon in Moravia.

A. Matalová, E. Matalová, *Gregor Mendel - The Scientist*, Springer Biographies,
https://doi.org/10.1007/978-3-030-98923-1_14

In 1878, Mendel supported the introduction of weather forecasts in Moravia in a newspaper article (Orel 1996, p. 249).

In 1882, Mendel extended his meteorological work to astronomy. Through his new Fritsch telescope, he observed and recorded the intensity and positions of sunspots (Iltis 1924).

On the title page of August Kunzek's *Textbook of Meteorology* (*Lehrbuch der Meteorologie, 1850*), which Mendel studied and annotated, he wrote his conviction: *He who cannot be alone, is not in harmony with himself. (Wer nicht einsam sein kann, ist auch nicht versöhnt mit sich.)*

Most of Mendel's printed publications are in meteorology, published in the scientific annuals of the *Naturforschender Verein in Brünn:*

Bemerkungen zu der graphisch-tabellarischen Uebersicht der meteorologischen Verhältnisse von Brünn. *Verhandlungen des naturforschenden Vereines in Brünn*, vol. 1(1862) 1863, Abhandlungen p. 246–249.

Meteorologische Beobachtungen aus Mähren und Schlesien für das Jahr 1863. *Verhandlungen des naturforschenden Vereines in Brünn*, vol. 2(1863) 1864, p. 99–121.

Meteorologische Beobachtungen aus Mähren und Schlesien für das Jahr 1864. *Verhandlungen des naturforschenden Vereines in Brünn*, vol. 3(1864) 1865, p. 209–220.

Meteorologische Beobachtungen aus Mähren und Schlesien für das Jahr 1865. *Verhandlungen des naturforschenden Vereines in Brünn*, vol. 4(1865) 1866, p. 318–330.

Meteorologische Beobachtungen aus Mähren und Schlesien für das Jahr 1866. *Verhandlungen des naturforschenden Vereines in Brünn*, vol. 5(1866) 1867, p. 160–172.

Meteorologische Beobachtungen aus Mähren und Schlesien für das Jahr 1869. *Verhandlungen des naturforschenden Vereines in Brünn*, vol. 8(1869) 1870, p. 131–143.

Die Windhose vom 13. October 1870. *Verhandlungen des naturforschenden Vereines in Brünn*, vol. 9(1870) 1871, p. 229–246.

Regenfall und Gewitter zu Brünn im Juni 1879. *Zeitschrift der österreichischen Gesellschaft für Meteorologie*, vol. 14, 1879, p. 315–316.

Gewitter in Brünn und Blansko am 15. August. *Zeitschrift der österreichischen Gesellschaft für Meteorologie,* vol. 17, 1882, p. 407–408.

Die Grundlage der Wetterprognosen. *Mittheilungen der k. k. Mährisch-Schlesischen Gesellschaft zur Beförderung des Ackerbaues, der Natur—und Landeskunde Brünn* 59, 1879, p. 29–31.

J. Liznar, an outstanding meteorologist and Mendel's former pupil, was in personal contact with Abbot Mendel and reported on Mendel's meteorological instruments installed in the frontal area of the prelate quarters facing the square. Mendel's meteorological station included a telescope, barometer, sundials, thermometer and ombrometer. Liznar got Mendel's farewell letter 2 weeks before Mendel's death. Translation by Eden and Cedar Paul in Iltis (1966, p. 275): *Dear*

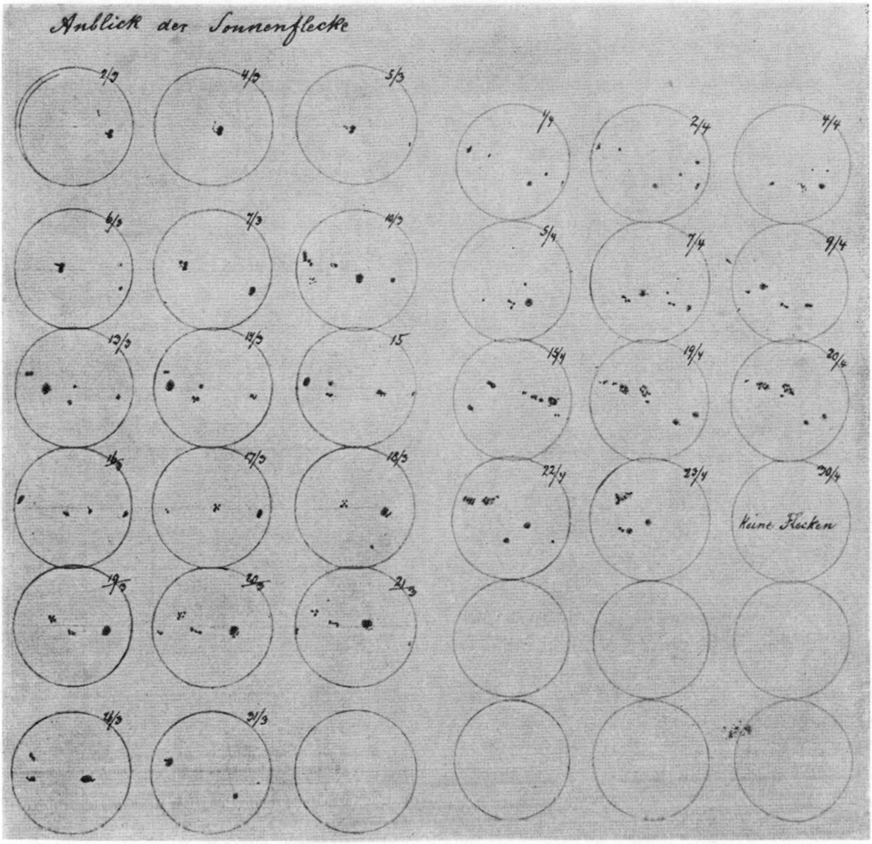

Mendel's drawing of sunspots. The circles are annotated with the day of observation, from 01 March to 30 April. (From Iltis 1924, p. 157.) The original at the University of Illinois in Urbana-Champaign

Friend,—Above all please accept my heartfelt thanks for the various writings of your own you have sent me from time to time, and for which I have not always expressed my gratitude in particular instances.—You are now entering upon the years of most active work, whereas I must be said to be in the opposite condition. Today I have found it necessary to ask to be completely excused from further meteorological observations, for since last May I have been suffering from heart trouble, which is now so severe that I can no longer take the readings of the meteorological instruments without assistance.—Since we are not likely to meet again in this world, let me take this opportunity of wishing you farewell, and of invoking upon your head all the blessings of the meteorological deities.—Best wishes to yourself and to your wife.—Gregor Mendel.

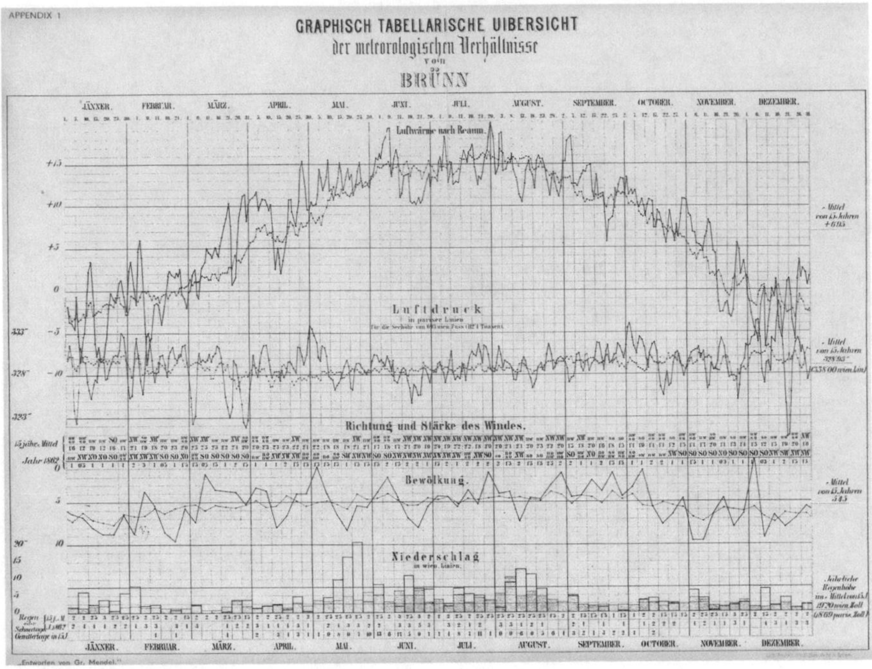

Mendel's first meteorological graphic and tabular overview of meteorological observations in Brno for the year 1862. It was republished appended to Alexander Makowsky's Die Flora des Brünner Kreises. Nach Pflanzengeographischen Prinzipien. Mit einer meteorologischen Tabelle von Prof. G. Mendel in the first annual volume of the *Naturforschender Verein* in Brünn/Brno 1863

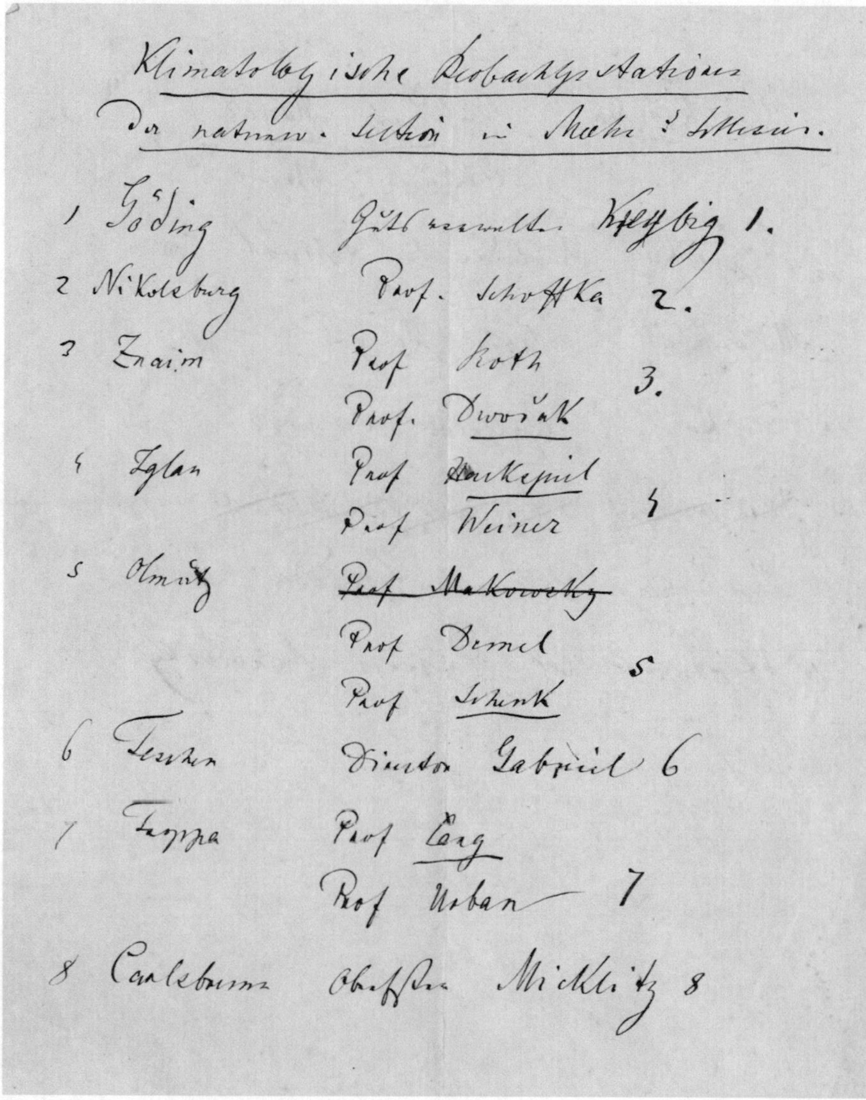
Klimatologische Beobachtungsstationen
der naturw. Sektion in Mähren u. Schlesien.

1 Göding Gutsverwalter Kellig 1.
2 Nikolsburg Prof. Schoffka 2.
3 Znaim Prof Roth
Prof. Dvořák 3.
4 Iglau Prof Hackspiel
Prof Weiner 4
5 Olmütz ~~Prof Makowsky~~
Prof Demel
Prof Schenk 5
6 Teschen Director Gabriel 6
7 Troppau Prof Lang
Prof Urban 7
8 Carlsbrunn Oberförster Micklitz 8

D675 The list of climatological observation stations in Moravia and Silesia registered by the Natural Science Section of the Agricultural Society. Two pages. In July 1862, it was decided to create a network of meteorological observers in Moravia and Silesia. Mendel was made responsible for their activity

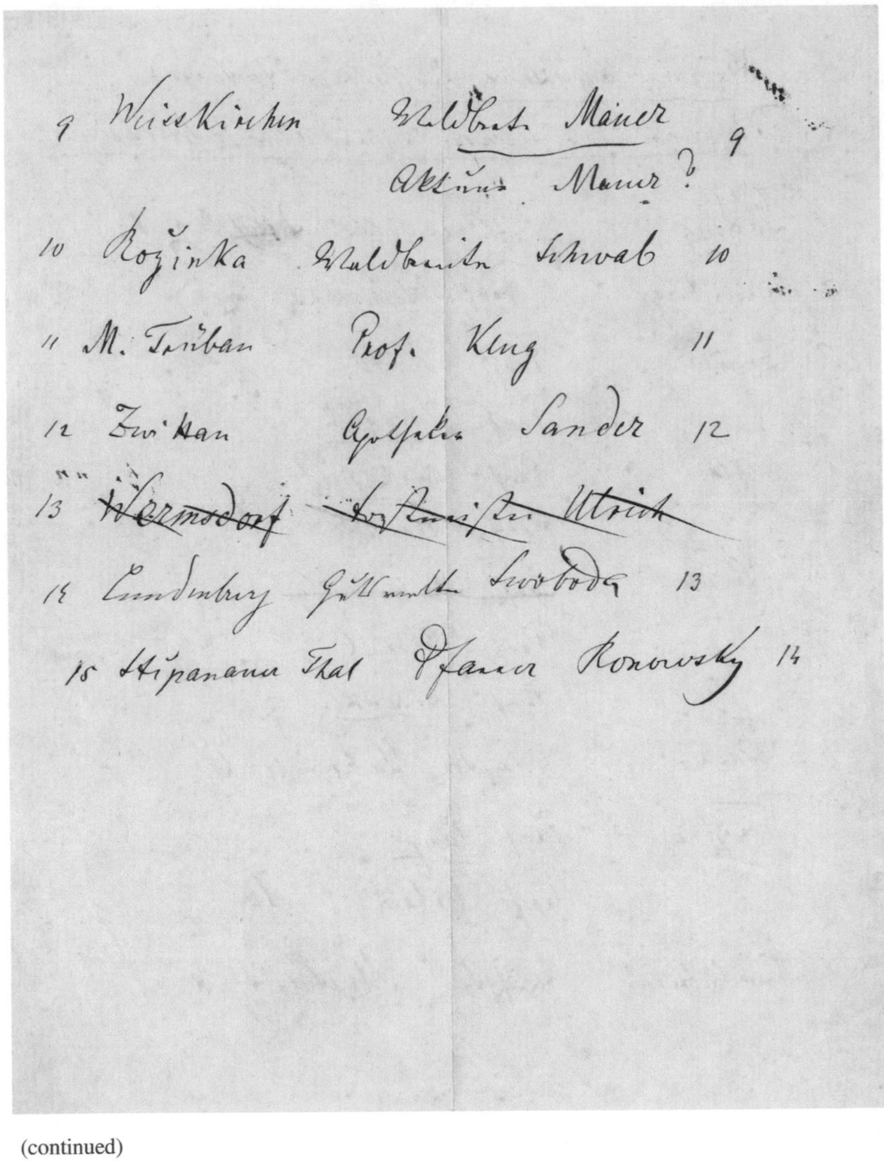

9	Weisskirchen	Waldbereiter Mauer	9
		Aktuar Mauer?	
10	Rožinka	Waldbereiter Schwab	10
11	M. Trübau	Prof. Kny	11
12	Zwittau	Apotheker Sander	12
13	~~Wermsdorf~~	~~Postmeister Ulrich~~	
14	Lundenburg	Gutsverwalter Swoboda	13
15	[illegible] Thal	Pfarrer Konowsky	14

(continued)

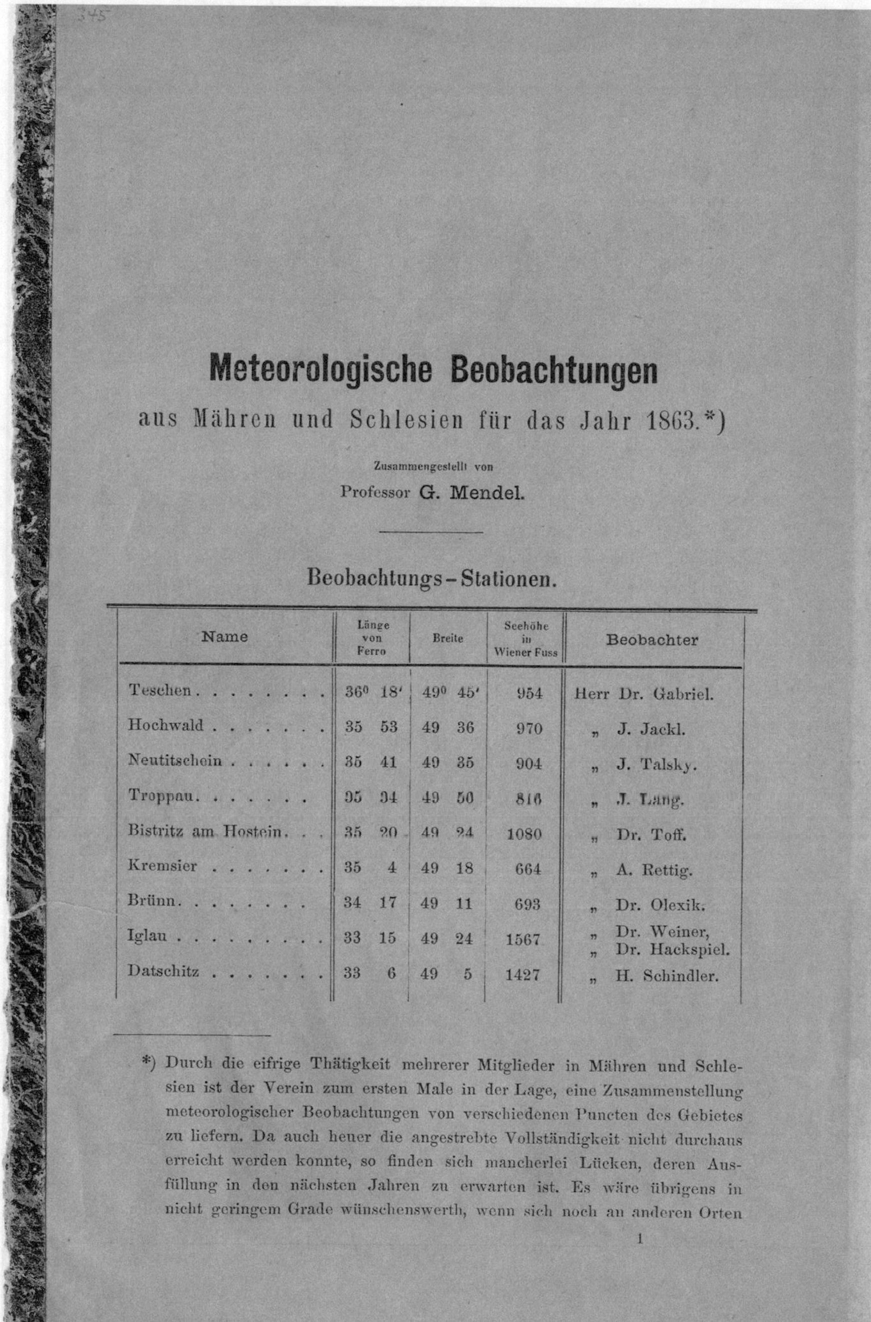

345

Meteorologische Beobachtungen

aus Mähren und Schlesien für das Jahr 1863.*)

Zusammengestellt von

Professor G. Mendel.

Beobachtungs-Stationen.

Name	Länge von Ferro		Breite		Seehöhe in Wiener Fuss	Beobachter
Teschen	36°	18′	49°	45′	954	Herr Dr. Gabriel.
Hochwald	35	53	49	36	970	„ J. Jackl.
Neutitschein	35	41	49	35	904	„ J. Talsky.
Troppau	35	34	49	50	816	„ J. Lang.
Bistritz am Hostein	35	20	49	24	1080	„ Dr. Toff.
Kremsier	35	4	49	18	664	„ A. Rettig.
Brünn	34	17	49	11	693	„ Dr. Olexik.
Iglau	33	15	49	24	1567	„ Dr. Weiner, „ Dr. Hackspiel.
Datschitz	33	6	49	5	1427	„ H. Schindler.

*) Durch die eifrige Thätigkeit mehrerer Mitglieder in Mähren und Schlesien ist der Verein zum ersten Male in der Lage, eine Zusammenstellung meteorologischer Beobachtungen von verschiedenen Puncten des Gebietes zu liefern. Da auch heuer die angestrebte Vollständigkeit nicht durchaus erreicht werden konnte, so finden sich mancherlei Lücken, deren Ausfüllung in den nächsten Jahren zu erwarten ist. Es wäre übrigens in nicht geringem Grade wünschenswerth, wenn sich noch an anderen Orten

1

D345 The first page of Mendel's published report on meteorological observations in Moravia and Silesia in 1862 appended to *Die Flora des Brünner Kreises* by Alexander Makowsky in the *Verhandlungen des naturforschenden Vereines in Brünn,* Abhandlungen. Brünn/Brno, 1863

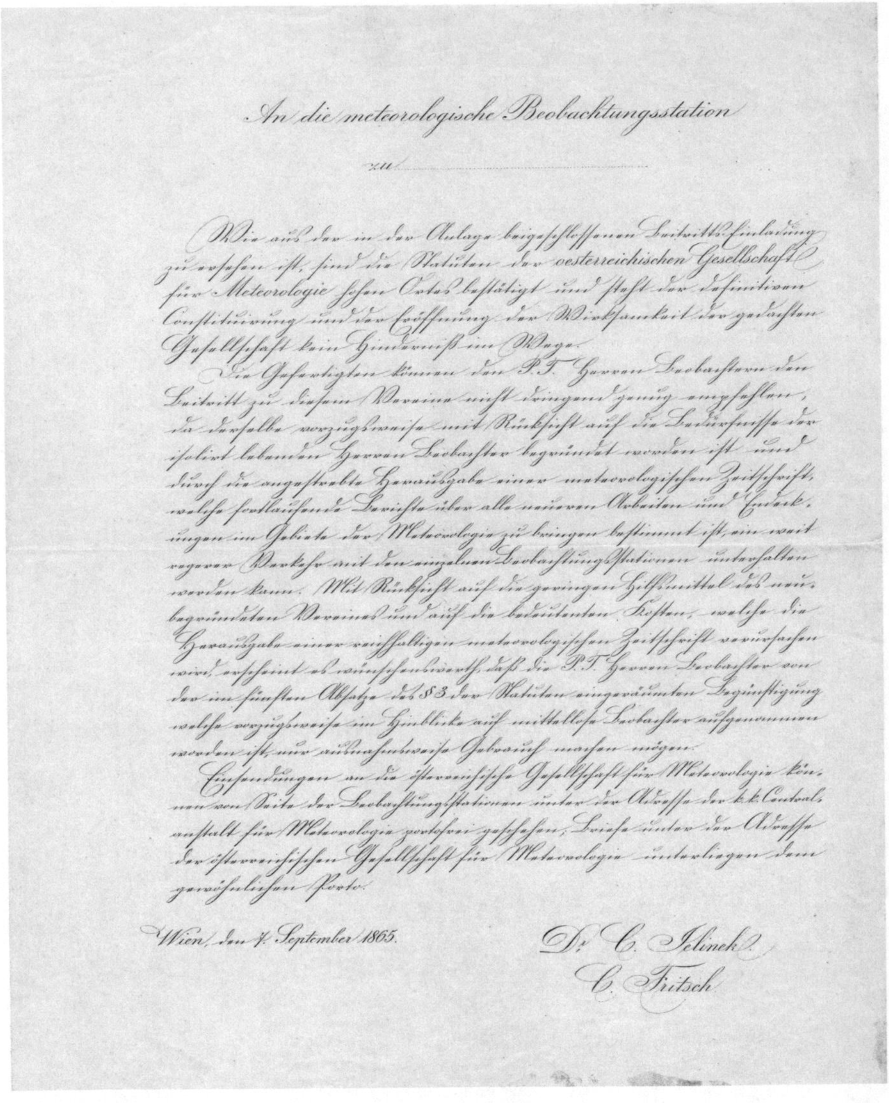

An die meteorologische Beobachtungsstation

zu ....

Wie aus der in der Anlage beigeschlossenen Beitritts-Einladung zu ersehen ist, sind die Statuten der oesterreichischen Gesellschaft für Meteorologie hohen Ortes bestätigt und steht der definitiven Constituirung und der Eröffnung der Wirksamkeit der gedachten Gesellschaft kein Hinderniß im Wege.

Die Gefertigten können den P.T. Herren Beobachtern den Beitritt zu diesem Vereine nicht dringend genug empfehlen, da derselbe vorzugsweise mit Rücksicht auf die Bedürfnisse der isolirt lebenden Herren Beobachter begründet worden ist und durch die angestrebte Herausgabe einer meteorologischen Zeitschrift, welche fortlaufende Berichte über alle neueren Arbeiten und Entdeckungen im Gebiete der Meteorologie zu bringen bestimmt ist, ein weit regerer Verkehr mit den einzelnen Beobachtungsstationen unterhalten werden kann. Mit Rücksicht auf die geringen Hilfsmittel des neu begründeten Vereines und auf die bedeutenden Kosten, welche die Herausgabe einer reichhaltigen meteorologischen Zeitschrift verursachen wird, erscheint es wünschenswerth, daß die P.T. Herren Beobachter von der im fünften Absatze des §3 der Statuten eingeräumten Begünstigung welche vorzugsweise im Hinblicke auf mittellose Beobachter aufgenommen worden ist, nur ausnahmsweise Gebrauch machen mögen.

Einsendungen an die österreichische Gesellschaft für Meteorologie können von Seite der Beobachtungsstationen unter der Adresse der k.k. Central-anstalt für Meteorologie portofrei geschehen, Briefe unter der Adresse der österreichischen Gesellschaft für Meteorologie unterliegen dem gewöhnlichen Porto.

Wien, den 7. September 1865.

Dr. C. Jelinek
C. Fritsch

D193 Dr. C. Jelinek and C. Fritsch announce the establishment of the Austrian Meteorological Society and the position of individual meteorological observers in the Statutes of the Austrian Meteorological Society. Dated in Wien/Vienna, September 7, 1865, signed by Dr. C. Jelinek and C. Fritsch

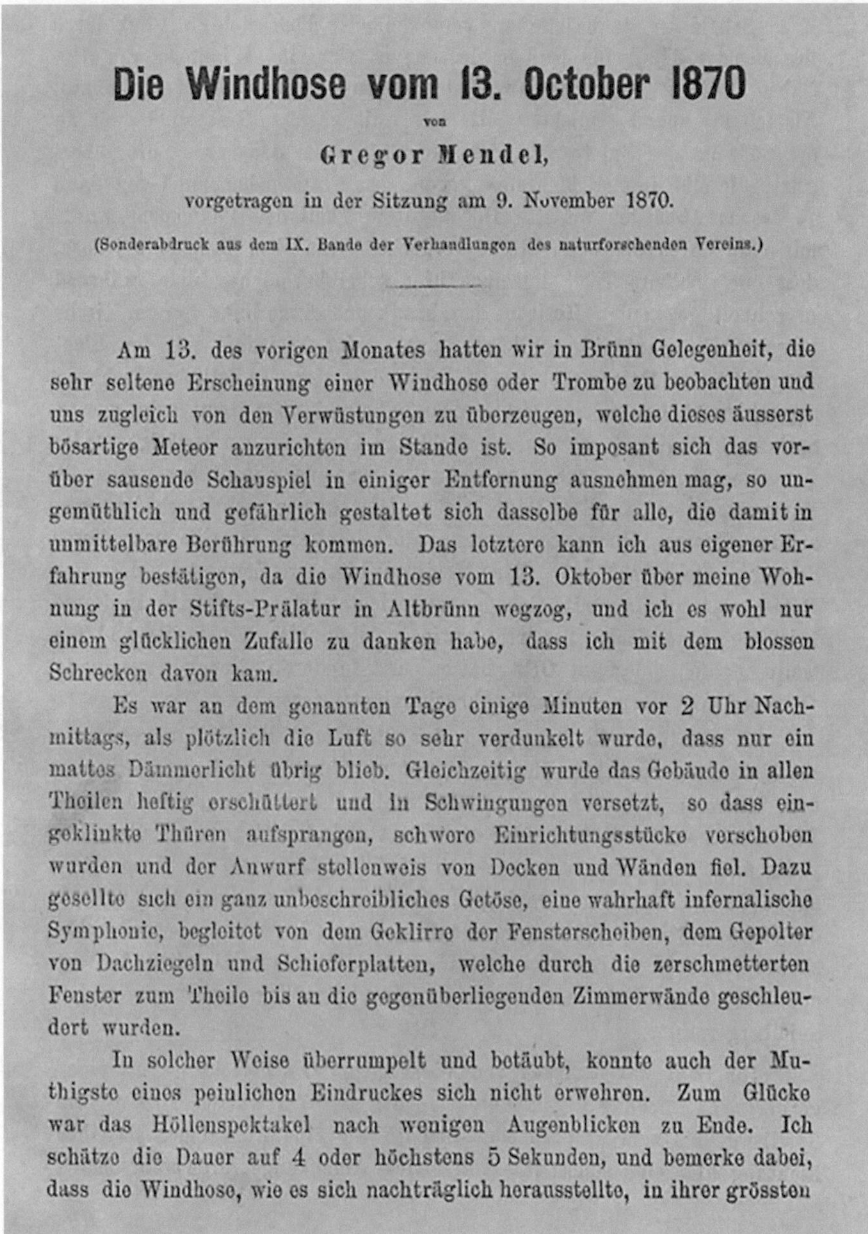

Die Windhose vom 13. October 1870

von

Gregor Mendel,

vorgetragen in der Sitzung am 9. November 1870.

(Sonderabdruck aus dem IX. Bande der Verhandlungen des naturforschenden Vereins.)

Am 13. des vorigen Monates hatten wir in Brünn Gelegenheit, die sehr seltene Erscheinung einer Windhose oder Trombe zu beobachten und uns zugleich von den Verwüstungen zu überzeugen, welche dieses äusserst bösartige Meteor anzurichten im Stande ist. So imposant sich das vorüber sausende Schauspiel in einiger Entfernung ausnehmen mag, so ungemüthlich und gefährlich gestaltet sich dasselbe für alle, die damit in unmittelbare Berührung kommen. Das letztere kann ich aus eigener Erfahrung bestätigen, da die Windhose vom 13. Oktober über meine Wohnung in der Stifts-Prälatur in Altbrünn wegzog, und ich es wohl nur einem glücklichen Zufalle zu danken habe, dass ich mit dem blossen Schrecken davon kam.

Es war an dem genannten Tage einige Minuten vor 2 Uhr Nachmittags, als plötzlich die Luft so sehr verdunkelt wurde, dass nur ein mattes Dämmerlicht übrig blieb. Gleichzeitig wurde das Gebäude in allen Theilen heftig erschüttert und in Schwingungen versetzt, so dass eingeklinkte Thüren aufsprangen, schwere Einrichtungsstücke verschoben wurden und der Anwurf stellenweis von Decken und Wänden fiel. Dazu gesellte sich ein ganz unbeschreibliches Getöse, eine wahrhaft infernalische Symphonie, begleitet von dem Geklirre der Fensterscheiben, dem Gepolter von Dachziegeln und Schieferplatten, welche durch die zerschmetterten Fenster zum Theile bis an die gegenüberliegenden Zimmerwände geschleudert wurden.

In solcher Weise überrumpelt und betäubt, konnte auch der Muthigste eines peinlichen Eindruckes sich nicht erwehren. Zum Glücke war das Höllenspektakel nach wenigen Augenblicken zu Ende. Ich schätze die Dauer auf 4 oder höchstens 5 Sekunden, und bemerke dabei, dass die Windhose, wie es sich nachträglich herausstellte, in ihrer grössten

D488 The first page of Mendel's paper on the tornado of 13 October 1870, read in the meeting of *Naturforschender Verein* on 09 November 1870. Reprint of the *Verhandlungen des naturforschenden Vereines in Brünn* (for the year 1870) 9: 229–246, 1871

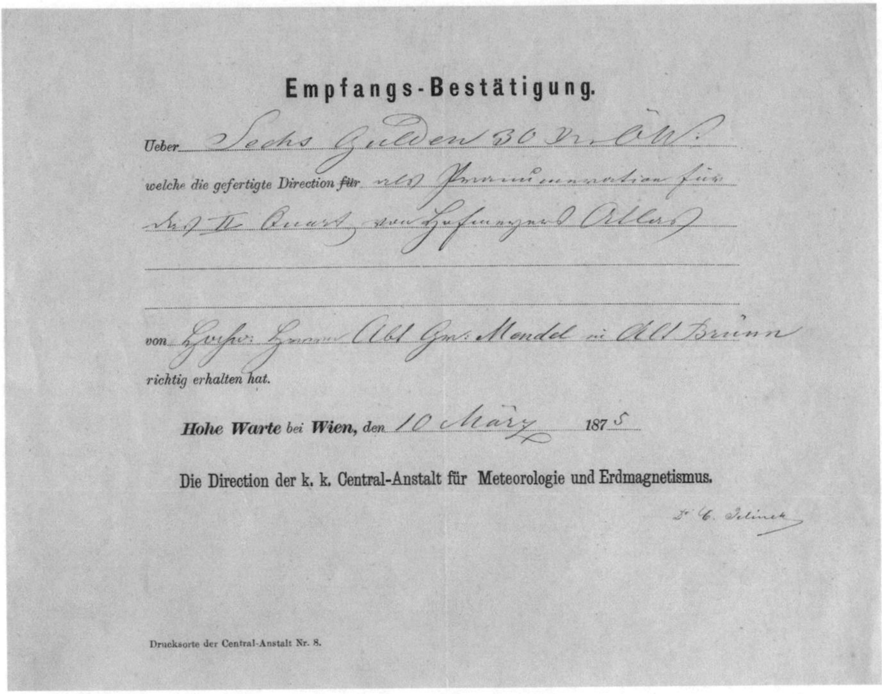

Empfangs-Bestätigung.

Ueber Sechs Gulden 30 kr. Ö.W.

welche die gefertigte Direction ~~für~~ als Pränumeration für das II Quart. von Hoffmeyers Atlas

von Hochw. Herrn Abt Gr. Mendel in Alt Brünn

richtig erhalten hat.

Hohe Warte bei *Wien*, den 10 März 1875

Die Direction der k. k. Central-Anstalt für Meteorologie und Erdmagnetismus.

Dr C. Jelinek

Drucksorte der Central-Anstalt Nr. 8.

D740 Mendel's subscription receipt for Hoffmeyer's Atlas for the second quarter of the year. Dated in Wien/Vienna, 10 March 1875, signed by Dr. C. Jelinek

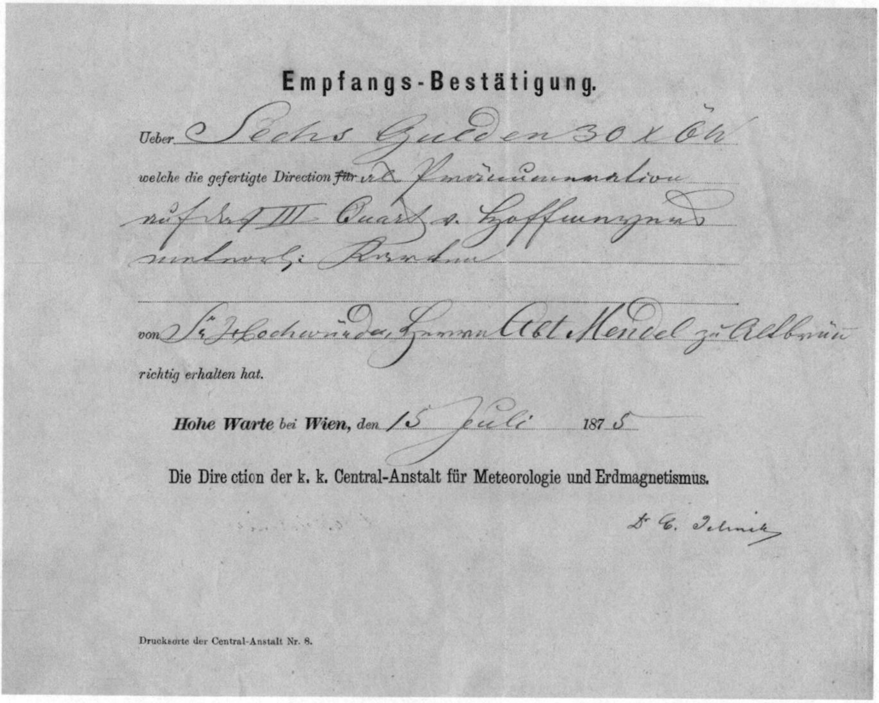

Empfangs-Bestätigung.

Ueber Sechs Gulden 30 x Öw

welche die gefertigte Direction ~~für~~ die Pränumeration

auf das III. Quart. d. Hoffmeyer'schen

[illegible] Karten

von Sr. Hochwürden [illegible] Abt Mendel zu Altbrünn

richtig erhalten hat.

Hohe Warte bei Wien, den 15 Juli 1875

Die Direction der k. k. Central-Anstalt für Meteorologie und Erdmagnetismus.

Dr C. Jelinek

Drucksorte der Central-Anstalt Nr. 8.

D741 Mendel's subscription receipt for Hoffmeyer's Atlas for the third quarter of the year. Dated in Wien/Vienna, 15 July 1875, signed by Dr. C. Jelinck

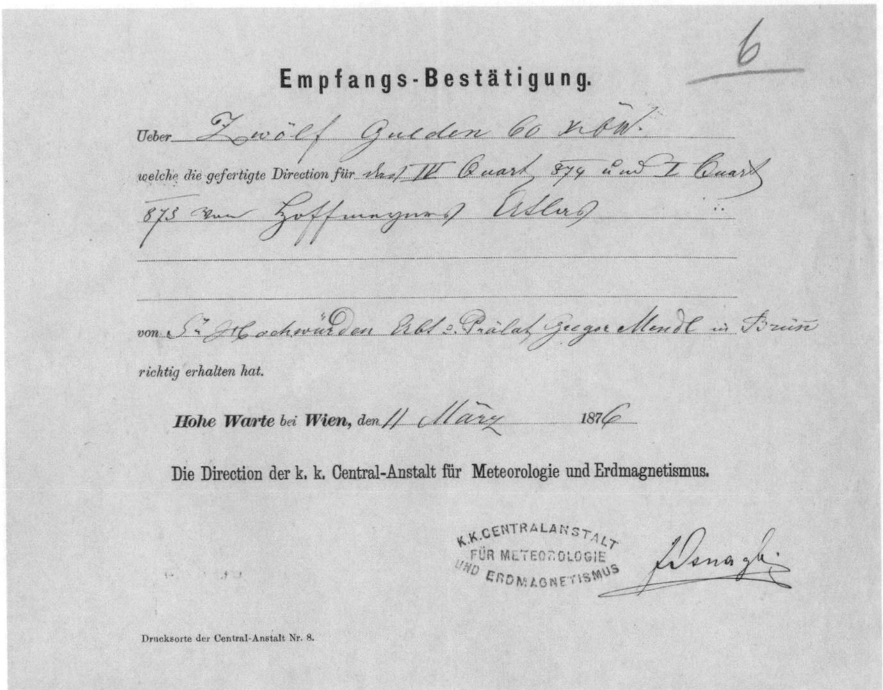

6

Empfangs-Bestätigung.

Ueber Zwölf Gulden 60 kr. ö.W.

welche die gefertigte Direction für das IV Quart. 874 und I Quart. 873 von Hoffmeyer's Atlas

von Sr. Hochwürden Abt u. Prälat Gregor Mendel in Brünn

richtig erhalten hat.

Hohe Warte bei Wien, den 11 März 1876

Die Direction der k. k. Central-Anstalt für Meteorologie und Erdmagnetismus.

K.K. CENTRALANSTALT FÜR METEOROLOGIE UND ERDMAGNETISMUS

Drucksorte der Central-Anstalt Nr. 8.

D742 Mendel's receipt of payment for Hoffmeyer's Atlas for the fourth quarter of 1874 and the first quarter of 1873. Dated in Wien/Vienna, 11 March 1876, signature not deciphered. Stamp of K. K. Centralanstalt für Meteorologie und Erdmagnetismus

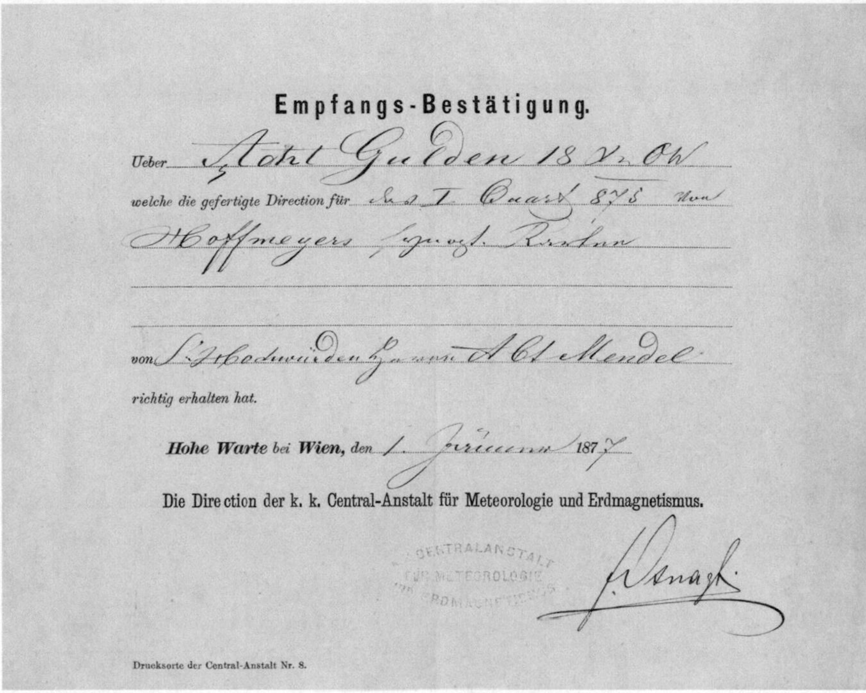

Empfangs-Bestätigung.

Ueber Acht Gulden 18 kr. Ö.W.

welche die gefertigte Direction für des I. Quart. 875 von

Hoffmeyers synopt. Karten

von Sr. Hochwürden Herrn Abt Mendel

richtig erhalten hat.

***Hohe Warte** bei **Wien**, den* 1. Jänner 187 7

Die Direction der k. k. Central-Anstalt für Meteorologie und Erdmagnetismus.

CENTRALANSTALT FÜR METEOROLOGIE UND ERDMAGNETISMUS

Drucksorte der Central-Anstalt Nr. 8.

D743 Mendel's receipt of payment for Hoffmeyer's Atlas for the first quarter of 1875. Dated in Wien/Vienna, 01 January 1877, signature not deciphered. Stamp of K. K. Centralanstalt für Meteorologie und Erdmagnetismus

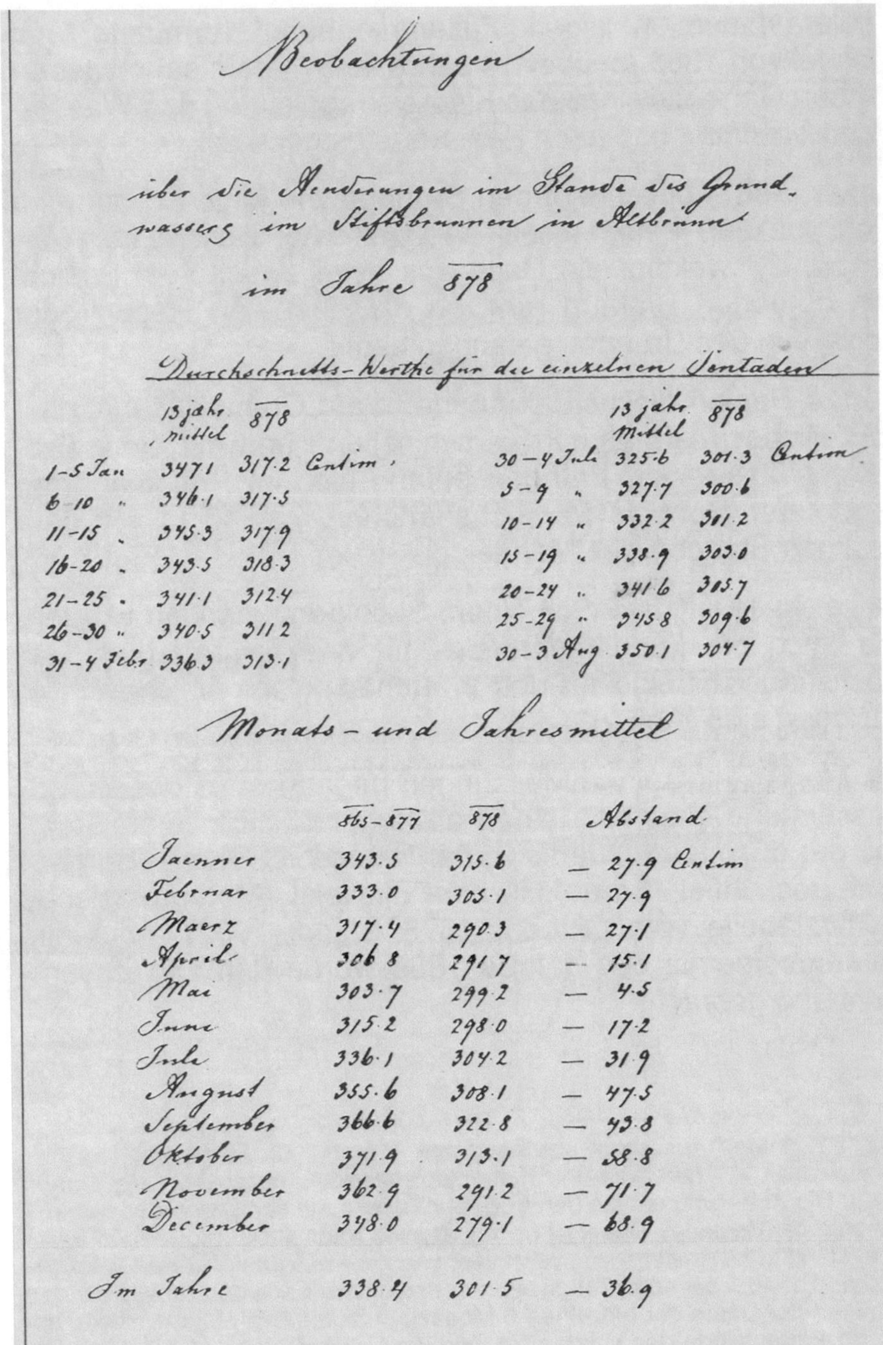

Beobachtungen

über die Aenderungen im Stande des Grund-
wassers im Stiftsbrunnen in Altbrünn

im Jahre 878

Durchschnitts-Werthe für die einzelnen Pentaden

	13 jahr. Mittel	878	
1–5 Jan	347.1	317.2	Centim.
6–10 „	346.1	317.5	
11–15 „	345.3	317.9	
16–20 „	343.5	318.3	
21–25 „	341.1	312.4	
26–30 „	340.5	311.2	
31–4 Febr	336.3	313.1	

	13 jahr. Mittel	878	
30–4 Juli	325.6	301.3	Centim.
5–9 „	327.7	300.6	
10–14 „	332.2	301.2	
15–19 „	338.9	303.0	
20–24 „	341.6	305.7	
25–29 „	345.8	309.6	
30–3 Aug	350.1	304.7	

Monats- und Jahresmittel

	865–877	878	Abstand	
Jaenner	343.5	315.6	– 27.9	Centim
Februar	333.0	305.1	– 27.9	
Maerz	317.4	290.3	– 27.1	
April	306.8	291.7	– 15.1	
Mai	303.7	299.2	– 4.5	
Juni	315.2	298.0	– 17.2	
Juli	336.1	304.2	– 31.9	
August	355.6	308.1	– 47.5	
September	366.6	322.8	– 43.8	
Oktober	371.9	313.1	– 58.8	
November	362.9	291.2	– 71.7	
December	348.0	279.1	– 68.9	
Im Jahre	338.4	301.5	– 36.9	

The first page of Mendel's measurements of the subsoil water-level fluctuations in the monastery well in the year 1878 (Weiling 1993, p. 386) analyzed and published by Liznar (1902)

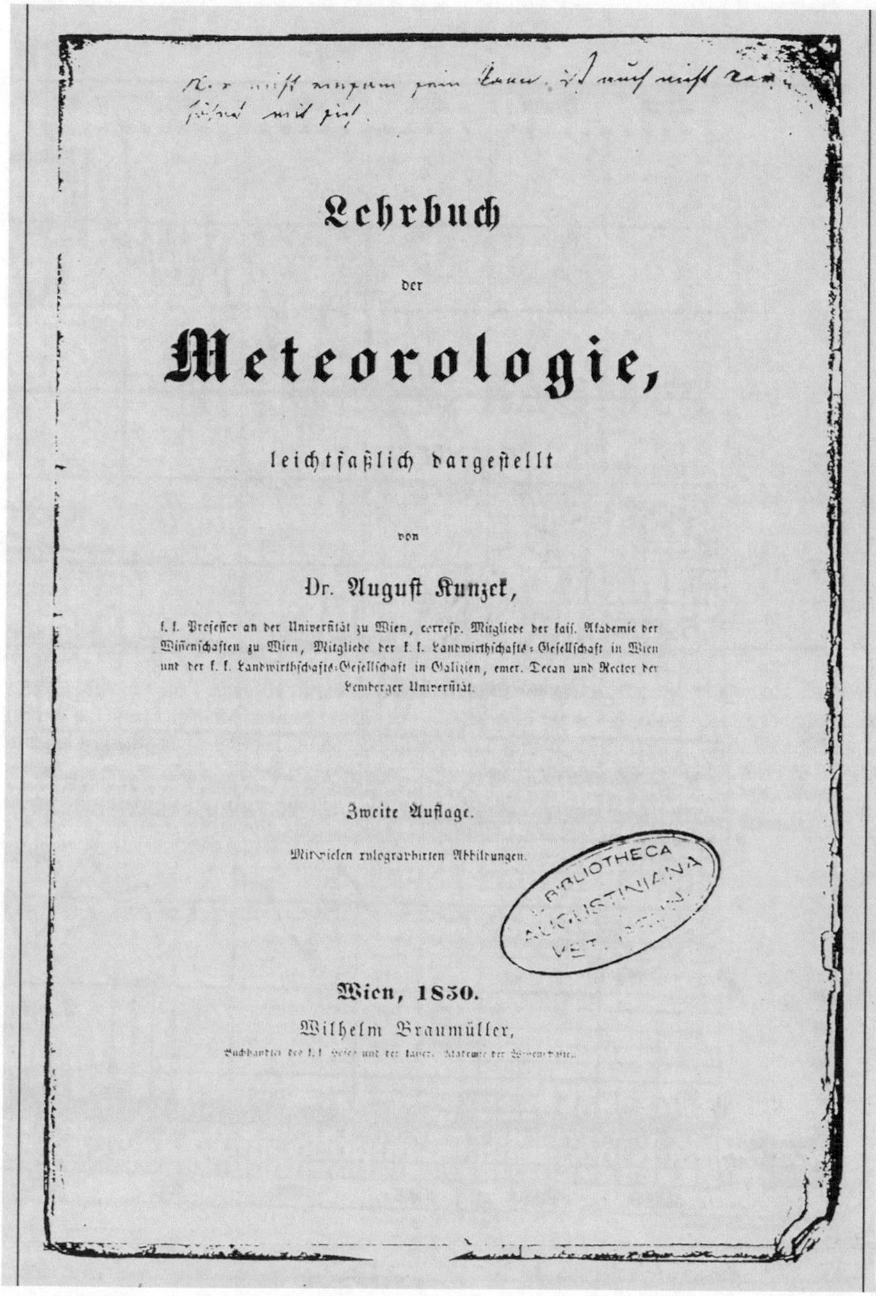

Lehrbuch

der

Meteorologie,

leichtfaßlich dargestellt

von

Dr. August Kunzek,

k. k. Professor an der Universität zu Wien, corresp. Mitgliede der kais. Akademie der Wissenschaften zu Wien, Mitgliede der k. k. Landwirthschafts-Gesellschaft in Wien und der k. k. Landwirthschafts-Gesellschaft in Galizien, emer. Decan und Rector der Lemberger Universität.

Zweite Auflage.

Mit vielen xylographirten Abbildungen.

Wien, 1850.

Wilhelm Braumüller,

Buchhändler des k. k. Hofes und der kaiser. Akademie der Wissenschaften.

On the title page of Dr. August Kunzek's *Lehrbuch der Meteorologie, leichtfasslich dargestellt* Mendel noted *Wer nicht einsam sein kann, ist auch nicht versöhnt mit sich.* (Weiling 1993, p. 383)

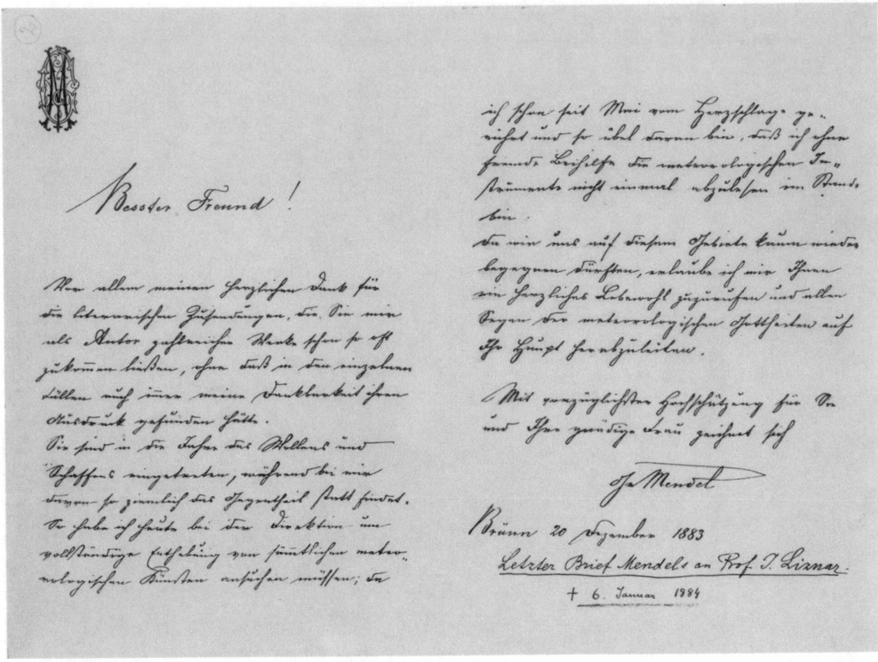
Bester Freund!

Mendel

Brünn 20 Dezember 1883

Letzter Brief Mendels an Prof. J. Liznar.

+ 6. Januar 1884

Mendel's farewell letter to J. Liznar, his former student at the *Realschule*. Dated in Brünn/Brno, 20 December 1883, signed by Gregor Mendel. On the second page of Mendel's personal writing paper adorned with his monogram there is a note added by an unknown person: "the last letter of Mendel to Prof. J. Liznar. + 6 January 1884"

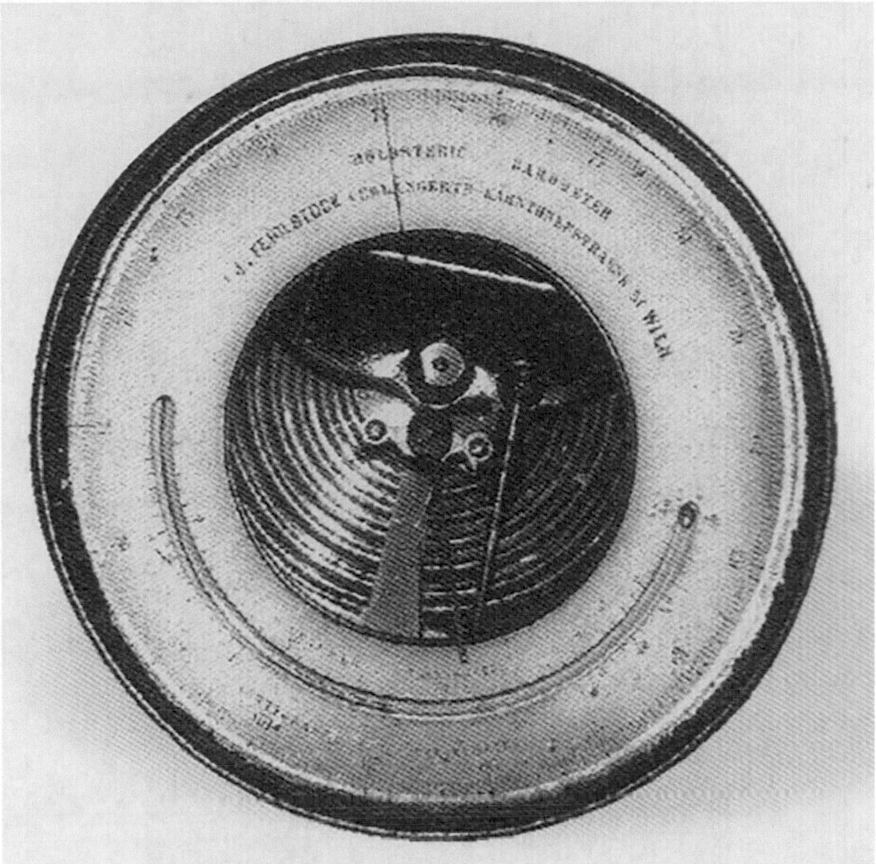

Barometer from Mendel's meteorological station

Mendel's telescope (coronograph) that Mendel used for the registration of sunspots after he extended his meteorological measurements to astronomy. The telescope was produced by Carl Fritsch in Vienna

References and Historical Printed Sources

Iltis H (1924) Gregor Johann Mendel. Leben, Werk und Wirkung. Julius Springer, Berlin. Life of Mendel. English translation by Eden and Cedar Paul. First edition. Norton, New York 1932; Second edition Allen and Unwin London, 1966

Liznar J (1902) Über die Änderungen des Grundwasserstandes nach den vom Prälaten Gregor Mendel in den Jahren 1865-1880 in Brünn ausgeführten Messungen. Festschrift zur Erinnerung an die Feier des 50jährigen Bestandes der deutschen Staatsoberrealschule in Brünn. C. Winiker, brno, pp 225–233

Makowsky A (1863) Die Flora des Brünner Kreizes. Verhandlungen des naturforsch. Vereines. Abhandlungen, Brünn 1, 43:210

Orel V (1996) Gregor Mendel the first geneticist. Translated by Stephen Finn. Oxford University Press. ISBN: 0 19 854774 9

Weiling F (1993/1994) Johann Gregor Mendel. Der Mensch und Forscher. Forscher in der Kontroverse. J. G. Mendel im Urteil der Zeitgenossen. Medizinische Genetik 5:35–51, 208–222, 274–289, 379–393, 6:35–51, 241–255.

Chapter 15
Experimentalist in Apiculture

The symbol of apiculture in the Great Chapter Hall ceiling paintings counterpoints the old straw beehive with Dzierzon's modern double hive.

Mendel joined the apiculturists in the Agricultural Society in 1870 at the invitation of Žiwansky, who organized the 14th Congress of Austrian and German Apiculturists in Brno. Napp was president of the congress, which took place from the 12th to the 14th of September 1865. Three-hundred European beekeepers attended, including the Silesian J. Dzierzon, author of the theory of parthenogenesis, and G. Dathe, an expert on heather bee. Hybridization and fertilization were at the centre of attention of the congress attendees. Segregation ratios of imported light Italian bees hybridized with the domestic dark drones were 3:1 in colour. Beekeepers "italianized" their bee colonies for aesthetic reasons. Mendel made no experiments in this respect. His name is not entered in the list of participants of the Congress of European beekeepers in Brno in 1865.

In 1865, Mendel reached the peak of his scientific endeavour. He presented his lectures on artificial pea fertilization and hybridization in the February and March meetings of the *Naturforschender Verein*. Napp did not mention Mendel's achievements in the field of plant hybridization in connection with the bee hybridization discussed at the congress. Whether Napp understood Mendel's conclusions of his pea experiments remains open.

After the European Apicultural Congress in Brno, Žiwansky, who was the chair of the Apicultural Association of the Agricultural Society, began to publish a scientific journal, *The Honeybee of Brno*. It appeared in German (*Die Honigbiene von Brünn*) and Czech (*Včela brněnská*). Žiwansky hybridized various bee races, published on the importance of growing plants useful for honeybee pasture and propagated breeding bees in movable beehives.

Mendel joined the Apicultural Association in 1869. A beehouse on the southern slope of the monastery orchard was built in 1871. It was a brick building in an L shape, surrounded by acacias, fruit trees, linden trees and nectar-producing plants. Mendel had a cellar excavated in the steep slope behind the beehouse, intended for

A. Matalová, E. Matalová, *Gregor Mendel - The Scientist*, Springer Biographies,
https://doi.org/10.1007/978-3-030-98923-1_15

the wintering of the bee colonies. His experiments were noted in Žiwansky's bee journal.

In 1871, Žiwansky and Mendel participated in the 17th Congress of German Apiculturists in Kiel (Dittmar 1972). Mendel visited the apiculturist Dathe in Eystrup.

In Brno, Alexander Makowsky, Mendel's colleague in the *Realschule and Naturforschender Verein*, gave Mendel a colony of the South American stingless bees imported by chance in an imported Pernambuco tree trunk used for colouring in a textile factory in Brno. Mendel's acclimatization experiment with the tropical bee was reported in *Zoologischer Anzeiger* in 1879 and 1880. The Hungarian Bee journal *Ungarische Biene* published Kühne's account of Mendel's beehouse, which Kühne visited in 1879, and Mendel's letter on the acclimatization experiment in 1880. The Russian *Zoological Garden and Acclimatisation* (*Zoologicheskyi sad i aklimatisatsiya*) included information about Mendel's acclimatization experiments with tropical bees in 1885 (Orel 1969, p. 240).

In Brno, Mendel planted the southern slopes of the Spielberg Hill near the monastery with trees and plants for foraging his bees. He also wrote a number of short communications for Žiwansky's bee journal. In 1875, Mendel urged bee-keepers to experiment as a means for making progress. In the meeting of the Committee of the Apicultural Association in Brno on 12 July 1877, a committee member named Kment enthusiastically reported about his visit to Mendel's beehouse in the Augustinian monastery garden: *I found the bees in the most splendid condition that any beekeeper could wish. Swarm after swarm, and the continuous rearing of new queens, so that I could wish everyone to come and look at this model beekeeping, for I may express here my complete conviction that everyone would take away with him some good and useful information*. Mendel kept many different races of bees in his apiary: Cyprian bees, heather bees, Carnollian bees, Italian bees and Egyptian bees (Orel, Rozman, Veselý 1965, p. 13). The beehouse consisted of thirty beehives, and another twenty were movable. Mendel studied bee characteristics, and their advantages best suited for the Brno region from the viewpoint of improving the bee colonies for honey production. In 1877, he came to the conclusion that *the Cyprian bee has certain advantages, and would be suitable for cultured breeds by crossing with other bees, the goal for which the whole apicultural world has been striving* (Včela brněnská 1877, p. 82). In 1877, Mendel was elected an honorary member of the Apicultural Association after he had declined to be its chairman.

Mendel's research into bee heredity had to confront several obstacles, which could be overcome by experiments in controlled mating of the queen mother in confinement. Mendel's colleague Žiwansky excluded any possibility of controlled mating of the bee queen with a selected drone (selected drones) in confinement (Žiwansky 1873). Despite Žiwansky's statement, Mendel designed a cage with walls of wire mesh for his experiments on controlled mating of the bee mother. He described his experiments in 1880 in a letter to Kühne, an apiculturist in Temeswar (then Timisuara), who visited Mendel in Brno. Kühne published Mendel's report in *Ungarische Biene* in 1881 (p. 16). Translation by Stephen Finn in Beránek and Orel

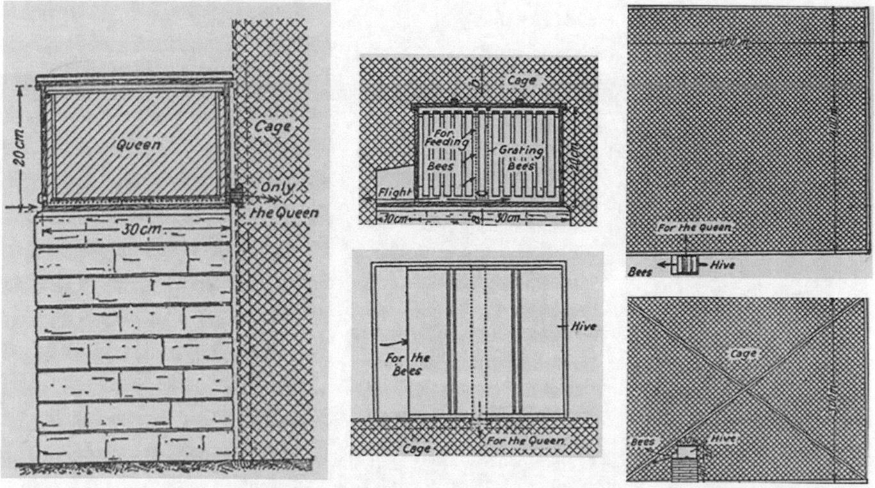

The cage for mating the queen bee with selected drones in confinement, made by carpenter Ludwig from Jungdorf/Jundrov according to Mendel's proposal (Iltis 1924, p. 147)

1988, p. 8. *As regards fertilisation of the queen, this year I did not obtain any result, but the experience I have acquired leads me to suppose that the experiment will eventually succeed. Last year the fault laid in the drones, this year in the queen, which, as I had the opportunity to observe, in the course of nine flights into the mating cage remained quite indifferent to her passionate lovers and repeatedly tried to escape from the cage. As an after-swarm queen she had already known free flight, whereas the drones had been out in the cage (the upper extension) soon after they left the hive and did not know what an open space was. The queen used last year was caught on an alighting board before taking off, and the drones all caught on the alighting boards. In the coming year I want to repeat the experiment with a quacking queen and drones which have not experienced free flight. I will also adapt the apparatus so that the queen and the drones will be separated into two small beehives, whence they will be able to fly into the mating cage.*

Die
Honigbiene von Brünn.
Organ der Bienenfreunde in Mähren und Schlesien.
Nr. 5 und 6. | Brünn am 1. Mai 1867. | Jahrgang I.

The title page of the German monthly journal of the Moravian Apicultural Association, *The Honeybee of Brno*, April 1872

VČELA BRNĚNSKÁ.
Časopis přátel včelařství na Moravě a Slezsku.
Číslo 7. | V Brně dne 16. srpna 1867. | Ročník I.

The title page of the Czech monthly journal of the Moravian Apicultural Association titled *The Honeybee of Brno*, May 1873

The title page of the Journal of the South Hungarian Bee Keepers´ Association titled *Hungarian Honeybee*

Auf meine Anfrage erhielt ich folgende gütige Antwort:

„Die stachellosen Miniaturbienchen Trigona linenta haben in einem geheizten Zimmer, wo am Tage die Temperatur 13—14° + R. betrug, aber in der Nacht oft auf nur 10° herabsank, ganz gut überwintert und sich bedeutend vermehrt. — Gefüttert wurden sie während des Winters mit Zuckerlösung und mit dem von den Bienen eingetragenen Pollen. Von letzteren haben sie ein bedeutendes Quanium verbraucht. Mit Anfang Mai fliegen sie wieder aus einem Fenster des Gartenhauses, wo es ihnen ganz gut zu gefallen scheint. Durch volle 35 Wochen, vom 20. September bis 25 Mai (1879/80) blieben sie ohne Ausfluge in ihrer Winterruhe. Als sie an dem genannten Maitage ihren ersten Ausflug hielten, war es ein Vergnügen anzusehen, wie sie sich eiligst durch das Flugloch drängten und mit einem hell klingenden Freudenton vor dem Stöckchen hin und her schwärmten. Schon nach 10 Minuten kam die erste mit Höschen zurück. Am Abende des nächsten Tages waren zwei Krüge mehr als zur Hälfte ausgebaut, der eine mit frischem Honig, der andere mit Blüthenstaub gefüllt. Das Völkchen dürfte gegenwärtig bei 300 Individuen zählen. Ein Abschwärmen, wie bei der Honigbiene oder anderweitiges Ausschicken von Colonien habe ich bis jetzt nicht bemerken können. Ihr Fleiß ist musterhaft und man muß geradezu staunen über die Größe der Pollenhöschen, welche diese kaum 1½ Linien messenden Bienchen nach Hause schleppen, sowie über den Umfang und die Höhe der Honigkrüge, welche sie in kurzer Zeit aus braunem Wachse zu bauen und zu füllen verstehen. Ihr Flug beginnt jedoch erst bei 16—17° + R. und wird bei 19° und darüber zum Vollfluge. Aus diesem Grunde ist an eine Aklimatisirung dieser kleinen Thierchen kaum zu denken. Als Curiosität will ich sie indessen so lange als möglich zu erhalten suchen."

Mendel's report on his acclimatization experiments with the stingless bee *Trigona lineata*, published by Kühne in the journal *Ungarische Biene* in 1881

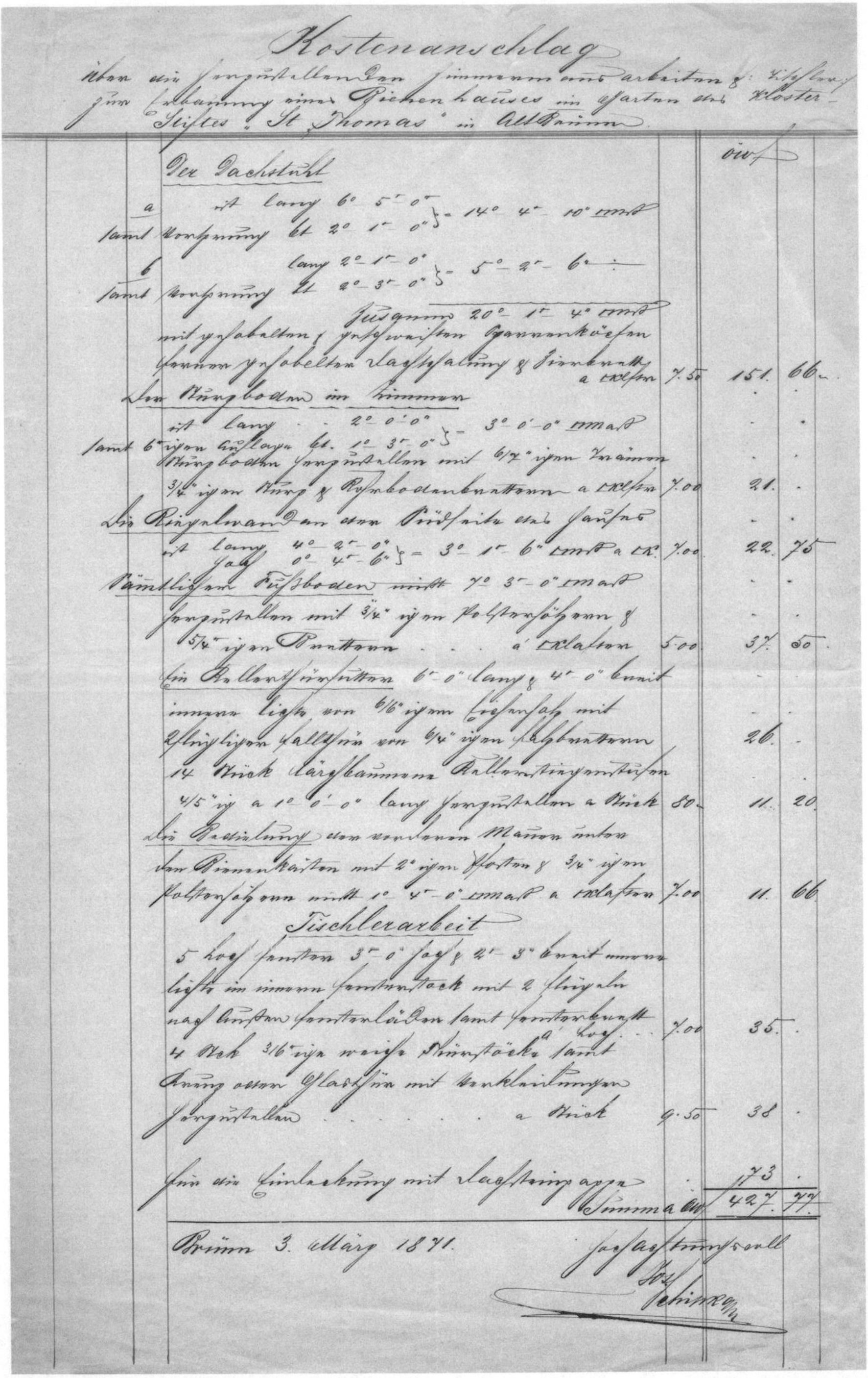
Kostenanschlag

St. Thomas

Tischlerarbeit

Brünn 3. März 1871.

D261 Josef Schipka's budget for building a beehouse in the garden of the Augustinian Monastery of St. Thomas in Old Brno. Dated in Brünn/Brno, 03 March 1871

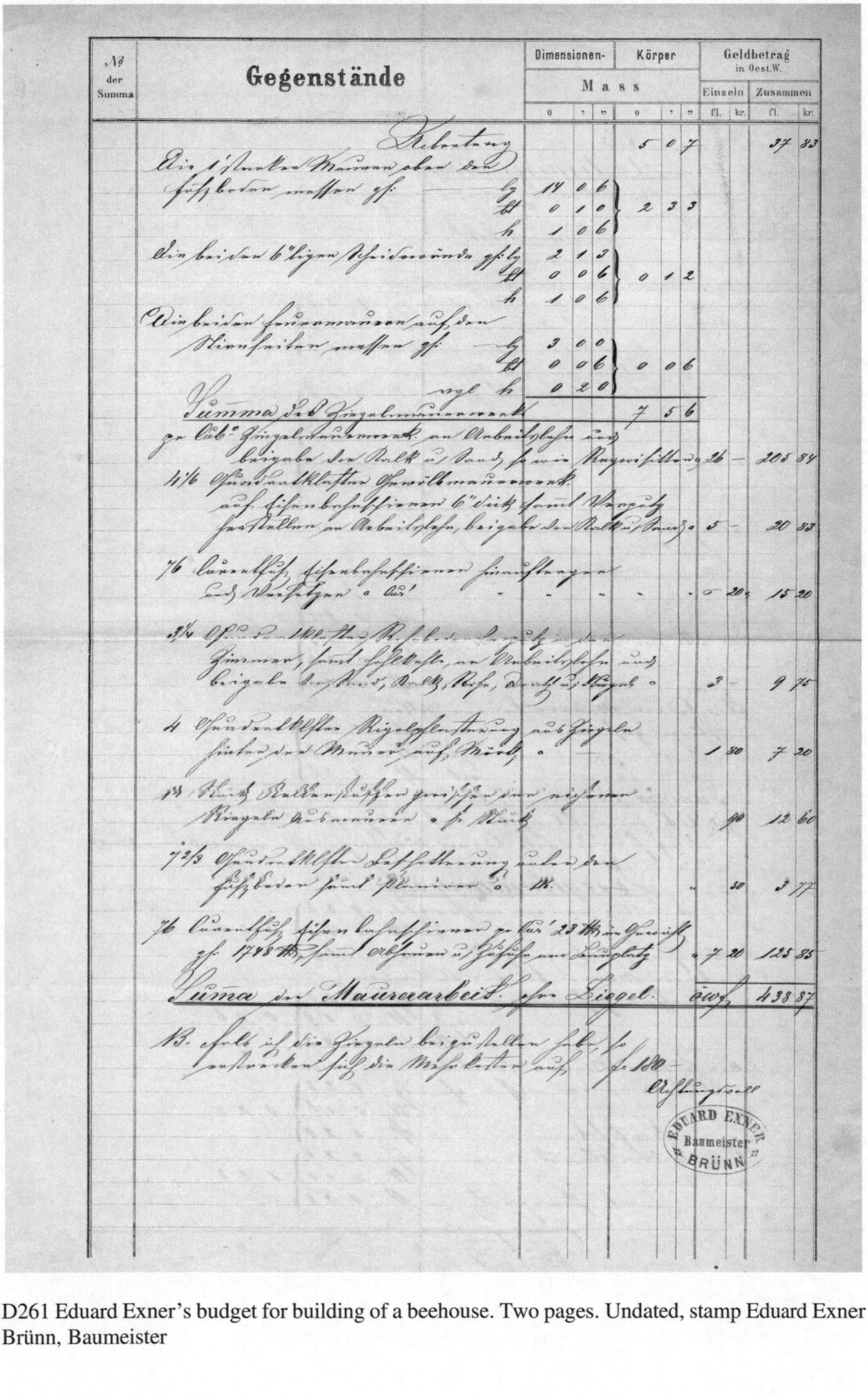

№ der Summa	Gegenstände	Dimensionen- Körper Mass	Geldbetrag in Oest.W. Einzeln	Zusammen
				37 83
				205 84
				20 83
				15 20
				9 75
				7 20
				12 60
				3 77
				125 85
	Summa der Mauerarbeit ohne Ziegel			438 87

EDUARD EXNER
Baumeister
BRÜNN

D261 Eduard Exner's budget for building of a beehouse. Two pages. Undated, stamp Eduard Exner Brünn, Baumeister

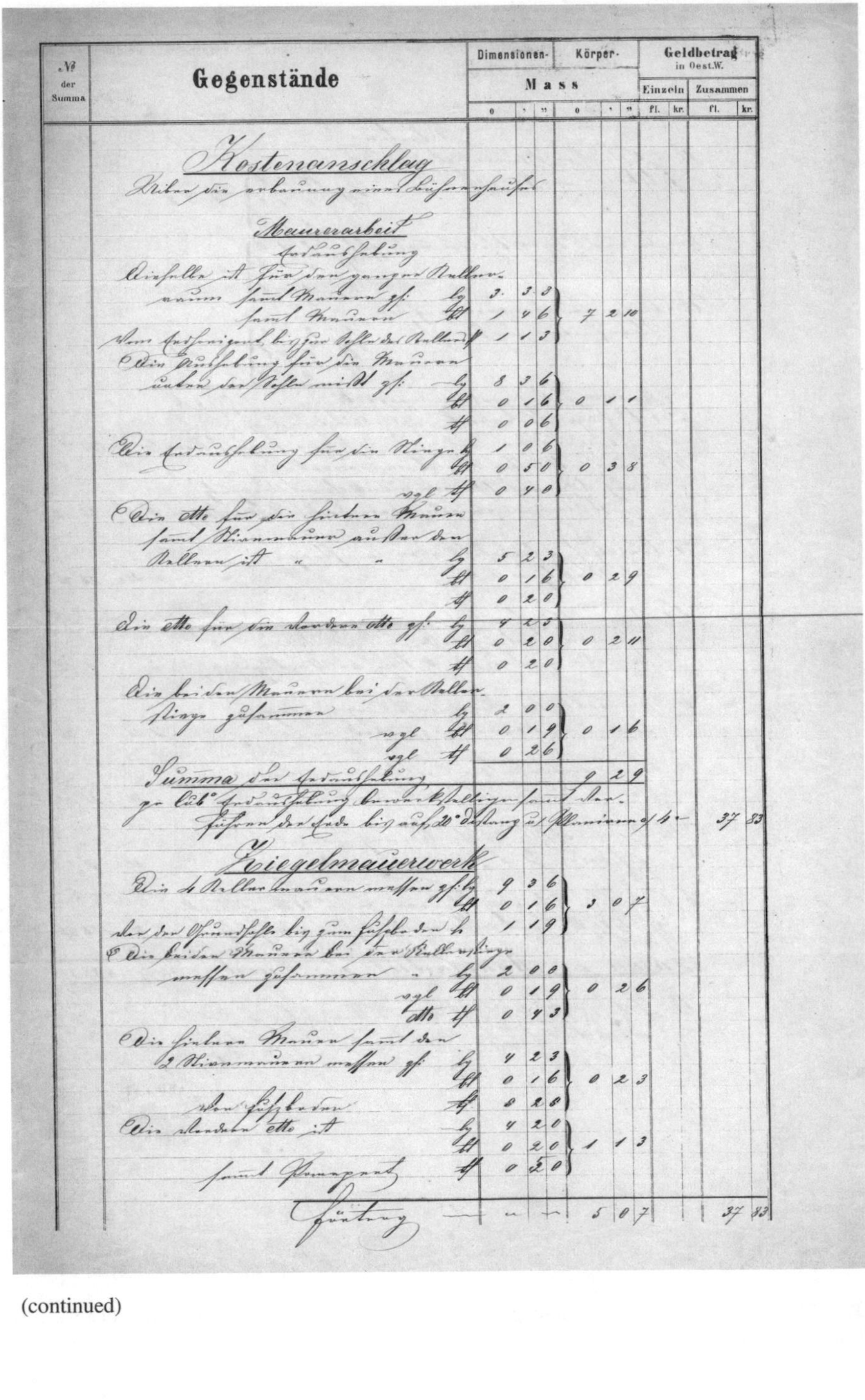

№ der Summa	Gegenstände	Dimensionen-Mass			Körper-Mass			Geldbetrag in Oest.W. Einzeln		Zusammen	
		°	'	"	°	'	"	fl.	kr.	fl.	kr.
	Kostenanschlag										
	Maurerarbeit										
	Summa der Grundaushebung				9	2	9			37	83
	Ziegelmauerwerk										
	Fürtrag				5	0	7			37	83

(continued)

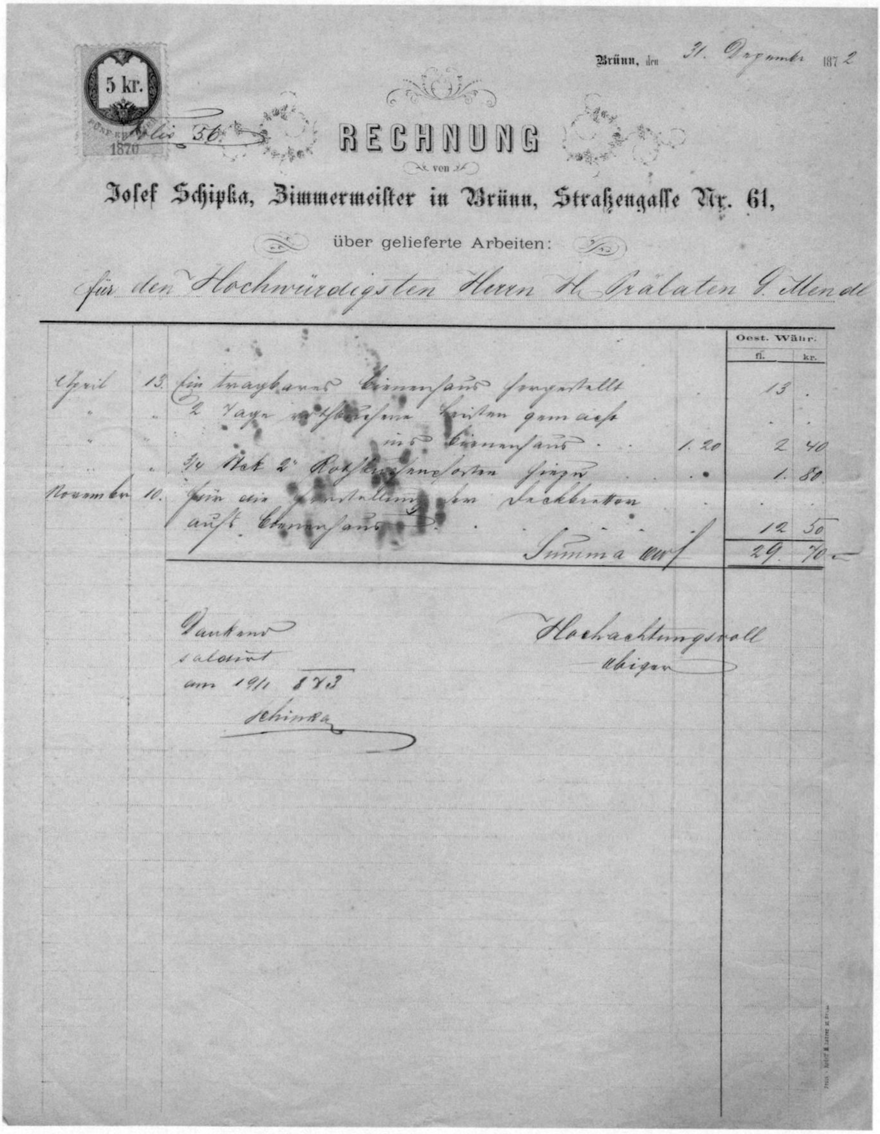

Brünn, den 31. Dezember 1872

5 kr. 1870

Folio 56.

RECHNUNG
von
Josef Schipka, Zimmermeister in Brünn, Straßengasse Nr. 61,
über gelieferte Arbeiten:
für den Hochwürdigsten Herrn H. Prälaten G. Mendl

				fl.	kr.
April	13.	für tragbares Bienenhaus hergestellt		13	.
.	.	2 Tage [illegible]		.	.
.	.	[illegible] Bienenhaus	1.20	2	40
.	.	3/4 [illegible]	.	1	80
November	10.	für die [illegible]	.	.	.
		[illegible] Bienenhaus		12	50
		Summa		29	70

Dankend saldirt am 19/1 873 Schipka

Hochachtungsvoll obiger

D201 Invoice of Josef Schipka to "Prelate G. Mendl" for a movable bee house. Dated 19 January 1873

References and Historical Printed Sources

Beránek V, Orel V (1988) New documents to Mendel's experiments with beas. Folia Mendeliana 23:9–16

Dittmar G (1972) Zur Reise von Gregor Mendel von Brünn nach Kiel im September 1871. FM 7: 37–42

Iltis H (1924) Gregor Johann Mendel. Leben, Werk und Wirkung. Julius Springer, Berlin. Life of Mendel. English translation by Eden and Cedar Paul. First edition. Norton, New York 1932; Second edition Allen and Unwin London, 1966

Orel V (1969) Abbot Mendel's expert opinion on the first weather forecasts for agriculture. FM 4: 37–40

Orel V, Rozman J, Veselý V (1965) Mendel as a beekeeper. Moravian Museum, Brno

Žiwansky F (1873) Kurze Anleitung zum Betriebe der vernunftgemessenen Bienenzucht, Brno

Chapter 16
Accredited Examiner in Pomology

The symbol of pomiculture in the whitewashed Great Chapter Hall showed Father Schreiber and Mendel's father Anton at work with a fruit tree grown in a pot. In the background is the church of Sts. Peter and Paul in Vražné and the Pohoř Hill, a good place for botanical excursions. Schreiber, a priest of the Vražné parish, followed in the footsteps of Jan Amos Komenský, known as Johann Amos Comenius (1592–1670) and included fruit tree cultivation and beekeeping in the school curriculum. Fruit tree cultivation was taught at Comenius's school in Fulnek and at the philanthropinum school in Kunín. The school in the Kunín castle was founded according to the model of Philanthropinum in Schnepfenthal in Saxony. In that tradition, agriculture, and cultivation of fruit trees and vines were compulsory subjects, taught in two semesters at the philosophical institutes. They were compulsory for priest candidates who were expected to teach their parishioners to cultivate fruit trees in their orchards, along the roads and paths, on the hill slopes and in the fields. The fruits, fresh, dried or cooked, diminished hunger of poor people and helped them to survive the long and cold winter months.

Schreiber had originally been a teacher in the philanthropinum school for boys and girls of Countess Maria Walpurga-Truchsess-Waldburg-Zeil (1762–1828) at her castle in Kunín. The school was considered to be an alternative to the schools sponsored by the church (Schreiber 1801). In 1802, Schreiber was forced to leave the Kunín school and was transferred to the parish in Vražné. In 1809, he published a book supporting education in useful subjects (*Lesebuch zur Beförderung des Unterrichts in gemeinnützigen Gegenständen*). Schreiber influenced the teaching programmes at schools of his parish, including Hynčice. He contributed to newspapers founded and published by Carl Christian André, co-operator of Salzmann of Schnepfenthal. André was invited to Brno to organize the evangelical school system for the children of protestant factory owners. In 1816, he became a founding member of the Pomological Association of the Agricultural Society in Brno. At that time, he was the only pomologist from Silesia in the Moravia-Silesian Agricultural Society. His practical knowledge and philanthropic ideas were well known.

A. Matalová, E. Matalová, *Gregor Mendel - The Scientist*, Springer Biographies,
https://doi.org/10.1007/978-3-030-98923-1_16

Carl Jurende was Schreiber's successor at the philanthropic school in Kunín, continuing the tradition of fruit tree growing. Jurende encouraged the teachers and villagers with their children to collect seeds from fruit trees. From one million collected seeds, thousands of fruit trees were raised and distributed to interested people, especially landholders and gardeners. Mendel's family was also active in this respect. These extraordinary pomiculture activities did not escape the attention of the daily press and popular calendars, which promoted fruit tree breeding. In 1809, Jurende published the *Moravian Pilgrim*, which was the first calendar on the scientific basis of field work and agricultural economy. The calendars contained articles on weather forecasts, improvement of the school system in Silesia, possibilities of increasing salaries of the village teachers, and advice on how to transform the region into a cultural landscape.

Fruit tree cultivation flourished in the suburbs of Brno, which were famous for their fertile gardens, vineyards and orchards. In the sixteenth century, travellers Thomas Jordan of Clausenburg and Johann Sporisch from Ottenthal admired the beauty of Brno precincts with gardens full of ornamental flowers of diverse colours, various sorts of vegetables and fields abounding in crops and fruits (Jordánková and Sulitková 2008, pp. 7–8). The exceptionally fertile soil of Brno, a town between two rivers, yielded great quantities of quality products. That tradition of economically prospering gardens lasted until Mendel's arrival in Brno in 1843.

Mendel had a feeling for nature, which he derived from his childhood home. In his early years as a friar, he attended lectures and passed examinations in agriculture, pomiculture, viniculture and horticulture by F. Diebl, author of a book on fruit tree and vine growing (1844), at the Philosophical Institute in Brno. Mendel drew special information on fruit tree improvement from the extensive illustrated handbook by F. Jahn, E. Lucas and G. G. Oberdieck on pomiculture, published in two volumes in Stuttgart in 1850 and 1860, pomological journals, books and leaflets. Napp supported fruit tree cultivation and improvement in the monastery's estates, changing vineyards to orchards. An abundance of pomological literature is deposited in the monastic library. Some books contain Mendel's notes pertaining to fruit tree hybridization. Through such hybridization, Mendel hoped to achieve better flavour of fruit, durability of the pulp, resistance of trees to diseases, increased fruit yield and tree sustainability in the regional soil and local climatic conditions.

As abbot Mendel was elected into the Central Board of the Agricultural Society (Orel 1970a, b). In 1868, Olexik and Mendel became accredited examiners of the Agricultural Society, testing the practical and theoretical knowledge of the candidates specialized in fruit and forest tree cultivation. A collection of microtomic sections of wood cuttings has been preserved along with a collection of fruit tree leaves that Mendel used in the examinations of fruit tree breeders. After Mendel's death, more than 300 fruit trees were prepared in the monastic garden for distribution in the regular Agricultural Society exhibits.

On 04 April 1883, Mendel asked his nephews Alois and Ferdinand Schindler in a letter to send him some grafts from specific fruit trees in the garden of his brother-in-law Alois Sturm in Hynčice, and from the former garden of his retired parents (Iltis 1924, p. 193). Translation by Eden and Cedar Paul in Iltis 1965, p. 275–6: *My dear doctors in spe!—Anticipating that you will so soon be returning home by way of Brünn, I am not this time sending you a postal remittance to Vienna. You see, then, that I am confidentially expecting you, but it would be as well that you should let me have a postcard the day before your arrival.—I suppose you have both been in Arcady during these lovely March days! The weather has been just as bad in Brünn. My servant Josef declares that he has been watching the March violets blooming on our noses, which have been purple with the cold. Only once during the last thirty-nine years, namely in 1845, was the average temperature lower, so this year 1883 ranks as the thirty-eighth.—I need some grafting shoots. Would you be good enough to ask Alois Sturm for them in my name? I want him to send me one from the Günsbirne [a pear tree], two from the Quaglich [another pear tree], and three from the good apple-tree in the reservation garden. I shall be glad to make returns in kind. – With warmest wishes and kisses,—Your affectionate uncle—Gregor.*

Günsbirne should be corrected to *Gansbirne (goose pear).*

In the Hynčice region the green "goose pears" grew ripe in the late autumn when geese were killed and baked in the oven with the green "goose pears" as a side-dish, traditionally on the feast of St. Martin on 11 November.

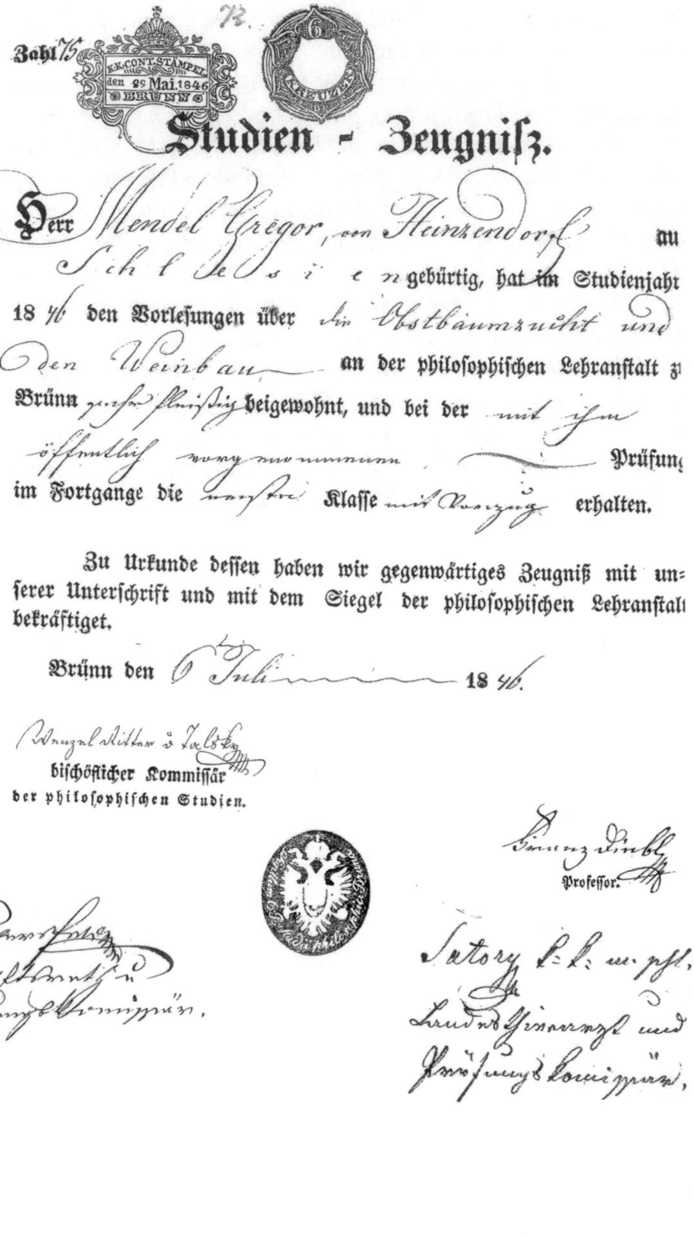

Zahl 75

K.K. CONT. STÄMPEL den 29 Mai 1846 BRÜNN

6 KREUZER

Studien - Zeugniß.

Herr Mendel Gregor, von Heinzendorf aus Schlesien gebürtig, hat im Studienjahr 18 46 den Vorlesungen über die Obstbaumzucht und den Weinbau an der philosophischen Lehranstalt zu Brünn sehr fleißig beigewohnt, und bei der mit ihm öffentlich vorgenommenen Prüfung im Fortgange die erste Klasse mit Vorzug erhalten.

Zu Urkunde dessen haben wir gegenwärtiges Zeugniß mit unserer Unterschrift und mit dem Siegel der philosophischen Lehranstalt bekräftiget.

Brünn den 6 Juli 1846.

Wenzel Ritter v. Talsky
bischöflicher Kommissär
der philosophischen Studien.

Franz Diebl
Professor.

Sutory

D63 Mendel's certificate from the lectures on pomiculture and viniculture which he attended at the Philosophical Institute in Brno during his studies at the Theological Institute. Dated in Brünn/Brno, 06 June 1846, signed by Franz Diebl, Wenzel Ritter v. Talsky, Sutory, the name of the fourth examiner undeciphered

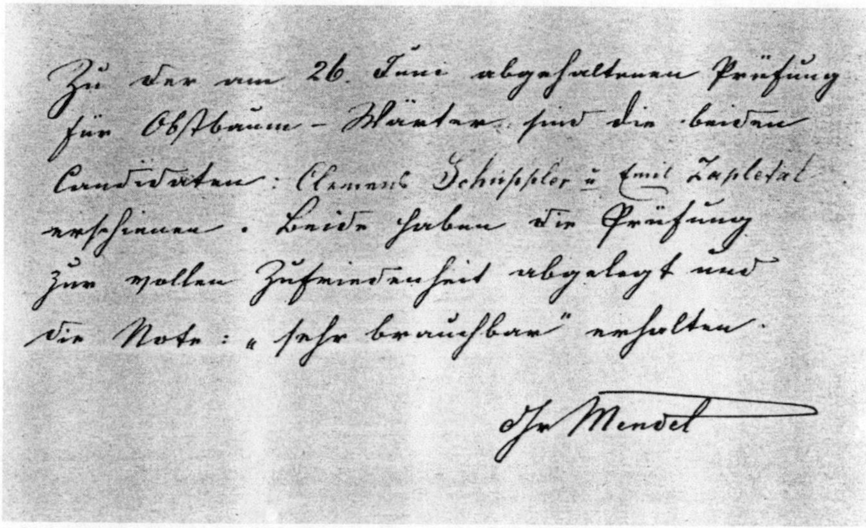

Zu der am 26. Juni abgehaltenen Prüfung für Obstbaum-Wärter sind die beiden Candidaten: Clemens Schuppler u. Emil Zapletal erschienen. Beide haben die Prüfung zur vollen Zufriedenheit abgelegt und die Note: „sehr brauchbar" erhalten.

Gr Mendel

D605 Mendel's evaluation of the examination of Clemens Schuppler and Emil Zapletal from fruit tree breeding. Undated, signed by Gr. Mendel

Der pomologisch-oenologische Vereins-Ausschuss

FÜR MAEHREN UND SCHLESIEN

hat Seiner Hochwürden den Herrn Herrn

GREGOR MENDL

Augustiner Priester, supplirender Professor an der k.k. Ober-Realschule zu Brünn

zur Förderung seiner gemeinnützigen patriotischen Zwecke zum wirklichen Mitgliede erwählt.

Brünn den 9 Mai 1863.

Praeses

D802 Mendel's decree appointing him a member of the Pomological-Oenological Association of Moravia and Silesia. Dated in Brünn/Brno, 09 May 1863

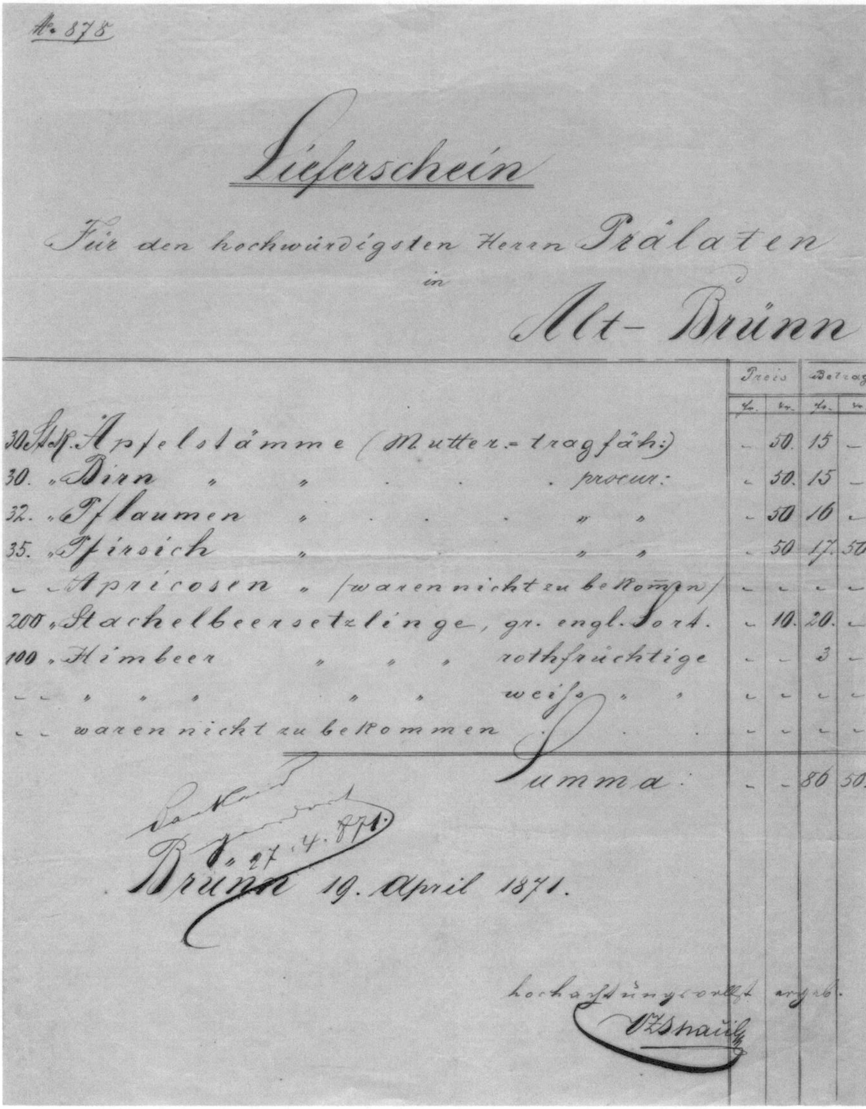

No. 878

Lieferschein

Für den hochwürdigsten Herrn Prälaten
in
Alt-Brünn

	Preis fl.	kr.	Betrag fl.	kr.
30 Stk. Apfelstämme (Mutter-tragfäh:)	–	50.	15	–
30. " Birn " " . . . . procur:	–	50.	15	–
32. " Pflaumen " . . . . " "	–	50	16	–
35. " Pfirsich " " "	.	50	17.	50.
– " Apricosen " (waren nicht zu bekommen)	–	–	–	–
200 " Stachelbeersetzlinge, gr. engl. Sort.	–	10.	20.	–
100 " Himbeer " " " rothfrüchtige	–	–	3	–
– – " " " " " weiße " "	–	–	–	–
– – waren nicht zu bekommen	–	–	–	–
Summa:	–	–	86	50.

Bezahlt [illegible] d. 27.4.871.

Brünn 19. April 1871.

hochachtungsvollst ergeb.
V. Z. Spačil

D729 Bill of delivery of apple trees, pear trees, plum trees, peach trees, apricot trees, blackberries and strawberries. Dated in Brünn/Brno, 19 April 1871, signed by V. Z. Spačil

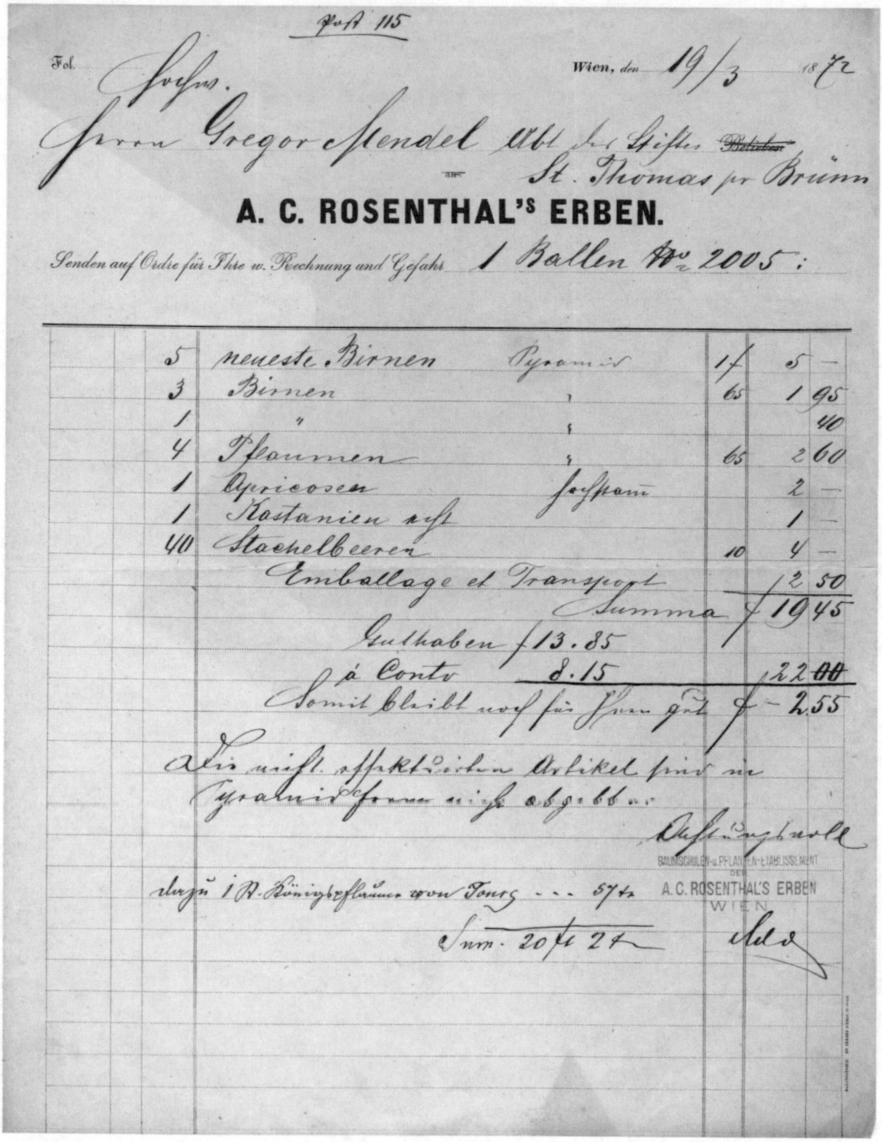

Post 115

Fol. Wien, den 19/3 1872

Hochw.
Herrn Gregor Mendel Abt des Stiftes ~~Betrieben~~
St. Thomas per Brünn

A. C. ROSENTHAL's ERBEN.

Senden auf Ordre für Ihre w. Rechnung und Gefahr 1 Ballen No 2005:

5	neueste Birnen	Pyramide	1	5 —
3	Birnen	"	65	1 95
1	"	"		40
4	Pflaumen	"	65	2 60
1	Apricosen	Hochstamm		2 —
1	Kastanien echt			1 —
40	Stachelbeeren		10	4 —
	Emballage et Transport			2 50
	Summa			f 19 45

Guthaben f 13.85
à Conto 8.15 — f 22 ~~00~~

Somit bleibt noch für Ihnen gut f — 2.55

Die nicht effektuirten Artikel sind in Pyramidenform nicht abgebbar.

Hochachtungsvolle

BAUMSCHULEN- u. PFLANZEN-ETABLISSEMENT DER A. C. ROSENTHAL'S ERBEN WIEN

[illegible]

dazu 1 St. Königspflaumen von Tours . . . 57 kr

Sum. 20 fl 2 kr

D730 A. C. Rosenthal's Erben bill of delivery addressed to Gregor Mendel for pear trees, plum trees, apricot tree, chestnut tree and blackberries. Dated in Wien/Vienna, 19 March 1872, signature undeciphered

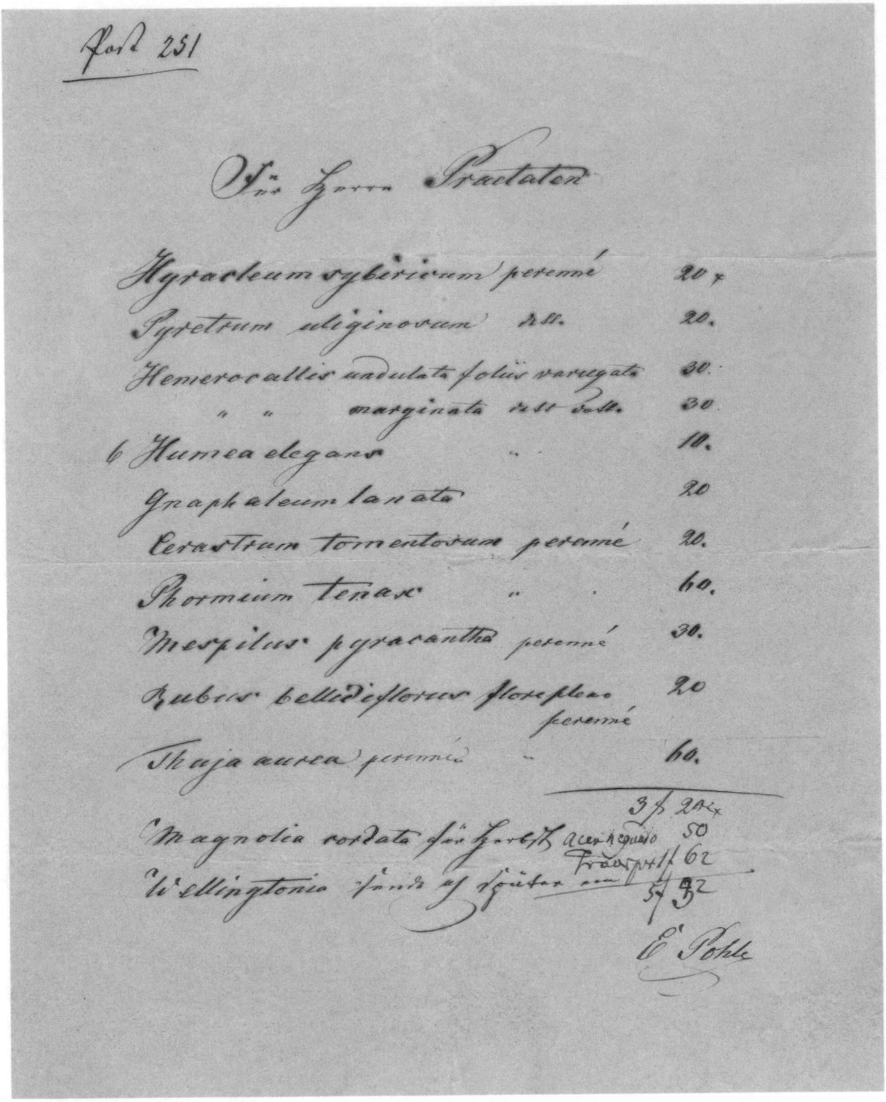

Fol. 251

Für Herrn Praelaten

Hyracleum sybiricum perenne 20 x
Pyrethrum uliginosum ditto 20.
Hemerocallis undulata foliis variegata 30.
" " marginata ditto ditto 30
6 Humea elegans 10.
Gnaphaleum lanata 20
Cerastrum tomentosum perenne 20.
Phormium tenax " 60.
Mespilus pyracantha perenne 30.
Rubus bellidiflorus flore pleno perenne 20
Thuja aurea perenne 60.

3 fl 20 x
Magnolia cordata für Herbst 50
Wellingtonia [illegible] Transport 1 fl 62
5 fl 32

E. Pohle

D359 Bill of delivery addressed "to Reverend Prelate" by gardener E. Pohle. Undated, signed by E. Pohle

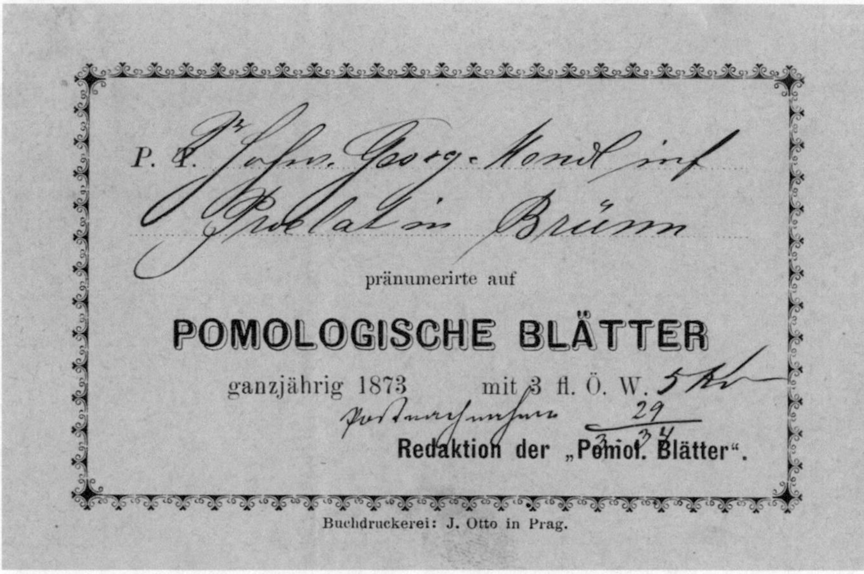

P. T. [illegible] Georg Mendl inf. Prälat in Brünn

pränumerirte auf

POMOLOGISCHE BLÄTTER

ganzjährig 1873 mit 3 fl. Ö. W. 5 kr

[illegible] 29

3 34

Redaktion der „Pomol. Blätter".

Buchdruckerei: J. Otto in Prag.

D739 Receipt of prepayment from "Georg Mendl mitred prelate in Brünn/Brno" for Pomological Journal/*Pomologische Blätter* for the year 1873, undated, without signature

Lehre

von der

Baum-Zucht

überhaupt, und von der

Obstbaumzucht, dem Weinbaue

und der

wilden- oder Waldbaumzucht

insbesondere.

Zum Behufe der öffentlichen und Privat-Belehrung.

von

Franz Diebl,

k. k. Professor der Landwirthschaftslehre und allgemeinen Naturgeschichte an der philosophischen Lehranstalt zu Brünn und Mitgliede mehrerer landwirthschaftlichen Gesellschaften des Inn- und Auslandes.

Brünn 1844,

Gedruckt bei Rudolph Rohrer's sel. Wittwe.

The title page of F. Diebl's textbook on tree breeding with special attention to fruit trees, viniculture and wild and forest tree breeding. Brünn/Brno 1844

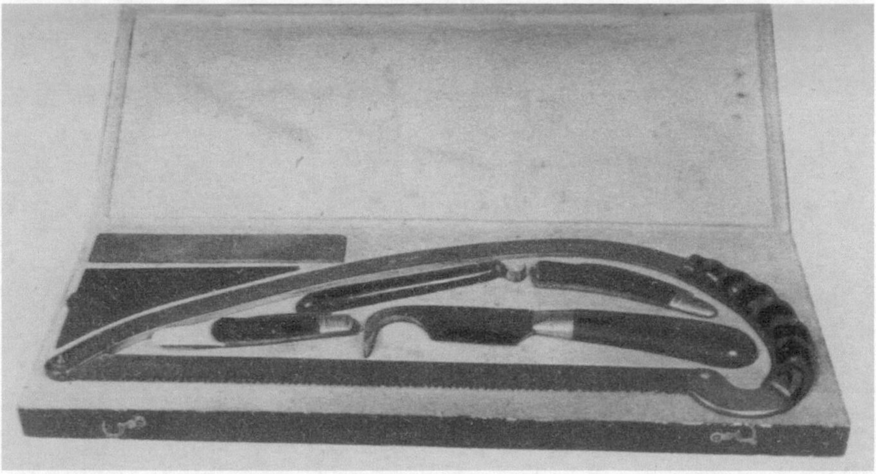

IM85 Mendel's garden tools in their original case. Photo by Josef Tichý, 1962

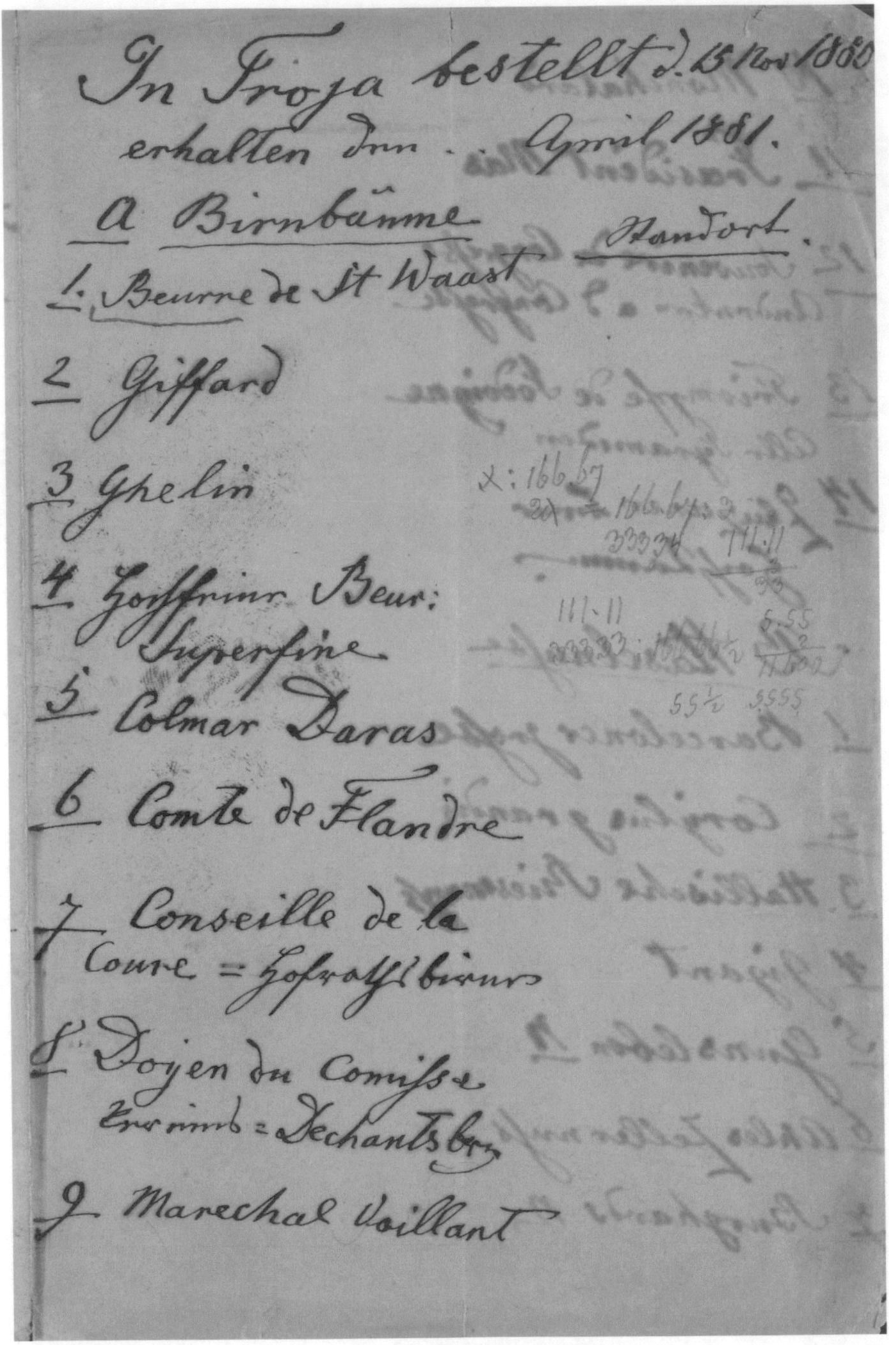

In Troja bestellt d. 15 Nov 1880
erhalten den .. April 1881.

a Birnbäume Standort.

1. Beurre de St Waast
2 Giffard
3 Ghelin
4 Hoffrims Beur: Superfine
5 Colmar Daras
6 Comte de Flandre
7 Conseille de la Coure = Hofrathsbirne
8 Doyen du Comisse = Dechantsbirn
9 Marechal Voillant

The list of pear and walnut trees ordered in Troja on 15 November 1880, received in April 1881. Two sheets. Item No. 159, Box 5, Record Series 15/24/56, courtesy of the *University of Illinois at Urbana-Champaign Archives*. Reproduction by Christina Laukaitis, 2021

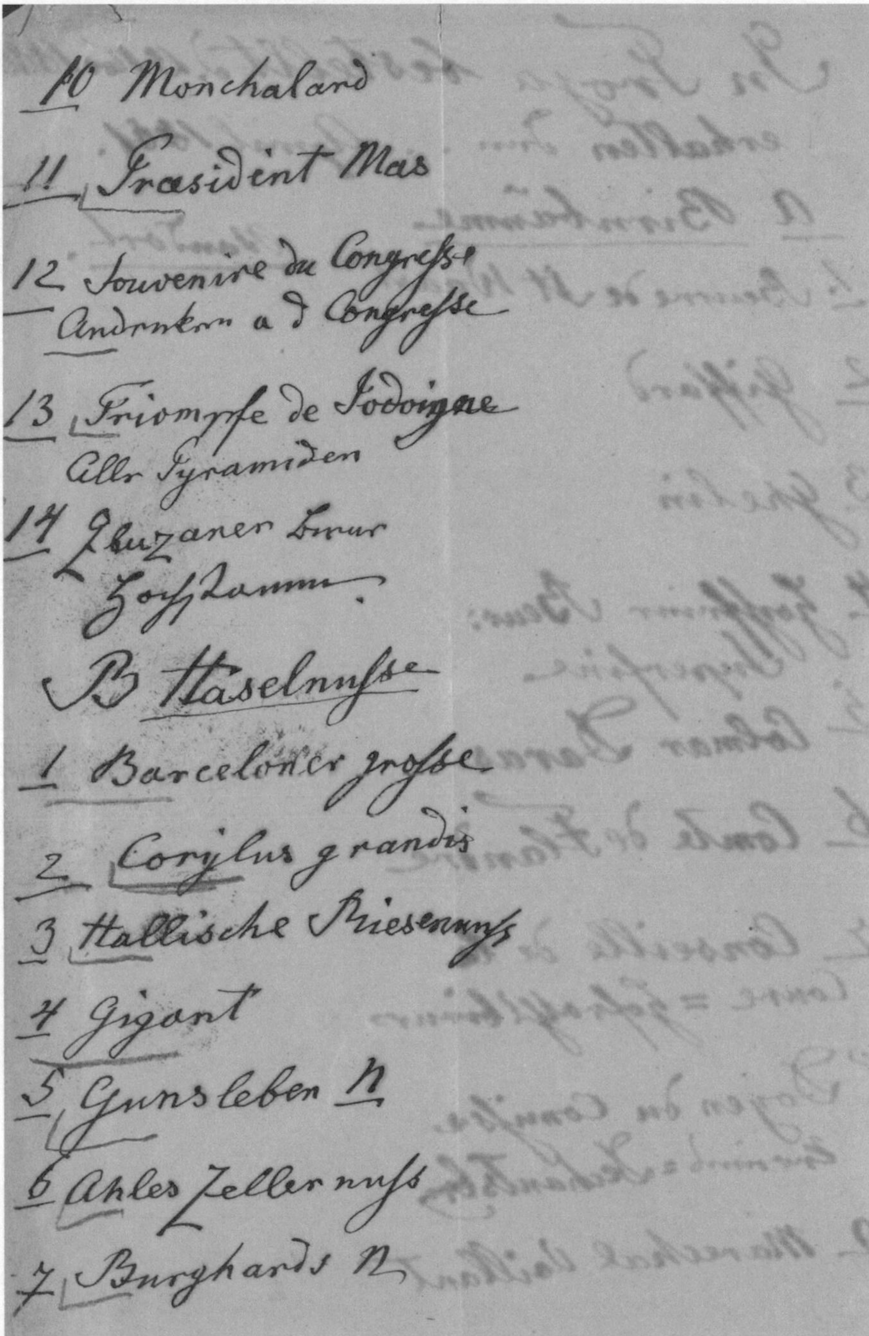

10 Monchalard

11 Präsident Mas

12 Souvenire du Congresse
Andenken a d Congresse

13 Triomphe de Jodoigne
Aller Pyramiden

14 Bozener Birne
Hochstamm.

B Haselnüsse

1 Barcelonner grosse

2 Corylus grandis

3 Hallische Riesennuss

4 Gigant

5 Gunsleben N

6 Ahles Zellernuss

7 Burghards N

(continued)

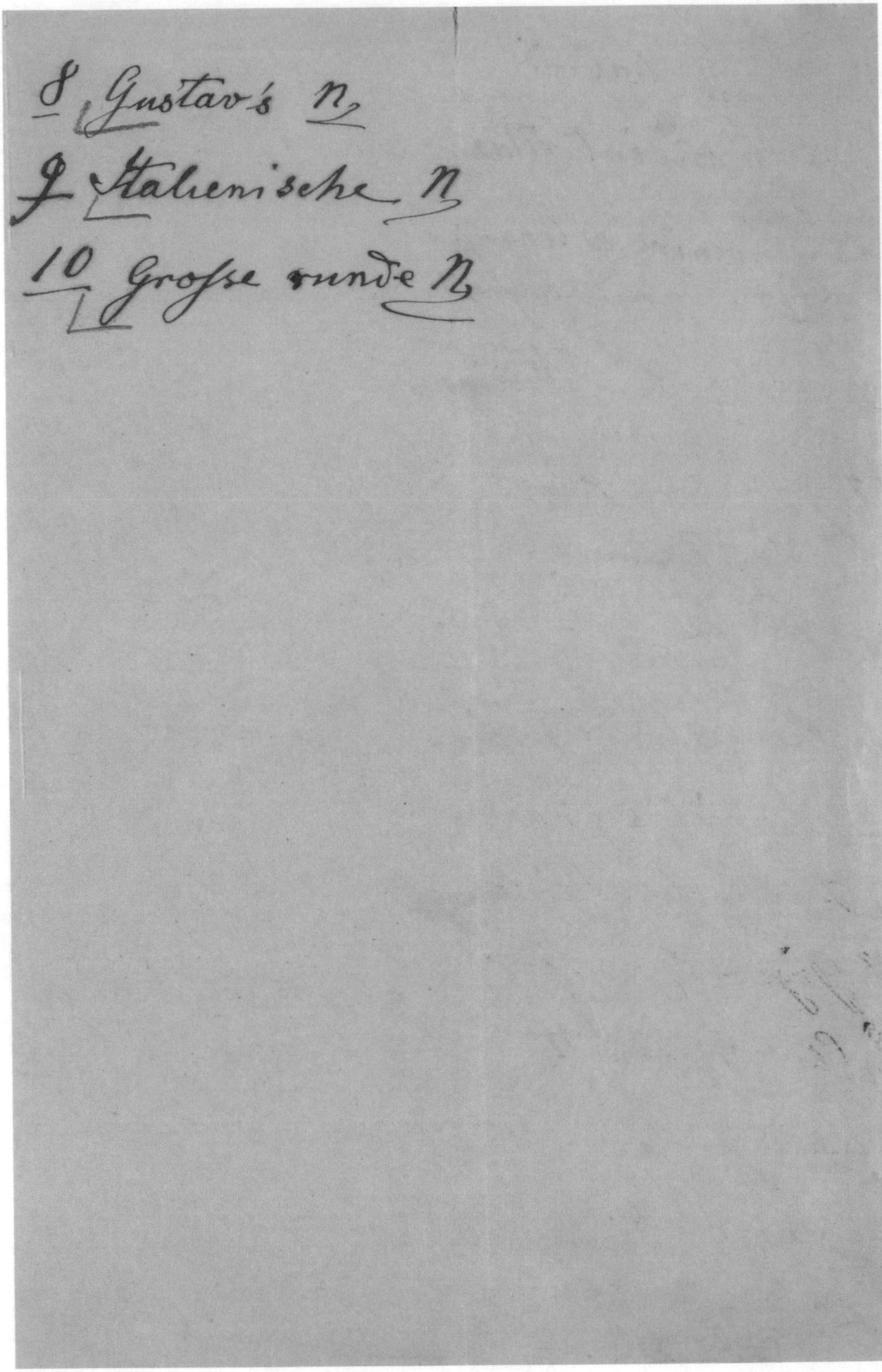

8 Gustav's N

9 Italienische N

10 Grosse runde N

(continued)

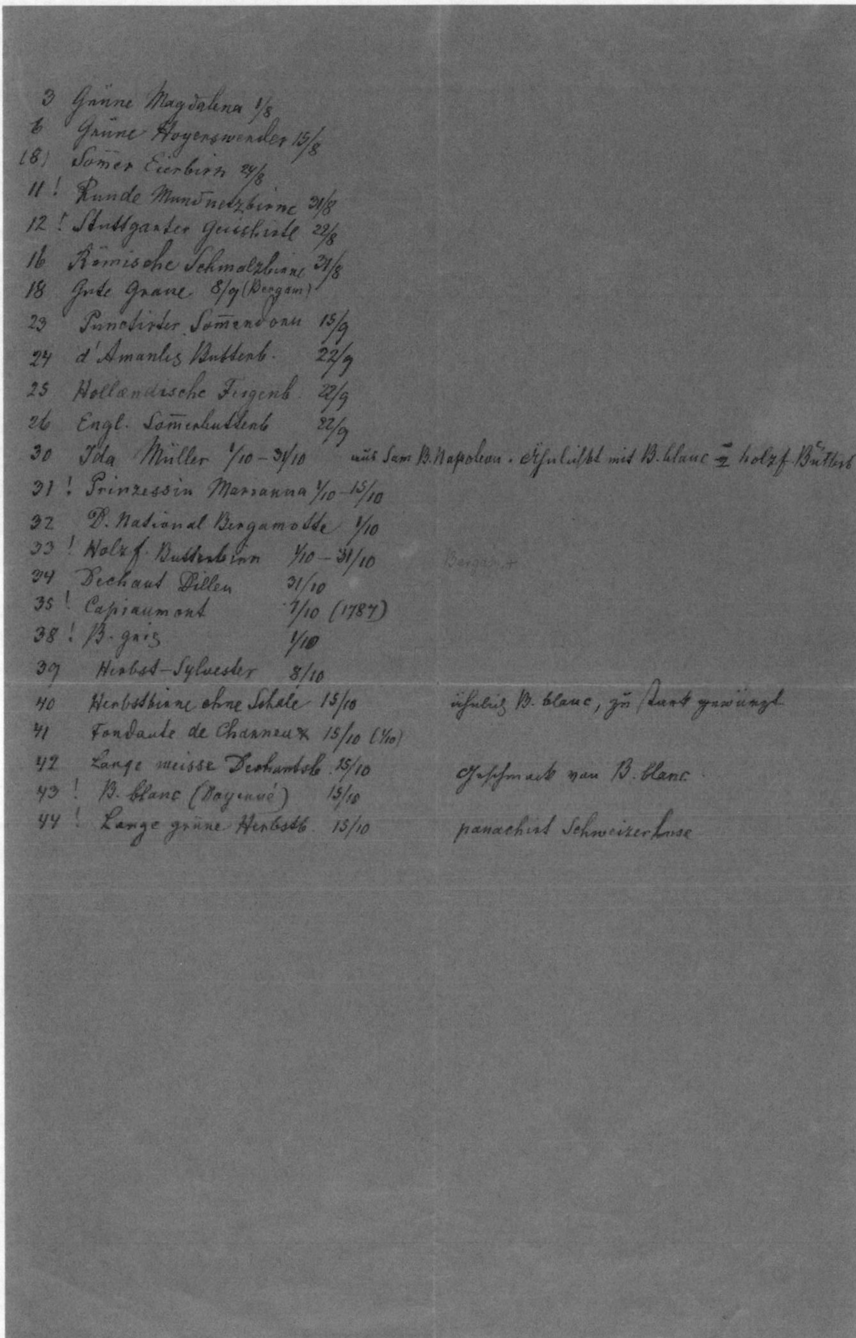

Mendel's notes on fruit trees. One sheet. Item No. 159 Box 5, Record Series 15/24/56, courtesy of the *University of Illinois at Urbana-Champaign Archives*. Reproduction by Christina Laukaitis, 2021

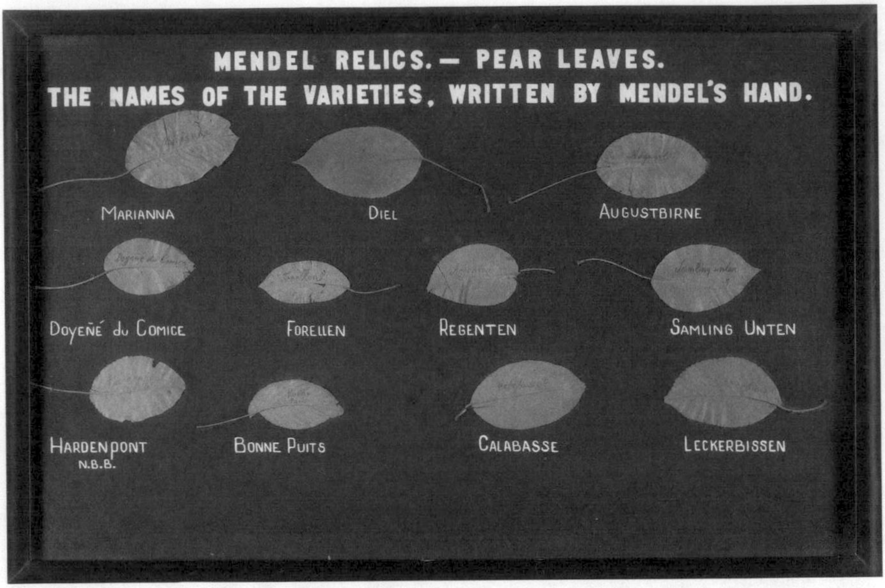

Eleven pear leaves. The names of the varieties are written by Mendel's hand: Marianna, Diel, Augustbirne, Doyene du Comice, Forellen, Regenten, Samling Unten, Hardenpont N.B.B., Nonne Puits, Calabasse, Leckerbissen. Item No. 161, Box 5, Record Series 15/24/56, courtesy of the *University of Illinois at Urbana-Champaign Archives*. Reproduction by Christina Laukaitis, 2021

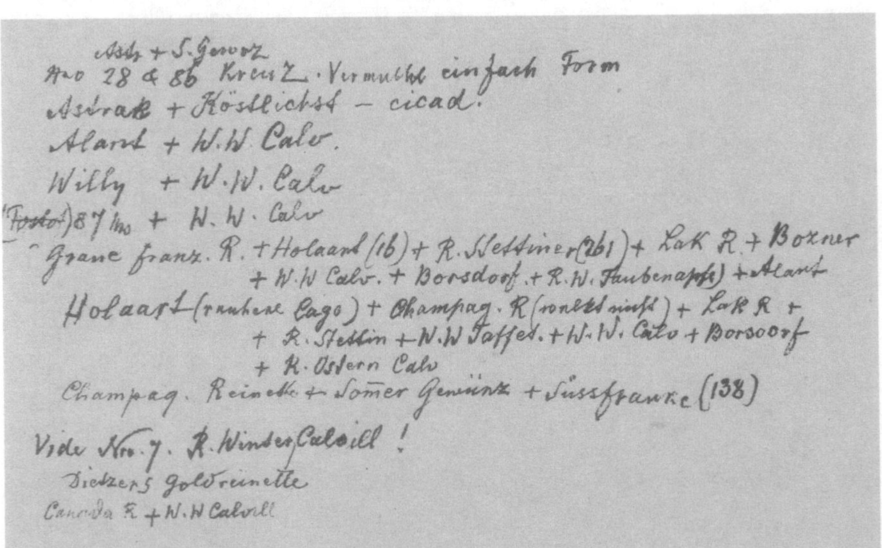
Astr. + S. Gervoz
Aro 28 & 86 Kreuz. Vermuthl. einfach Form
Astrak + Köstlichst – cicad.
Alant + W.W. Calv.
Willy + W.W. Calv
87 No + W.W. Calv
Grane franz. R. + Holaart (16) + R. Stettiner (261) + Lak R. + Bozner
+ W.W Calv. + Borsdorf. + R.W. Taubenapfel) + Alant
Holaart (rauhere Cago) + Champag. R (vielleicht nicht) + Lak R +
+ R. Stettin + W.W Taffet. + W.W. Calv + Borsdorf
+ R. Ostern Calv
Champag. Reinette + Somer Gewürz + Süssfranke (138)
Vide No. 7. R. Winter Calvill !
Dietzers Goldreinette
Canada R + W.W Calvill

IM86 Mendel's pencilled notes on the inner side of the cover of the first volume of handbook on pomology by Jahn, Lucas and Oberdieck 1860

References and Historical Printed Sources

Iltis H (1924) Gregor Johann Mendel. Leben, Werk und Wirkung. Julius Springer, Berlin. Life of Mendel. English translation by Eden and Cedar Paul. First edition. Norton, New York 1932; Second edition Allen and Unwin London, 1966

Jordánková H, Sulitková L (2008) Brno uvnitř hradeb /Brno inside the City Walls/. Paseka, Praha - Litomyšl

Orel V (1970a) Die Auseinandersetzung um die Organisation der Brünner Naturforscher in der Zeit, da G. Mendel seine Pisum-Versuche durchführte. FM 5:55–72

Orel V (1970b) Mendel and the Central Board of the Agricultural Society. FM 5:39–54

Schreiber JAE (1801) Nachricht von der Industrialschule zu Kunewald. Patriotisches Tagesblatt 42-5 Brno

Chapter 17
Organizer of Agricultural Society Exhibits

Circular paintings showing arranged plants, apples, prunes and vine grapes on the ceiling of the Augustinian Library Hall recall the exhibits organized by the Agricultural Society each year. Abbot Napp and Professor Diebl were the first organizers of the exhibits. Olexik and Mendel continued in their organizational work, announced and awarded prizes for new plant varieties achieved by means of hybridization. The exhibits were held in the Redouta Theatre, the museum and the new pavilion in Lužánky Park.

The Brno physician Olexik built glasshouses for growing rare ornamental plants imported from foreign plant breeders and showed them in the exhibits. Some exotic plants were supplied by the Augustinian gardener who grew exotic plants in pots in the orangery. Later, Mendel joined Olexik in purchasing bulbs of tulips and hyacinths and flower seedlings, which he grew for exhibits in the monastery garden plots and hothouses. The first exhibit of fruits, flowers and vegetables was held in Brno in April 1851 and was a great success. The founders were J. Patek, director of experimental fields of the Agricultural Society, F. Pluzar, director of the evangelical school, A. Illek, a professional gardener, P. Olexik, physician, meteorologist and gardener and L. Möser, an apothecarian. The participating Brno gardening firms were Molisch, Schebanek, Wittwe Rotwang and Absolon. In the spring, exhibitions highlighted camellias, tulips, hyacinths, rhododendrons and azaleas. In the fall, exquisite fruits, apples, pears, plums, selected vine grapes and vegetables of extraordinary forms, size or colour, In every exhibit, three prizes were awarded in the categories of best flower, best vegetable and best fruit. Mendel was an important member of the jury, appointing the winners, co-financing the prize money and formulating the objectives for the following years. If Mendel's exhibited products were chosen as the best, he never accepted the prize money that went to the second winner.

In 1868, Abbot Mendel announced a financial prize for breeding frost-resistant climbing roses obtained through hybridization. The roses had become popular in the aristocratic and feudal residential gardens, surviving winter without soil cover. The over-wintering roses were a great achievement of gardener Tvrdý. Count

A. Matalová, E. Matalová, *Gregor Mendel - The Scientist*, Springer Biographies,
https://doi.org/10.1007/978-3-030-98923-1_17

Mittrowsky planted them in great numbers in his castle gardens. The popularity of roses in Moravia was connected with the Rosencrucian movement tradition.

Mendel grew a series of new varieties of apricots and peaches, various sorts of vines, and plantations of strawberries and black currant bushes.

Johann N. Tvrdy (1806–1883) specialized in full-blooming fuchsias, verbenas and pelargonia, improved by means of artificial pollination (Vávra 1984). In the 1860s, Tvrdý was already known for his new varieties of fuchsias, which he announced in specialized European journals each year. He baptized his newly bred fuchsias with progressive names, such as Humboldt or Galilei. He named one of his new varieties Prelate Mendel. From 1850 onwards, Olexik played a significant role in the exhibits of flowers, fruits and vine grapes. In 1863, he was awarded the silver medal of the Agriculture Society in appreciation of his participation in the organization of exhibits of pomicultural, vinicultural and horticultural products, for his display of flowers and plants of exceptional beauty shown in many exhibits, and for increasing the interest of broad public in plant breeding.

Tvrdy was an expert in hybridization of fuchsias, verbenas, pelargonium, heliotropium, petunia and geranium. P. Olexik improved camellias, azaleas, rhododendrons, hyacinths, tulips, narcises, Jonquillas, tacetas, weigelias, ardisias, skimmias, hederas, spireas, lilies, cheiranthus, cinerium (cortoderia) and clematis. J. Absolon and F. Czastka specialized in breeding rhododendron, K. Dörrfeld begonias, J. Feigl azaleas and roses, A. Illek rhododendrons and calceolarias, K. Jelínek epacris and dianthus and V. Jusa rhododendron and azalea, all of them gardeners in Brno. A. Kubelka, also a gardener in Brno, improved rhododendron, azalea, begonia and pelargonium. J. Metzl, a gardener in Pernštejn, bred fuchsias and petunias. F. Molisch, a gardener in Brno bred verbenas, azaleas, pelargonium, cinerarias, violas tricolor, rhododendrons and heliotropium. J. Patek, a gardener in Rosenau/Rožínka bred gloxinias, calceolarias and violas tricolor. E. Pohle, a gardener in Eisgrub/Lednice, bred cinerarias, calceolarias, petunias, rhododendrons, azaleas, anemones, verbenas, amaryllis and dianthus. J. Reif, a gardener in Raiz/Rájec, bred primulas auricula, lophospermum (maurandia), begonias and cinerarias. J. Schwab, a gardener in Brno, bred rhododendrons. A. Schebanek, a gardener in Brno, bred juniperus, rhododendron, epacis, *Vitis quinquefolia*, *Amelopsis hederifolia* and *Primula veris*. J. Šeda, a gardener in Sokolnice, bred gloxinia. J. Zeitler, a physician in Brno, bred cheiranthus.

Brno gardeners produced also large amounts of vegetables and fruits. Surplus products were exported to Austria and neighbouring regions. The traditions of improving plant cultivation by means of import, hybridization and selection, and their protection against plant diseases, were intensively discussed in the Agricultural Society. In the second half of the nineteenth century, that dynamic industry offered new possibilities and opportunities and its commercial potential became institutionalized. The Moravian Mortgage Bank in Brno, with Mendel as its Director (Vybral 1968), is a good example. Economic interests prevailed over gardening.

In the ceiling paintings, Mendel expressed his satisfaction with his activity in the Agricultural Society as significant and meritorious. Presenting his emblem next to

The collections of plants and fruits are represented in circular arrangements in the ceiling paintings in the Library Hall in the Augustinian monastery. The paintings are a continuation of the series of paintings removed from the Great Chapter Hall depicting Mendel's fields of scientific interest: agriculture, meteorology, apiculture and pomology and signed with Mendel's monogram, GM, and dated 1875

Napp's shows that he thought of himself to be as successful as his predecessor. In the 1875 paintings Mendel was on an equal with Napp. However, he had no idea that his art of diplomacy when compared with the diplomatic Napp, equalled zero. His vain struggle against the increased contribution of the Augustinian monastery to the "religious fund" lasted ten years and made him prematurely an old man.

Names of plant genera mentioned in Mendel's works and in his letters to Nägeli (changes in the genus identification are given in brackets): *Antirrhinum*/snapdragon, *Aquilegia*/aquilegia, *Calceolaria*/lady's slipper, *Carex*/sedge, *Cirsium*/thistle, *Dianthus*/carnation, *Geum*/avens, *Hieracium*/hawkweed, *Cheiranthus*/wallflower (*Matthiola*/viola), *Ipomoea*/morning glory (*Pharbitis*/morning glory), *Lathyrus*/ sweetpea, *Lavatera*/mallow, *Lens*/lentils, *Linaria*/strawberry, *Geranium*/geranium, *Lychnis*/catchfly (*Melandium*/rose champion), *Malva*/mallow, *Matthiola*/viola, *Mirabilis*/ambrella-wort, *Nicotiana*/tobacco, *Oenothera*/evening primrose, *Phaseolus*/common bean, *Pisum*/garden pea, *Potentilla*/goose-grass *Salix*/willow, *Tropaeolum*/Indian cress, *Verbascum*/ Aarod's rod, *Veronica*/speedwell, *Viola*/viola and *Zea*/maize. The list of Mendel's plants was extended (Van Dijk et al. 2018, p. 349) on the basis of newspaper information from *Neuigkeiten* on 26 July 1861 (taken over from *Mährischer Korrespondent*): *Father Gregor Mendl, professor at the local k. k. Oberrealschule, is concerned with instructive experiments, which are*

The ceiling paintings in the Library Hall show flowers and fruits and abbatial emblems of Napp and Mendel, both placed on opposite sides of the ceiling

aimed at improving the vegetable and flower varieties cultivated in our region. Through artificial fertilization truly surprising results could be achieved. The vegetables grown by the professor, such as peas, fisols, cucumbers and beans ... Until now, the experiments carried out with potatoes were less successful. ... The carnations and fuchsias, of which the Professor grew several 100 pots ...

Van Dijk (2018, p. 349) adds a critical reaction to Mendel's hybridizing experiments published in the *Brünner Zeitung*, pointing to Mendel's small-scale experiments and their small economic importance.

The colour mix of flowers in the Library Hall paintings also recall Mendel's attempt at solving the colour scale in the development of hybrids as a composite trait that follows the same law as *Pisum.* Translated by Abbot and Fairbanks. 2016, pp. 407–422 (available also in FM 2016, 52/2).

Notwithstanding the many difficulties these observations had to confront, this experiment at least shows that the development of hybrids follows the same law as in Pisum in relation to those characters corresponding to the form of the plant. With respect to the colour characters, however, it seems difficult to find sufficient accordance. Laying aside the fact that a whole array of colours arises from the union of a white and purple-red colour, from purple to pale violet and white, it is a striking circumstance that of 31 plants that flowered, only one produced the recessive character of white colour, whereas with Pisum such is the case for every fourth plant on average.

But even these enigmatic phenomena might probably be explained according to the laws that are valid for Pisum if one could assume that the flower and seed colour

Frost-resistant climbing roses were achieved by means of artificial fertilization and got a prize with Mendel's financial support. In the background of the floral composition, pea leaves and tendrils are evident. The white pea flowers cannot be easily recognized in the white ceiling

of Ph. multiflorus are a complex of two or more completely independent colours that individually behave like other constant characters of a plant. If flower colour A were composed of the independent characters $A_1 + A_2 + \ldots$, *that create the total impression of purple-red colour, then through fertilization with the differing character of white colour the hybrid combinations* $A_1a + A_2a + \ldots$ *would be formed, and similar behaviour would be expected with the corresponding colour of the seed coat. According to the assumption stated above, each of these hybrid colour combinations would be self-sufficient and would thus develop completely independently from the others. One can easily see, then, by combining the individual developmental series, a complete colour series must arise. If, for example,* $A = A_1 + A_2$, *then the hybrids* A_1a *and* A_2a *would correspond to the developmental series*

$A_1 + 2A_1a + a$
$A_2 + 2A_2a + a$

The members of these series can occur in nine different combinations and each of them represents the designation for another colour:

Roses and ipomoea are depicted with pea leaves in the background

1 $A_1 \; A_2$	*2* $A_1a \; A_2$	*1* $A_2 \; a$
2 $A_1 \; A_2a$	*4* $A_1a \; A_2a$	*2* $A_2a \; a$
1 $A_1 \; a$	*2* $A_1a \; a$	*1* $a \; a$

The numbers assumed for the individual combinations simultaneously indicate how many plants with the corresponding colour belong to the series. Since that sum equals 16, all colours, on average, are distributed to each of 16 plants, although, as the series itself shows, in unequal proportions.

If the development of colours actually took place in this manner, the case noted above could be explained—that white flowers and seed colour occurred only once among 31 plants of the first generation. This colour is included only once in the series and could thus, on average, develop once for each 16; and with three colour characters, only once for each 64 plants.

It must not be forgotten, however, that the explanation proposed here is based only on a mere supposition that has no other support than the very imperfect result of the experiment just discussed. It would, of course, be a worthwhile labour to follow the development of colour in hybrids with similar experiments, since it is probable that in this way we would come to understand the extraordinary multitude of colours in our ornamental flowers.

Fully bloomed fuchsia, ipomoea and sorbus are arranged with pea leaves and tendrils

At this point, little more is known with certainty other than flower colour in most ornamental plants is an extremely variable character. The opinion has often been expressed that the stability of a species has been disrupted to a high degree or utterly broken through cultivation. There is a common inclination to refer to the development of cultivated forms as proceeding without rules and by chance; the colour of ornamental plants is generally cited as a pattern of instability. It is not apparent, however, why the mere placement in garden soil should result in such a drastic and persistent revolution in the plant organism. No one will seriously assert that the development of plants in a natural landscape is governed by different laws than in a garden bed. Here, just as there, typical variations must appear if the conditions of life are changed for a species, and it has the ability to adapt to the new conditions. It is freely admitted, through cultivation the production of new varieties is favoured, and by the hand of man many a variation is preserved that would have failed in the

Fully bloomed fuchsias and tansy with pea leaves in the background

wild state, but nothing gives us the right to assume that the tendency for new varieties to form is so extremely augmented that species soon lose all stability and that their offspring break up into an infinite array of highly variable forms. If the change in the conditions of vegetation were the sole cause of variability, then one would be justified in expecting that those domesticated plants cultivated under almost the same conditions for centuries would have acquired stability. As is well known, this is not the case, for especially among them not only the most different but also the most variable forms are found. Only the Leguminosae, like Pisum, Phaseolus, Lens, whose organs of fructification are protected by a keel, constitute an appreciable exception. Even for these, numerous varieties have arisen during cultivation for more than 1000 years under the most diversified conditions; however, under the same permanent conditions of life, they retain stability similar to that of species growing in the wild.

It remains more than probable that there is a factor in action for the variability of cultivated plants, which hitherto has received little attention. Different experiences urge us to the view that our ornamental plants, with few exceptions, are members of different hybrid series whose legitimate further development is modified and delayed through numerous intercrosses.

A collection of apples, white vine grapes, pears and plums are depicted near Mendel's emblem

In 1872, Mendel was awarded the Comthur Cross of the Emperor, on which occasion he introduced a change in his abbatial emblem. He substituted the shaking hands with a clergyman's arm holding a cross. The arm with the cross Mendel copied from Napp's emblem.

We do not know what prompted Mendel to change the emblem and why he replaced the symbol of human friendship with clergyman's arm holding a cross. Maybe he followed the command in Latin words from St. Augustine that is carved in marble in the Augustinian refectory: *Ante omnia fratres carissimi diligatur Deus deinde Proximus* (Above all, beloved brothers, love God and then the neighbour). The equation Alpha = Omega in the original emblem is replaced with a relation Alpha: Omega that may represent the algebraic expression of the origin of the hybrid A/a or A: a. The double-point colon expresses the hybrid connection. It is surprising why Mendel did not use the classic relation Alpha + Omega.

Antagonistic parental traits occur in mirror symmetry (Alpha = Omega) as discovered by Mendel:

The origin of the hybrid	A : a
In the hybrid, contrary types are complementary	Aa
1st generation from the hybrid	A Aa aA a
2nd generation from the hybrid	A A Aa aA a a
3rd generation from the hybrid	A A A Aa aA a a a

New varieties of fuchsias announced by Tvrdy in the flower catalogue for 1866: 1. Swan, 2. Majestica, 3. Spectabilis, 4. Goliath, 5. Remembrance of Humboldt, 6. Prince A. Schwarzenberg and 7. Princess of Dietrichstein

Mendel explains the equilibrium of dominant and recessive elements in his *Experiments on Plant Hybrids* (Translated by Sherwood in Stern and Sherwood 1966, p. 16):

> If one assumes, on the average, equal fertility for all plants in all generations, and if one considers, furthermore, that half of the seeds that each hybrid produces yield hybrids again while the other half the two traits become constant in equal proportions, then the numerical relationships for the progeny in each generation follow from the tabulation below, where A and a again denote the two parental traits and Aa the hybrid form. For brevity's sake one may assume that in each generation each plant supplied only four seeds.

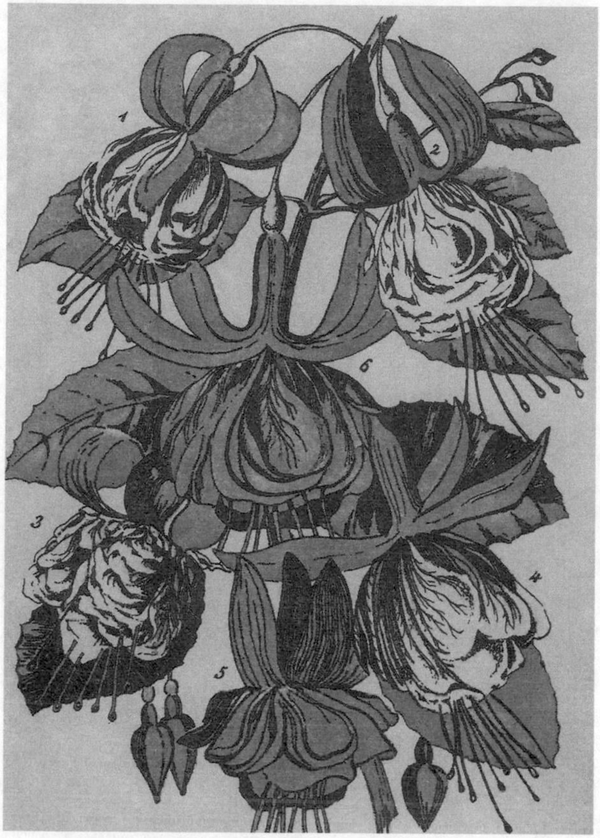

New varieties of fuchsias bred by Tvrdy for the 1867 flower catalogue: 1. Fata Morgana, 2. Custozza, 3. Count C. of Sternberg, 4. Madame Rambousek, 5. Vice Admiral v. Tegetthoff and 6. Mr. Laurentius

Expressed in terms of ratios

Generation	*A*	*Aa*	*a*	*A : Aa : a*
1	1	2	1	1 : 2 : 1
2	6	4	6	3 : 2 : 3
3	28	8	28	7 : 2 : 7
4	120	16	120	15 : 2 : 15
5	496	32	496	31 : 2 : 31
n				$2^n - 1 : 2 : 2^n - 1$

This is the first model in biology, calculated by Mendel the physicist.

Mendel obtained different results in *Hieracium* than in *Pisum*. In his letter to Nägeli dated 03 July 1870 he writes (in Stern and Sherwood 1966, p. 90): *On this occasion I cannot resist remarking how striking it is that the hybrids of Hieracium show a behaviour exactly opposite to those of Pisum. Evidently, we are here dealing*

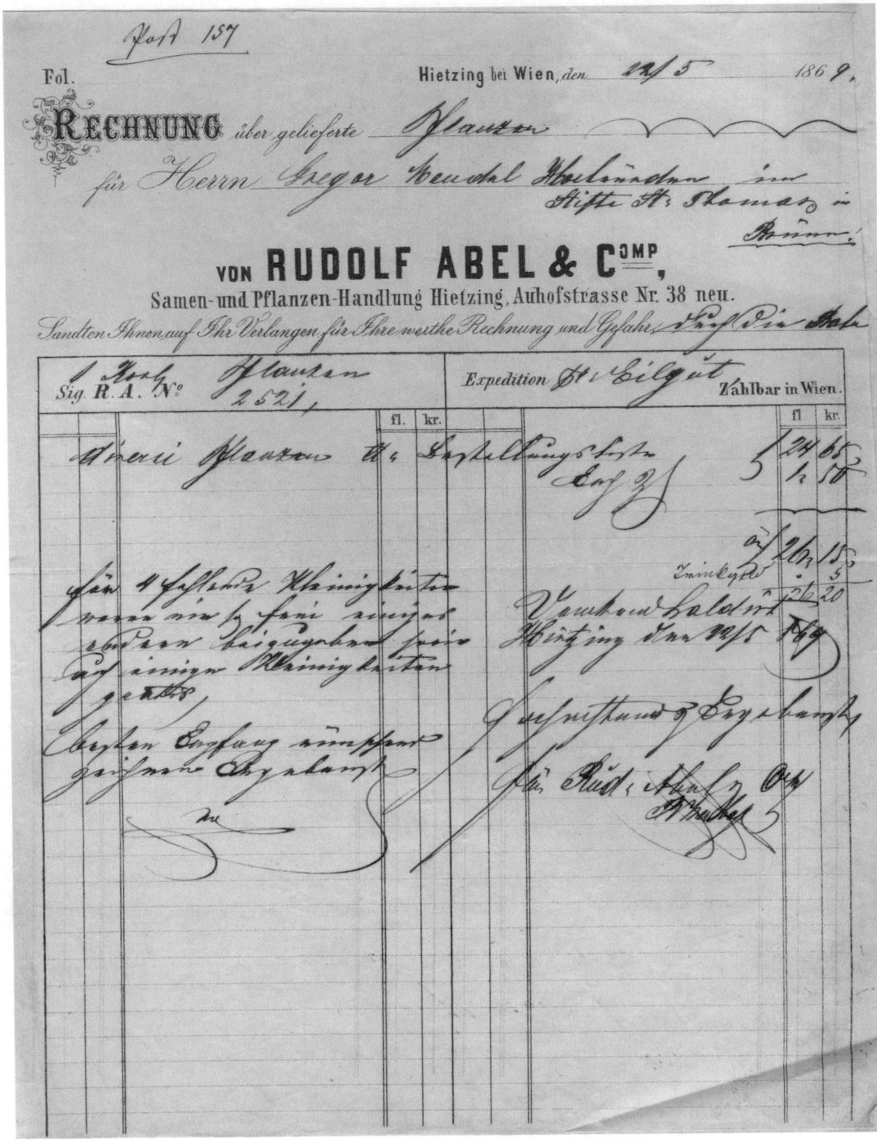

Fol. Hietzing bei Wien, den 1869.

RECHNUNG über gelieferte

für Herrn

VON RUDOLF ABEL & C[OMP],
Samen-und Pflanzen-Handlung Hietzing, Auhofstrasse Nr. 38 neu.

Sandten Ihnen auf Ihr Verlangen für Ihre werthe Rechnung und Gefahr

Sig. R. A. Nº

Expedition

Zahlbar in Wien.

fl. kr.

D728 Mendel's account for plants delivered by Rudolf Abel, Hietzing, dated 23 May 1868

only with individual phenomena, that are the manifestation of a higher, more fundamental law.

When Mendel compared the Hieracium results with those obtained from crosses of Pisum he found *a very real distinction: In Pisum the hybrids, obtained from the intermediate crossing of two forms, all have the same type, but their posterity, on the*

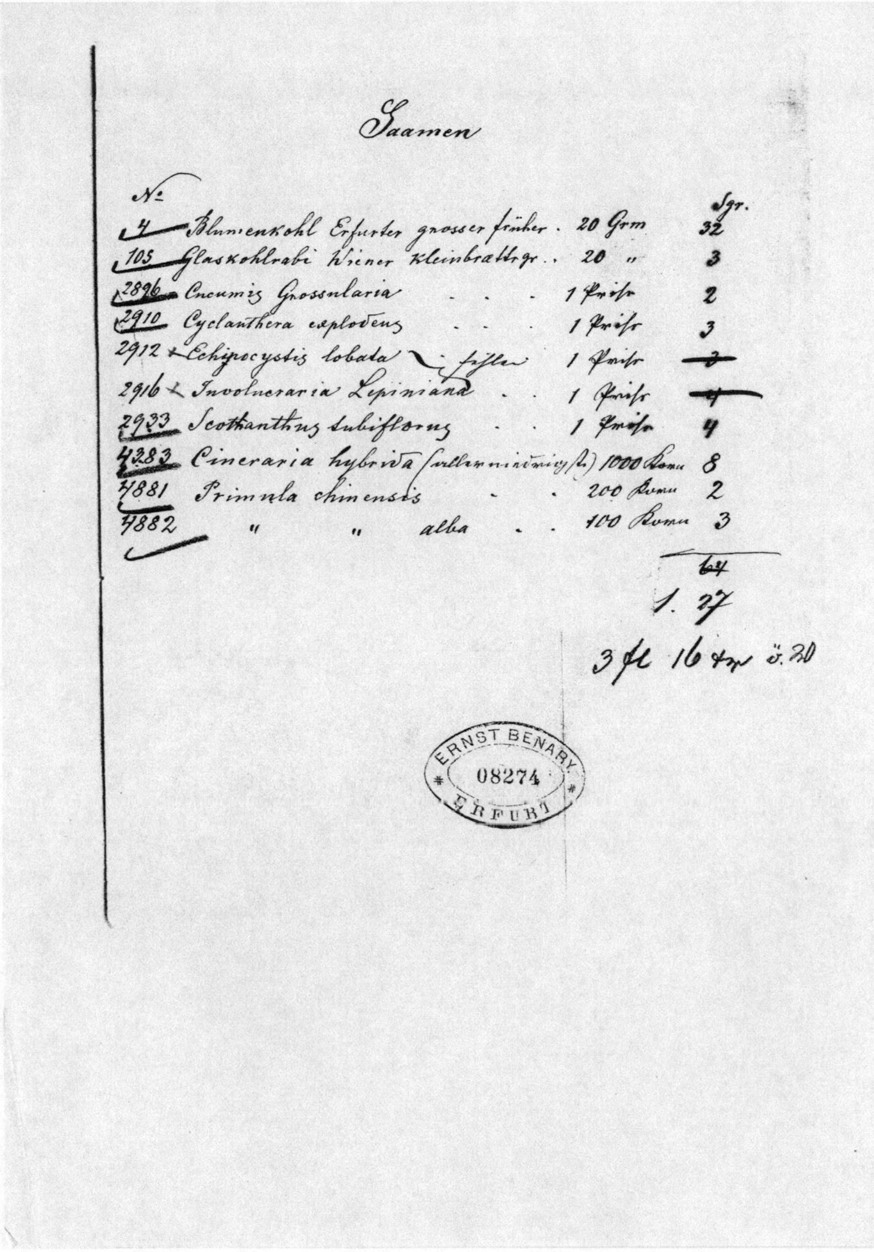

Saamen

No.			Sgr.
4	Blumenkohl Erfurter grosser früher .	20 Grm	32
105	Glaskohlrabi Wiener kleinblattr. gr. .	20 "	3
2896	Cucumis Grossularia . . .	1 Prise	2
2910	Cyclanthera explodens . . .	1 Prise	3
2912	+ Echinocystis lobata } fehlen	1 Prise	~~3~~
2916	+ Involucraria Lepiniana . .	1 Prise	~~4~~
2933	Scotanthus tubiflorus . .	1 Prise	4
4283	Cineraria hybrida (selbst [illegible])	1000 Korn	8
4881	Primula chinensis . .	200 Korn	2
4882	" " alba . .	100 Korn	3
			64
			1. 27
			3 fl 16 kr ö.W.

D249 Mendel's order for seeds, bulbs and plants delivered by Ernst Benary. Two pages. Erfurt. 21 September 1873

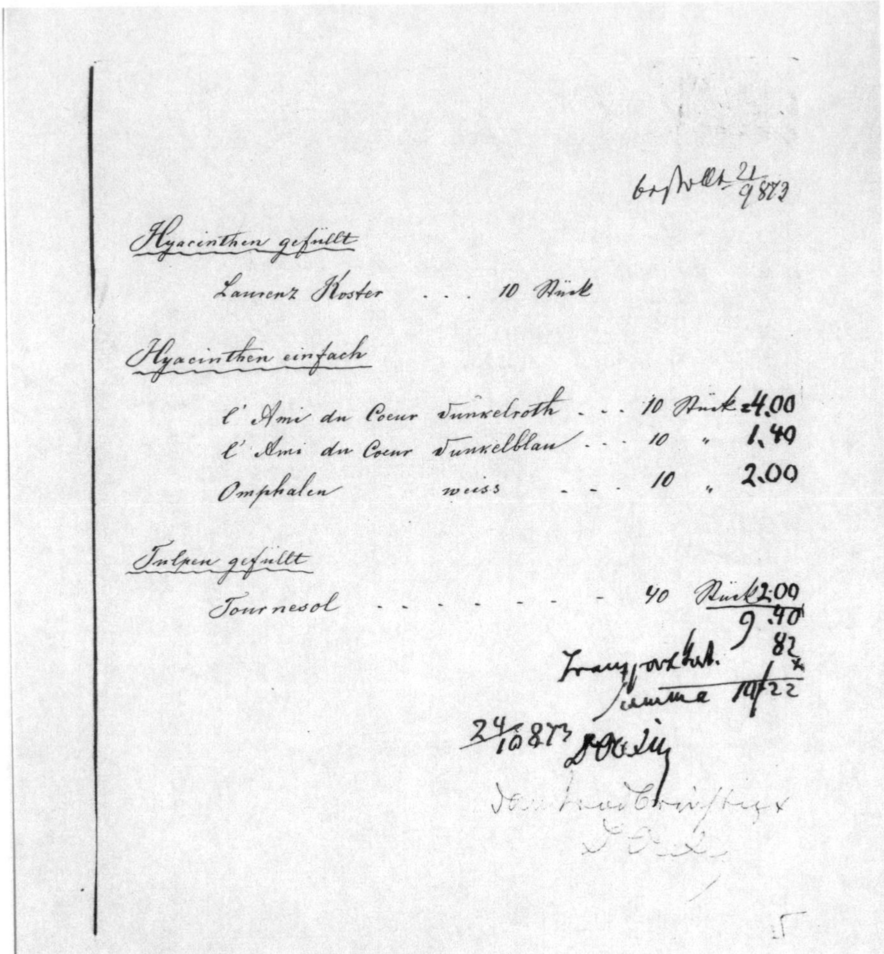

bestellt 21/9 873

Hyacinthen gefüllt

Laurenz Koster . . . 10 Stück

Hyacinthen einfach

l'Ami du Coeur dunkelroth . . . 10 Stück 4.00
l'Ami du Coeur dunkelblau . . . 10 " 1.40
Omphalen weiss . . . 10 " 2.00

Tulpen gefüllt

Tournesol . . . . 40 Stück 2.00
9.40
Transportkost. 82
Summe 10/22

24/10 873

(continued)

contrary, are variable and follow a definite law in their variations. In Hieracium according to the present experiments the exactly opposite phenomenon seems to be exhibited. In describing the Pisum experiments it was remarked that there are also hybrids whose posterity do not vary, and that, for example, according to Wichura

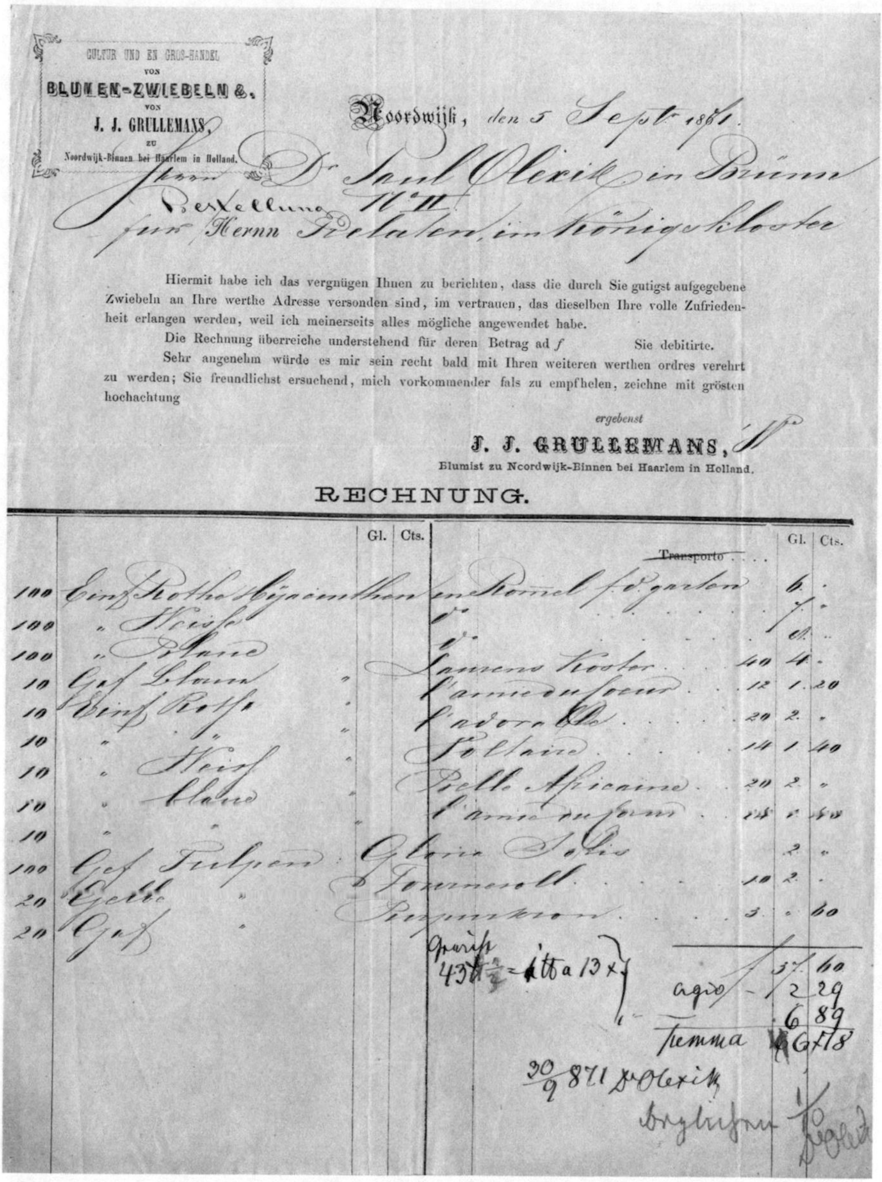

CULTUR UND EN GROS-HANDEL
VON
BLUMEN-ZWIEBELN &.
VON
J. J. GRULLEMANS
ZU
Noordwijk-Binnen bei Haarlem in Holland.

Noordwijk, den 5 Sept. 1871.

Hiermit habe ich das vergnügen Ihnen zu berichten, dass die durch Sie gutigst aufgegebene Zwiebeln an Ihre werthe Adresse versonden sind, im vertrauen, das dieselben Ihre volle Zufriedenheit erlangen werden, weil ich meinerseits alles mögliche angewendet habe.

Die Rechnung überreiche understehend für deren Betrag ad *f* Sie debitirte.

Sehr angenehm würde es mir sein recht bald mit Ihren weiteren werthen ordres verehrt zu werden; Sie freundlichst ersuchend, mich vorkommender fals zu empfhelen, zeichne mit grösten hochachtung

ergebenst

J. J. GRULLEMANS,

Blumist zu Noordwijk-Binnen bei Haarlem in Holland.

RECHNUNG.

	Gl.	Cts.		Gl.	Cts.
			Transporto . . .		

D691 Olexik's and Mendel's account for bulbs of hyacinths delivered by J. J. Grullemans, florist in Noordwijk-Binnen near Haarlem. Dated in Noordwijk, September 1871

the hybrids of Salix reproduce themselves like pure species. In Hieracium we may take it we have a similar case. Whether from this circumstance we may venture to draw the conclusion that the polymorphism of the genera Salix and Hieracium is connected with the special condition of their hybrids is still an open question, which may well be raised but not as yet answered. (From Stern and Sherwood 1966, p. 55.)

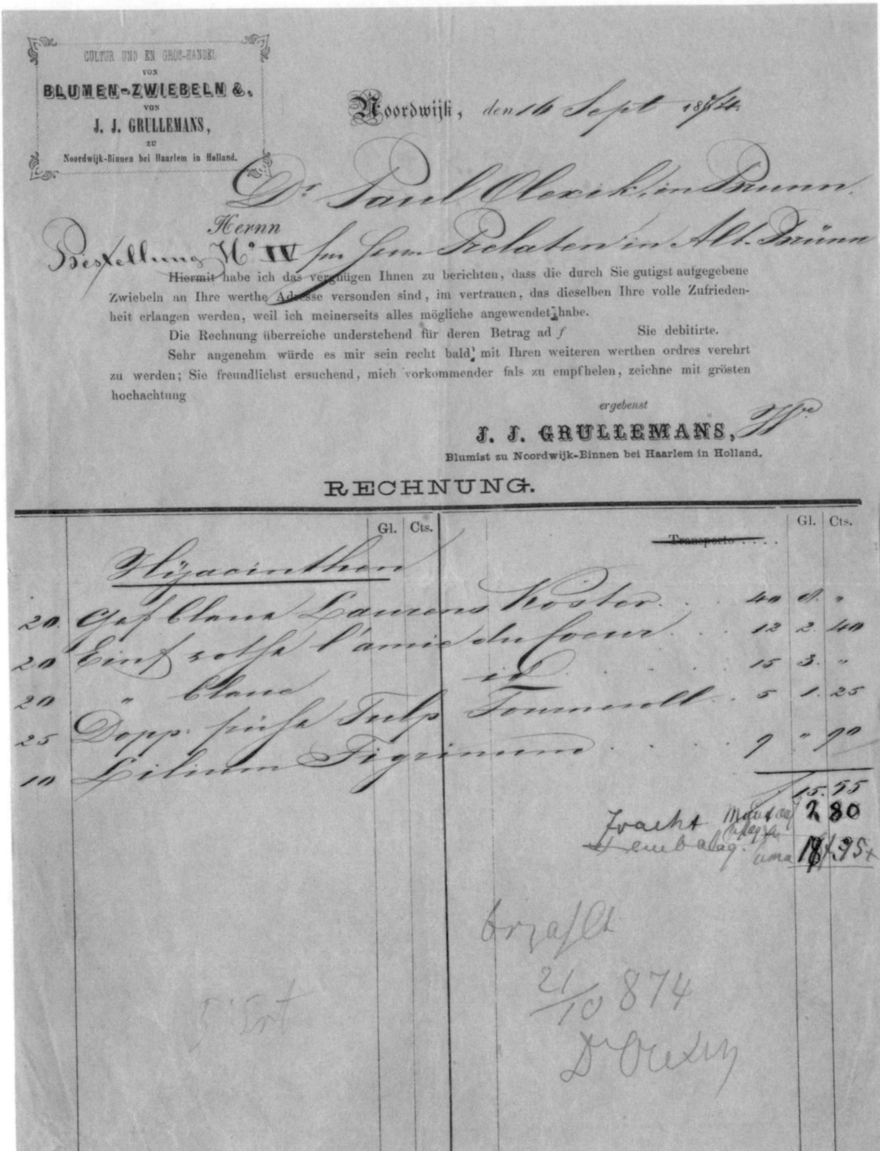

CULTUR UND EN GROS-HANDEL
VON
BLUMEN-ZWIEBELN &.
VON
J. J. GRULLEMANS,
ZU
Noordwijk-Binnen bei Haarlem in Holland.

Noordwijk, den 16 Sept 1874.

Herrn Dr. Paul Olexik in Brünn.

Bestellung No. IV für Herrn Prelaten in Alt-Brünn

Hiermit habe ich das vergnügen Ihnen zu berichten, dass die durch Sie gutigst aufgegebene Zwiebeln an Ihre werthe Adresse versonden sind, im vertrauen, das dieselben Ihre volle Zufriedenheit erlangen werden, weil ich meinerseits alles mögliche angewendet habe.

Die Rechnung überreiche understehend für deren Betrag ad *f* Sie debitirte.

Sehr angenehm würde es mir sein recht bald mit Ihren weiteren werthen ordres verehrt zu werden; Sie freundlichst ersuchend, mich vorkommender fals zu empfhelen, zeichne mit grösten hochachtung

ergebenst

J. J. GRULLEMANS,

Blumist zu Noordwijk-Binnen bei Haarlem in Holland.

RECHNUNG.

		Gl.	Cts.		Gl.	Cts.
	Hyacinthen			~~Transporto~~		
20	Gefbl. blau Laurens Koster			40	0.	"
20	Einf. rothe l'ami du Coeur			12	2.	40
20	" blau ...			15	3.	"
25	Dopp. frühe Tulp. ...			5	1.	25
10	Lilium Tigrinum			9	"	90
					15.	55
	Fracht				2	80
					18	35

bezahlt 21/10 874 Dr. Olexik

D691 Olexik's account for Mendel's delivery of hyacinth bulbs by J. J. Grullemans Noordwijk-Binnen near Haarlem. Dated in Noordwijk, 16 September 1874

At Mendel's death, no one highlighted his *Experiments on Plant Hybrids*. Only the Section of Pomiculture, Viticulture and Horticulture of the Agricultural Society recognized the "epoch-making" significance of Mendel's hybridization experiments in the obituary published in German in its monthly journal *Monatsberichte der*

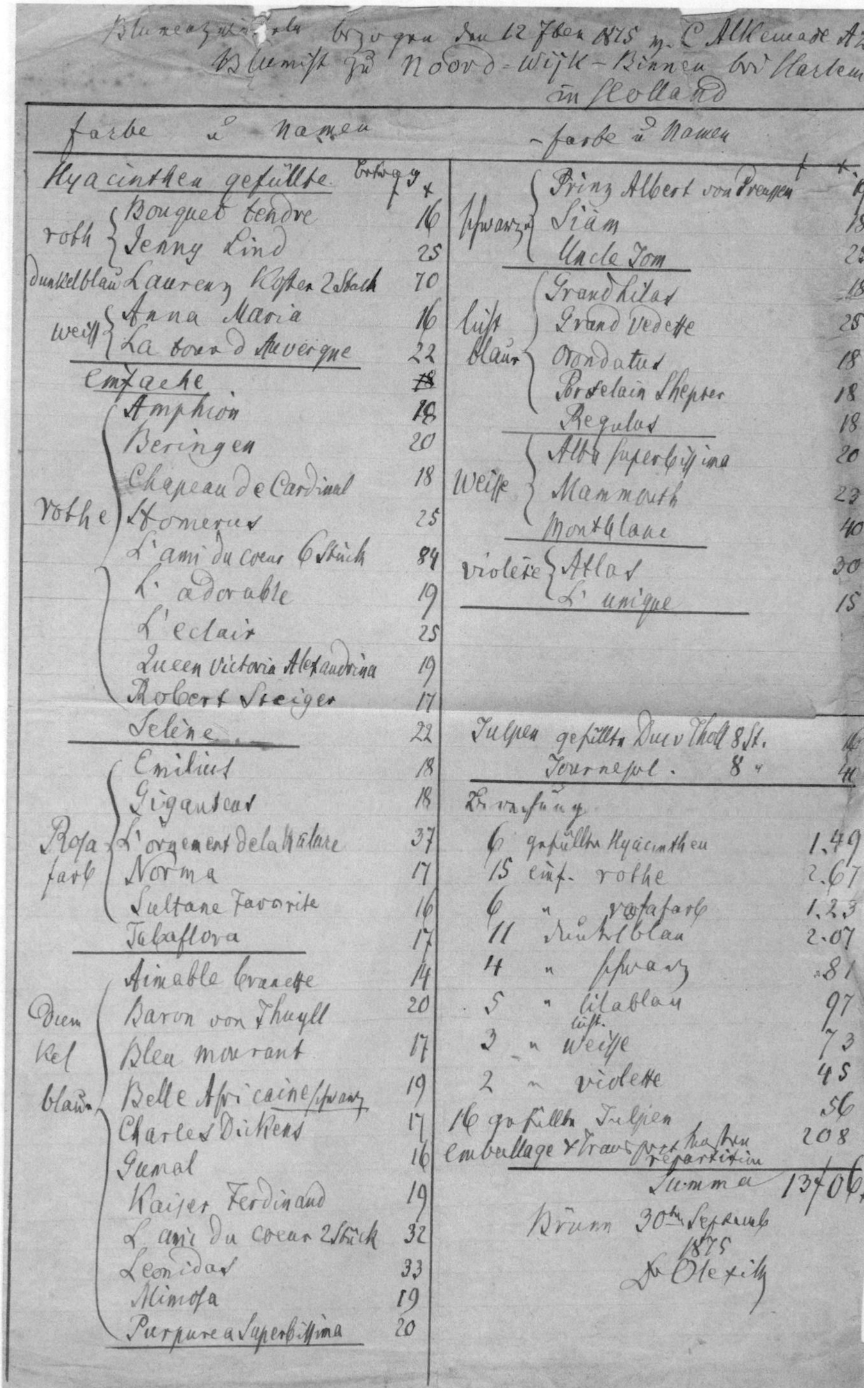

Blumenzwiebeln bezogen den 12 7ber 1875 v. C. Alkemade Az Blumist zu Noord-Wijk-Binnen bei Haarlem in Holland

farbe u. Namen		farbe u. Namen	
Hyacinthen gefüllte	Ertrag f. x	schwarz: Prinz Albert von Preussen	19
roth: Bouquet tendre	16	Siam	18
Jenny Lind	25	Uncle Tom	25
dunkelblau: Laurens Koster 2 Stück	70	licht blau: Grand Lilas	18
weiss: Anna Maria	16	Grand Vedette	28
La tour d Auvergne	22	Ondatus	18
Einfache	~~18~~	Porzelain Shepter	18
rothe: Amphion	18	Regulus	18
Beringen	20	weisse: Alba superbissima	20
Chapeau de Cardinal	18	Mammouth	23
Homerus	25	Montblanc	40
L'ami du coeur 6 Stück	84	violette: Atlas	30
L'adorable	19	L'unique	15
L'eclair	25		
Queen Victoria Alexandrina	19		
Robert Steiger	17		
Selene	22	Tulpen gefüllte Duc v Thol 8 St.	16
Rosa farb: Emilius	18	Tournesol. 8 "	40
Giganteas	18	Zusammenfassung	
L'orgement de la Nature	37	6 gefüllte Hyacinthen	1.49
Norma	17	15 einf. rothe	2.67
Sultane Favorite	16	6 " rosafarb	1.23
Tubiflora	17	11 dunkelblau	2.07
Dunkel blau: Aimable Rosette	14	4 " schwarz	.81
Baron von Thuyll	20	5 " lilablau licht	97
Bleu mourant	17	3 " weisse	73
Belle Africaine schwarz	19	2 " violette	45
Charles Dickens	17	16 gefüllte Tulpen	56
General	16	Emballage & Transport Kasten Reparation	208
Kaiser Ferdinand	19	Summa	13706 x
L'ami du coeur 2 Stück	32		
Leonidas	33		
Mimosa	19		
Purpurea Superbissima	20		

Brünn 30ten September 1875
Dr Olexik

D356 List of delivery by C. Alkemade in Noordwijk-Binnen near Haarlem, received by Dr. Olexik in Brno, 30 September 1875

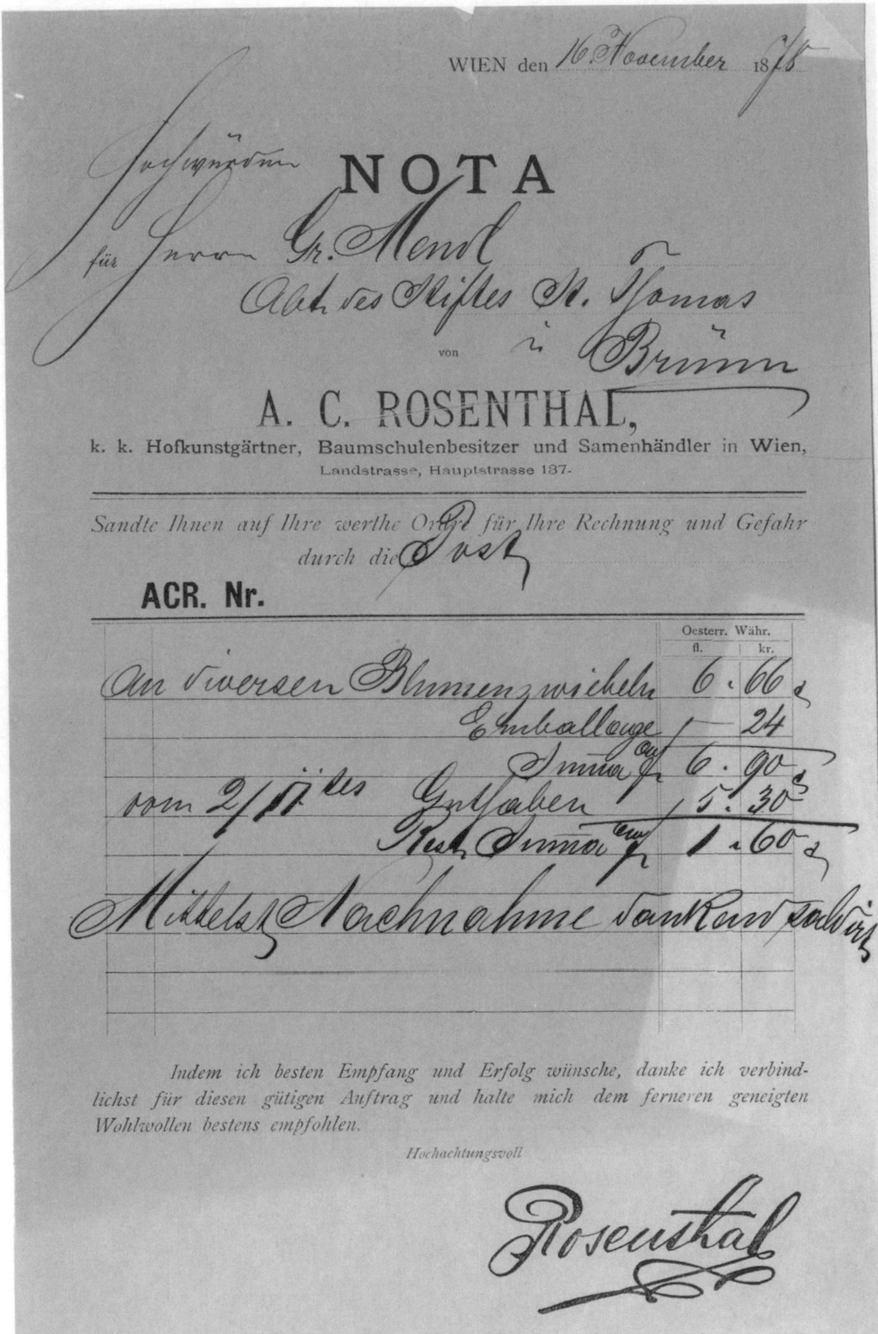

WIEN den 16. November 1878

Hochwürden

NOTA

für Herrn Gr. Mendel
Abt des Stiftes St. Thomas
von in Brünn

A. C. ROSENTHAL,

k. k. Hofkunstgärtner, Baumschulenbesitzer und Samenhändler in Wien,
Landstrasse, Hauptstrasse 137.

Sandte Ihnen auf Ihre werthe Ordre für Ihre Rechnung und Gefahr durch die Post

ACR. Nr.

	Oesterr. Währ. fl.	kr.
An diversen Blumenzwiebeln	6.	66
Emballage		24
Summa	6.	90
vom 2/11 ten Guthaben	5.	30
Rest Summa	1.	60
Mittelst Nachnahme dankend saldirt		

Indem ich besten Empfang und Erfolg wünsche, danke ich verbindlichst für diesen gütigen Auftrag und halte mich dem ferneren geneigten Wohlwollen bestens empfohlen.

Hochachtungsvoll

Rosenthal

D731 List of delivery of diverse plant bulbs to "Gr. Mendel" by A. C. Rosenthal. Dated in Vienna, 16 November 1878

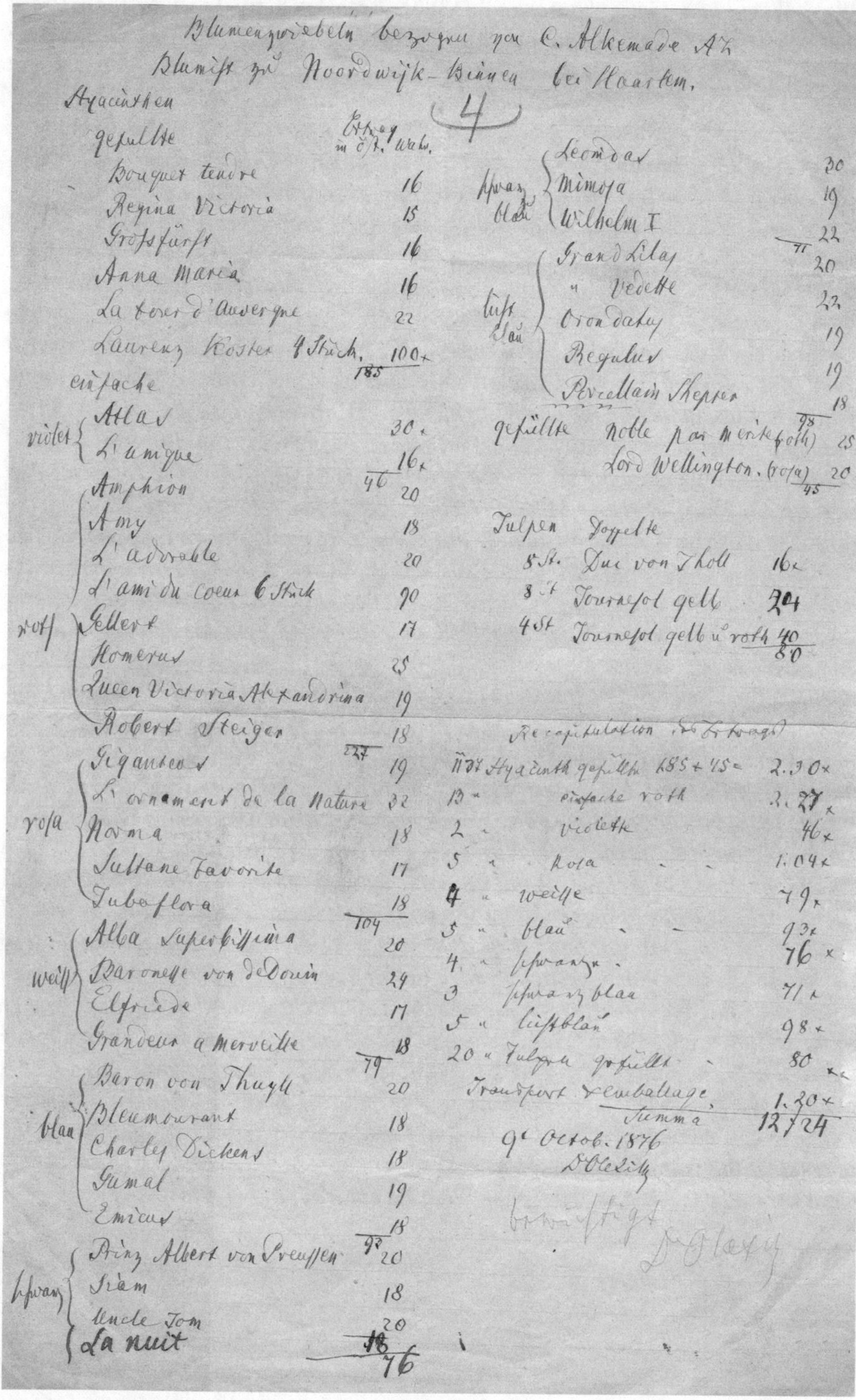

Blumenzwiebeln bezogen von C. Alkemade Az
Blumist zu Noordwijk-Binnen bei Haarlem.

(4)

Hyacinthen

gefüllte — Ertrag in öst. Währ.

Bouquet tendre	16
Regina Victoria	15
Großfürst	16
Anna Maria	16
La tour d'Auvergne	22
Laurens Koster 4 Stück	100
	185

einfache

violet	Atlas	30
	L'unique	16
		46
roth	Amphion	20
	Amy	18
	L'adorable	20
	L'ami du coeur 6 Stück	90
	Gellert	17
	Homerus	25
	Queen Victoria Alexandrina	19
	Robert Steiger	18
		227
rosa	Giganteus	19
	L'ornament de la Nature	32
	Norma	18
	Sultane Favorite	17
	Tubaflora	18
		104
weiss	Alba superbissima	20
	Baronesse von de Donin	24
	Elfriede	17
	Grandeur a merveille	18
		79
blau	Baron von Thuyll	20
	Bleumourant	18
	Charles Dickens	18
	Gemal	19
	Emicus	18
		93
schwarz	Prinz Albert von Preussen	20
	Siam	18
	Uncle Tom	20
	La nuit	18
		76
schwarz blau	Leonidas	30
	Mimosa	19
	Wilhelm I	22
		71
licht blau	Grand Lilas	20
	" Vedette	22
	Orondates	19
	Regulus	19
	Porcellain Shepper	18
		98
gefüllte	noble par merite (roth)	25
	Lord Wellington (rosa)	20
		45

Tulpen doppelte

8 St. Duc von Thol	16
8 St. Tournesol gelb	24
4 St. Tournesol gelb u. roth	40
	80

Recapitulation des Ertrags

11 St. Hyacinth gefüllte 185 + 45 =	2.30
13 " einfache roth	2.27
2 " violette	46
5 " rosa	1.04
4 " weisse	79
5 " blau	93
4 " schwarze	76
3 schwarz blau	71
5 " lichtblau	98
20 " Tulpen gefüllte	80
Transport & emballage	1.30
Summa	12.24

9e Octob. 1876
D Olexik

bewilligt
Dr Olexik

D360 List of delivery of plant bulbs by C. Alkemade from Noordwijk-Binnen near Haarlem. Received by Dr. Olexik in Brno on 09 October 1876

Obst-, Wein- und Gartenbausektion (1884). Translated by Stephen Finn (from Matalová 1984, p. 219): *Besides his function as a priest, he carried out with pleasure natural scientific studies; from 1854 to 1867 he was active as teacher of physics and natural history at the Brno Realschule. In that position he acquired, due to his virtue and amiability, the affection of all his pupils, who preserved a grateful memory of him even in later years. After the death of the praiseworthy Prelate Cyrill Napp, 1867, his friars elected him as successor, and everywhere that election was welcomed with joy. In his activity as abbot, he gained the respect and admiration of everybody with his openhandedness, love and kindness, so that we can rightly claim he had no personal enemy.—Nobody seeking for help was dismissed without alms. Prelate Mendel had the rare gift of the ability to give alms without letting the beneficiary feel he was being given them.—His Majesty praised the highly merited abbot and awarded him the Cross of the Imperial-Royal Order of Franz Joseph. Many societies made him an honorary member. His lively participation in national economic societies, especially his achievements in pomology and apiculture, will secure him a lasting monument in the hearts of those who knew the reverend and ever-modest man.—Prelate Mendel also executed the office of President of the Moravian Mortgage Bank for nearly three years, resigning from that post due to poor health in October of last year. The Moravian Diet recognized his accomplishments at the above institution in a most distinctive way.—At his death, our country has lost a model of a noble and unselfish character, and his friars a careful father and cautious representative of the interests of the monastery.—The Section of Horticulture, however, has lost in Prelate Mendel a special promotor of its endeavours in which he always took part from his election as a member on April 2, 1863, and participated in all exhibitions of flowers, fruit and vegetables, not only as chairman of the jury, but also effectively helping as a richly experienced advisor. In May 1859, he participated at an exhibition arranged on premises in Lužánky with a presentation of exquisite vegetables grown by him, and remained faithful to those exhibitions until the end, when the itinerant meeting of the Austrian Pomological Society held in Brno in September 1883 unanimously awarded him the large gold medal of the Horticultural Society of Hietzing for a superior fruit seed collection of his own production.—His first-class varieties of flowers, of which we must mention especially a full beautiful fuchsia, were a successful achievement in domestic flower culture. How deeply the deceased was devoted to the plants has been shown in his abbatial emblem, one field of which is occupied by a flowering plant chosen as his symbol.—Many of his investigations in the field of natural science, especially meteorology, continued after Dr. Olexik's death, were published by him in the Journal of the Naturforschender Verein in Brünn.—In fact, his experiments with plant hybrids opened a new epoch.—What he did will never be forgotten.*

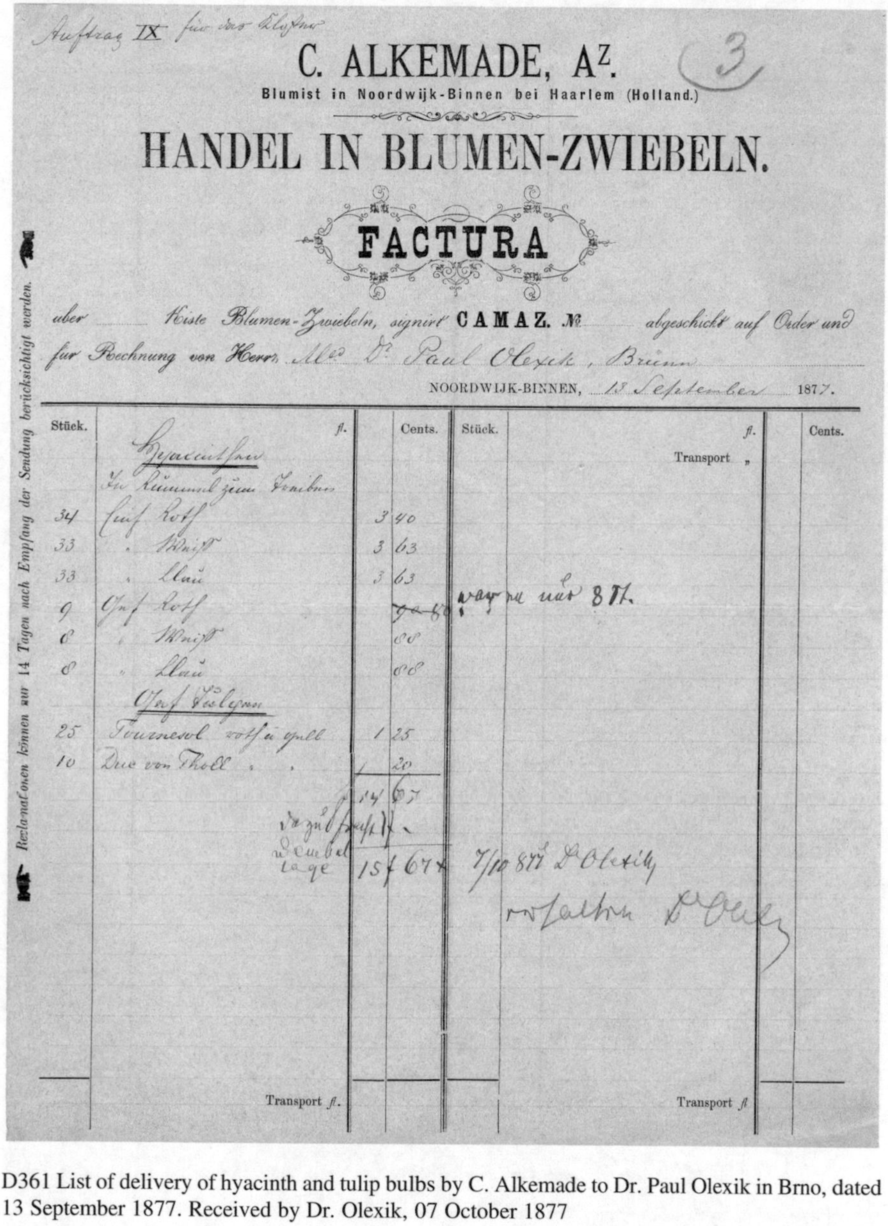

Auftrag IX für das Kloster

C. ALKEMADE, A^Z.

Blumist in Noordwijk-Binnen bei Haarlem (Holland.)

3

HANDEL IN BLUMEN-ZWIEBELN.

FACTURA

über Kiste Blumen-Zwiebeln, signirt CAMAZ. № abgeschickt auf Order und für Rechnung von Herrn Med. Dr. Paul Olexik, Brünn

NOORDWIJK-BINNEN, 13 September 1877.

Stück.		fl.	Cents.	Stück.		fl.	Cents.
	Hyacinthen				Transport	„	
	In Rummel zum Treiben						
34	Einf. Roth	3	40				
33	„ Weiss	3	63				
33	„ Blau	3	63				
9	Gef. Roth	~~9~~	~~60~~				
8	„ Weiss		88				
8	„ Blau		88				
	Einf. Tulpen						
25	Tournesol roth u. gelb	1	25				
10	Duc von Tholl „ „	1	20				
		14	67				
		15	67				
	Transport fl.				Transport fl.		

Kiste
15 f 67 x
7/10 877 Dr Olexik
erhalten Dr Olexik

Reclamationen können nur 14 Tagen nach Empfang der Sendung berücksichtigt werden.

D361 List of delivery of hyacinth and tulip bulbs by C. Alkemade to Dr. Paul Olexik in Brno, dated 13 September 1877. Received by Dr. Olexik, 07 October 1877

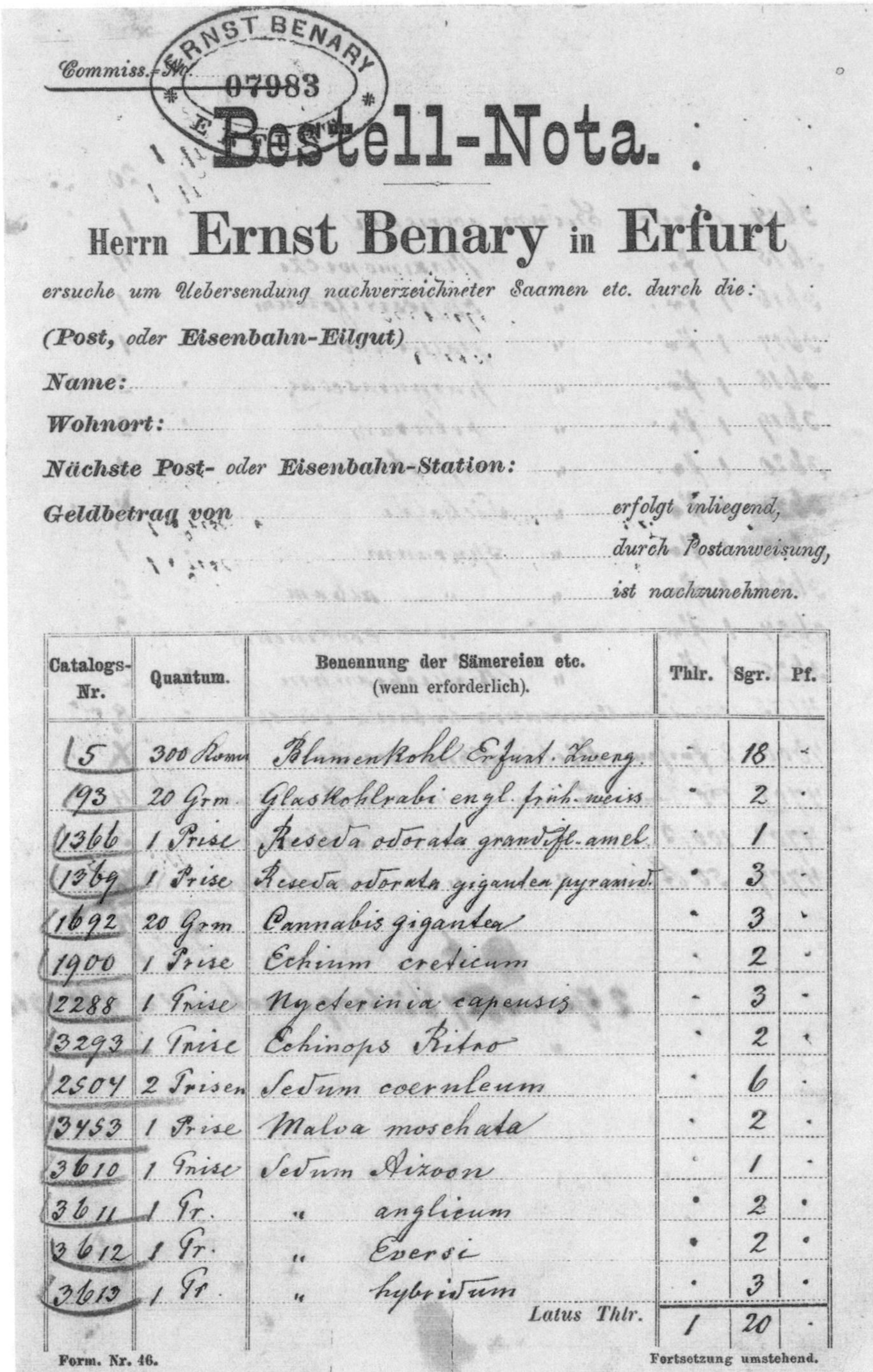

ERNST BENARY ERFURT

Commiss.-Nr. 07983

Bestell-Nota.

Herrn Ernst Benary in Erfurt

ersuche um Uebersendung nachverzeichneter Saamen etc. durch die:

(Post, *oder* **Eisenbahn-Eilgut)**

Name:

Wohnort:

Nächste Post- *oder* **Eisenbahn-Station:**

Geldbetrag von *erfolgt inliegend,*
durch Postanweisung,
ist nachzunehmen.

Catalogs-Nr.	Quantum.	Benennung der Sämereien etc. (wenn erforderlich).	Thlr.	Sgr.	Pf.
5	300 Korn	Blumenkohl Erfurt. Zwerg.	.	18	.
193	20 Grm.	Glaskohlrabi engl. früh-weiss	.	2	.
1366	1 Prise	Reseda odorata grandifl.-amel.	.	1	.
1369	1 Prise	Reseda odorata gigantea pyramid.	.	3	.
1692	20 Grm	Cannabis gigantea	.	3	.
1900	1 Prise	Echium creticum	.	2	.
2288	1 Prise	Nycterinia capensis	.	3	.
3293	1 Prise	Echinops Ritro	.	2	.
2504	2 Prisen	Sedum coeruleum	.	6	.
3453	1 Prise	Malva moschata	.	2	.
3610	1 Prise	Sedum Aizoon	.	1	.
3611	1 Pr.	" anglicum	.	2	.
3612	1 Pr.	" Eversi	.	2	.
3613	1 Pr.	" hybridum	.	3	.
		Latus Thlr.	1	20	.

Form. Nr. 46. Fortsetzung umstehend.

D249 Mendel's order of vegetable and plant seeds according to the catalogue numbers, to be delivered by Ernst Benary in Erfurt. Stamped Ernst Benary Erfurt No. 07983

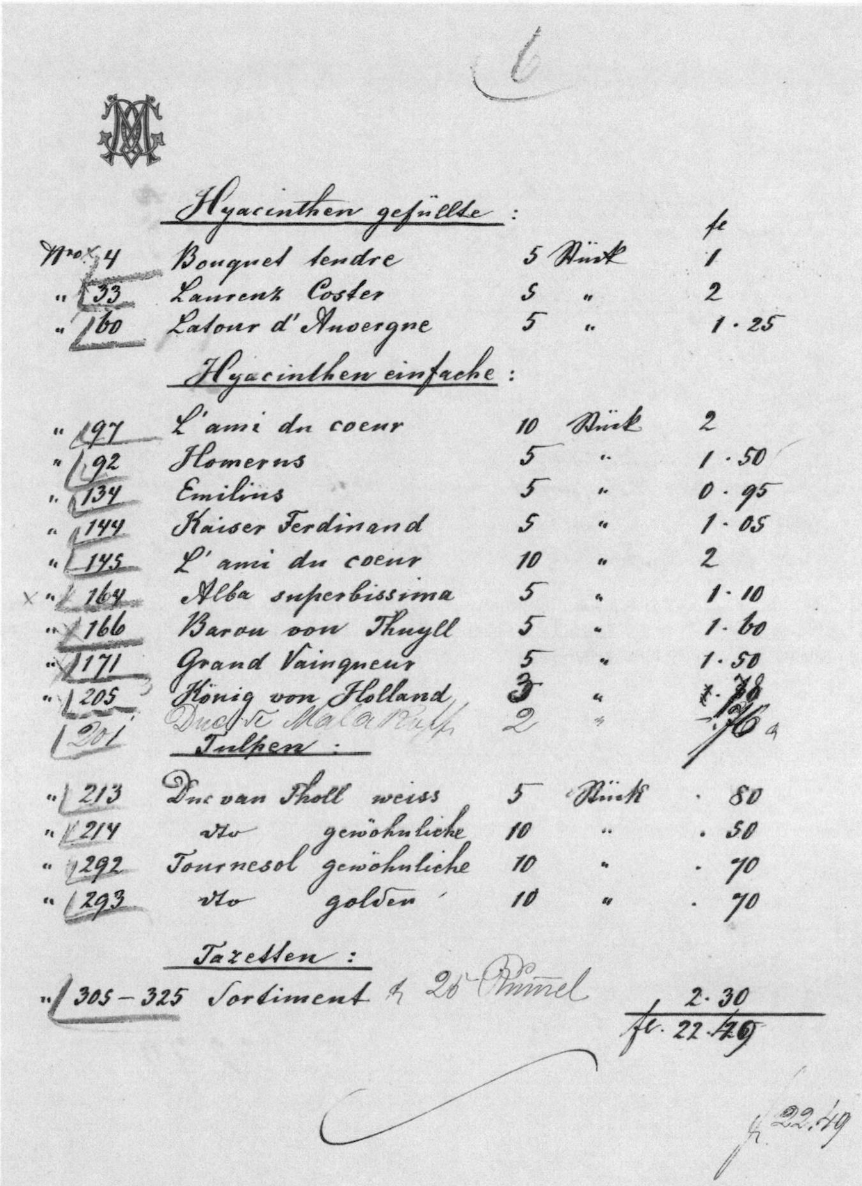

	Hyacinthen gefüllte:			fl
Nro 4	Bouquet tendre	5	Stück	1
" 33	Laurenz Coster	5	"	2
" 60	Latour d'Auvergne	5	"	1.25
	Hyacinthen einfache:			
" 97	L'ami du coeur	10	Stück	2
" 92	Homerus	5	"	1.50
" 134	Emilius	5	"	0.95
" 144	Kaiser Ferdinand	5	"	1.05
" 145	L'ami du coeur	10	"	2
" 164	Alba superbissima	5	"	1.10
" 166	Baron von Thuyll	5	"	1.60
" 171	Grand Vainqueur	5	"	1.50
" 205	König von Holland	3	"	1.78
201	Duc de Malakoff	2	"	.70
	Tulpen:			
" 213	Duc van Thol weiss	5	Stück	.80
" 214	dto gewöhnliche	10	"	.50
" 292	Tournesol gewöhnliche	10	"	.70
" 293	dto golden	10	"	.70
	Tazetten:			
" 305–325	Sortiment à 20 [illegible]			2.30
				fl. 22.49

fl 22.49

D249 Mendel's order of the bulbs of hyacinths and tulips. Mendel's specifications for the order were made according to the catalogue numbers, written on his private writing paper with his monogram GM

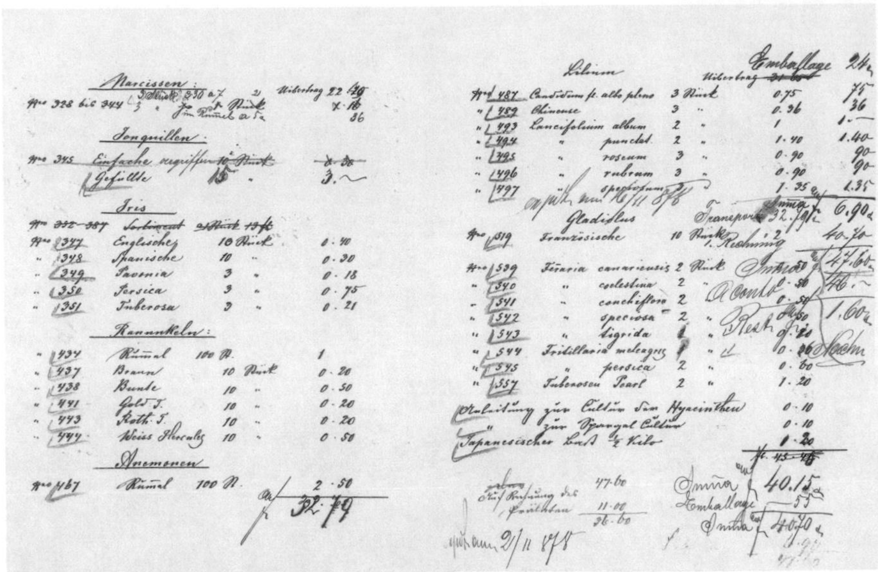

Narcissen:

No 328 bis 344 ... Uiberlag 22.49

Jonquillen:

No 345 Einfache vergriffen 10 Stück

Gefüllte 15 . 3.—

Iris

No 337 Englische 10 Stück 0.40
" 348 Spanische 10 " 0.30
" 349 Pavonia 3 " 0.18
" 350 Persica 3 " 0.75
" 351 Tuberosa 3 " 0.21

Ranunkeln:

" 434 Rümel 100 St. 1
" 437 Braun 10 Stück 0.20
" 438 Bunte 10 " 0.50
" 441 Gelb J. 10 " 0.20
" 443 Roth J. 10 " 0.20
" 444 Weiss ... 10 " 0.50

Anemonen

No 467 Rümel 100 St. 2.50

32.79

Lilium

Uiberlag ... Emballage 24

No 487 Candidum fl. albo pleno 3 Stück 0.75 75
" 489 Chinense 3 " 0.36 36
" 493 Lancifolium album 2 " 1 1.—
" 494 " punctat. 2 " 1.40 1.40
" 495 " roseum 3 " 0.90 90
" 496 " rubrum 3 " 0.90 90
" 497 " speciosum 3 " 1.35 1.35

Gladiolus

No 519 Französische 10 Stück Transport 32.79 6.90

No 539 ... canariensis 2 Stück
" 540 " coelestina 2
" 541 " conchiflora 2
" 542 " speciosa 2
" 543 " tigrida 2
" 544 Fritillaria meleagris 2
" 545 " persica 2 0.60
" 557 Tuberosen Pearl 2 1.20

Anleitung zur Cultur der Hyacinthen 0.10

" zur Spargel Cultur 0.10

Japanesischer Bast ½ Kilo 1.20

fl. 45.45

Auf Rechnung des Prälaten 47.60 11.00 36.60

Summa 40.15 Emballage 55 Summa 40.70

... 2/11 878

D249 Mendel’s order of narcissus, Jonquillas, irises, anemones, lilies and gladiolis. Above the date of 02 November 1878 is a note added by Mendel “auf Rechnung des Prälaten” (on the account of the prelate). Paid on 02 November 1878

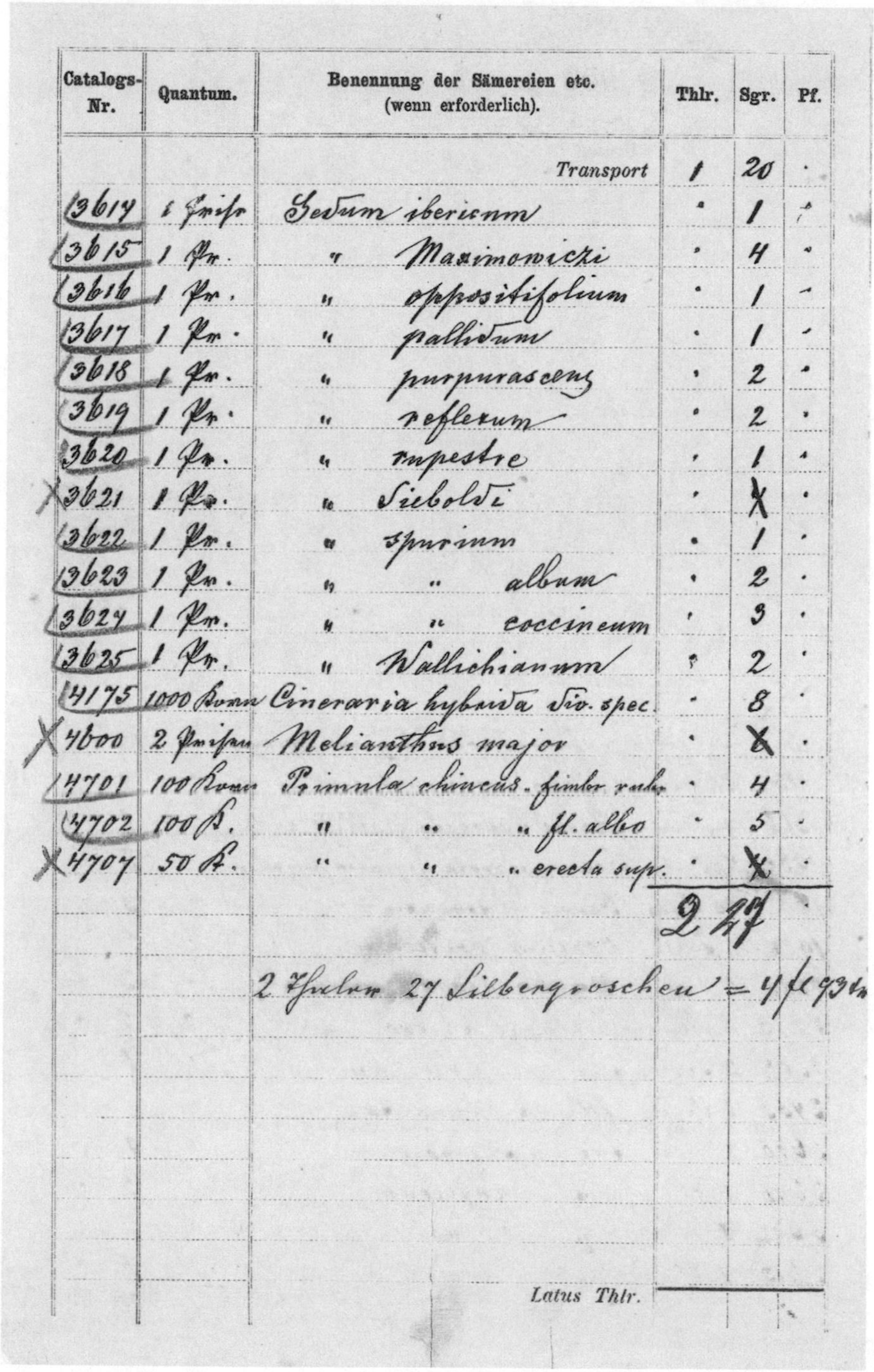

Catalogs-Nr.	Quantum.	Benennung der Sämereien etc. (wenn erforderlich).	Thlr.	Sgr.	Pf.
		Transport	1	20	.
3614	1 Prise	Sedum ibericum	.	1	.
3615	1 Pr.	" Maximowiczi	.	4	.
3616	1 Pr.	" oppositifolium	.	1	.
3617	1 Pr.	" pallidum	.	1	.
3618	1 Pr.	" purpurascens	.	2	.
3619	1 Pr.	" reflexum	.	2	.
3620	1 Pr.	" rupestre	.	1	.
X 3621	1 Pr.	" Sieboldi	.	X	.
3622	1 Pr.	" spurium	.	1	.
3623	1 Pr.	" " album	.	2	.
3624	1 Pr.	" " coccineum	.	3	.
3625	1 Pr.	" Wallichianum	.	2	.
4175	1000 Korn	Cineraria hybrida Vio. spec.	.	8	.
X 4600	2 Prisen	Melianthus major	.	X	.
4701	100 Korn	Primula chinens. fimbr. rubr.	.	4	.
4702	100 K.	" " " fl. albo	.	5	.
X 4707	50 K.	" " " erecta sup.	.	X	.
			2	27	

2 Thaler 27 Silbergroschen = 4 fl 93 kr

Latus Thlr.

D249 Mendel's order of sedum, cineraria, melianthus and primula. The total sum 2 Thaler 27 Silbergroschen. Without date

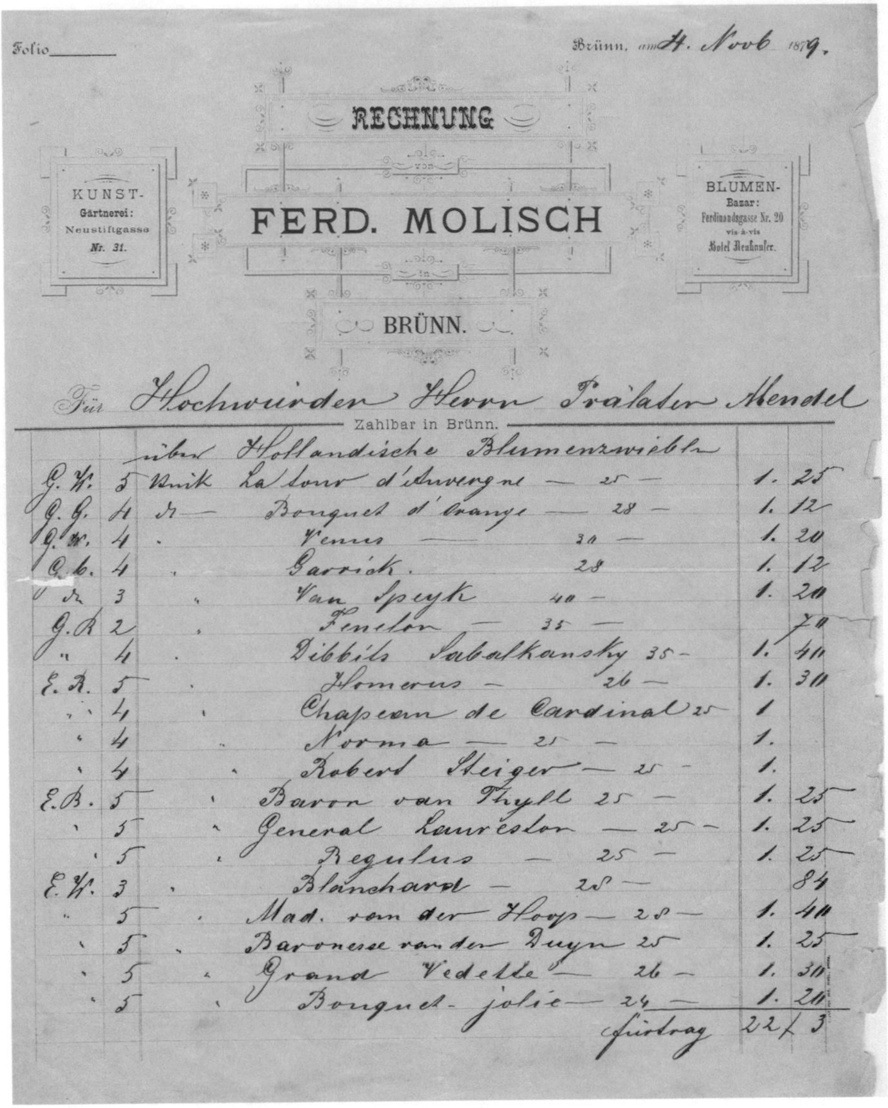

Folio ______ Brünn, am 4. Novb 1879.

RECHNUNG

von

FERD. MOLISCH

in

BRÜNN.

KUNST-Gärtnerei: Neustiftgasse Nr. 31.

BLUMEN-Bazar: Ferdinandsgasse Nr. 20 vis-à-vis Hotel Neuhauser.

Für Hochwürden Herrn Prälaten Mendel

Zahlbar in Brünn.

		über Holländische Blumenzwiebln			
G. W.	5	Stück La tour d'Auvergne	25	1.	25
G. G.	4	do Bouquet d'Orange	28	1.	12
G. W.	4	Venus	30	1.	20
G. C.	4	Garrick	28	1.	12
do	3	Van Speyk	40	1.	20
G. R.	2	Fenelon	35		70
"	4	Dibbits Sabalkansky	35	1.	40
E. R.	5	Homerus	26	1.	30
"	4	Chapeau de Cardinal	25	1	
"	4	Norma	25	1.	
"	4	Robert Steiger	25	1.	
E. B.	5	Baron van Thyll	25	1.	25
"	5	General Laurestor	25	1.	25
"	5	Regulus	25	1.	25
E. W.	3	Blanchard	28		84
"	5	Mad. van der Hoop	28	1.	40
"	5	Baronesse van der Duyn	25	1.	25
"	5	Grand Vedette	26	1.	30
"	5	Bouquet jolie	24	1.	20
		Fürtrag		22	3

D732 Account for tulip bulbs by Ferd. Molisch. Dated in Brno, 04 November 1879

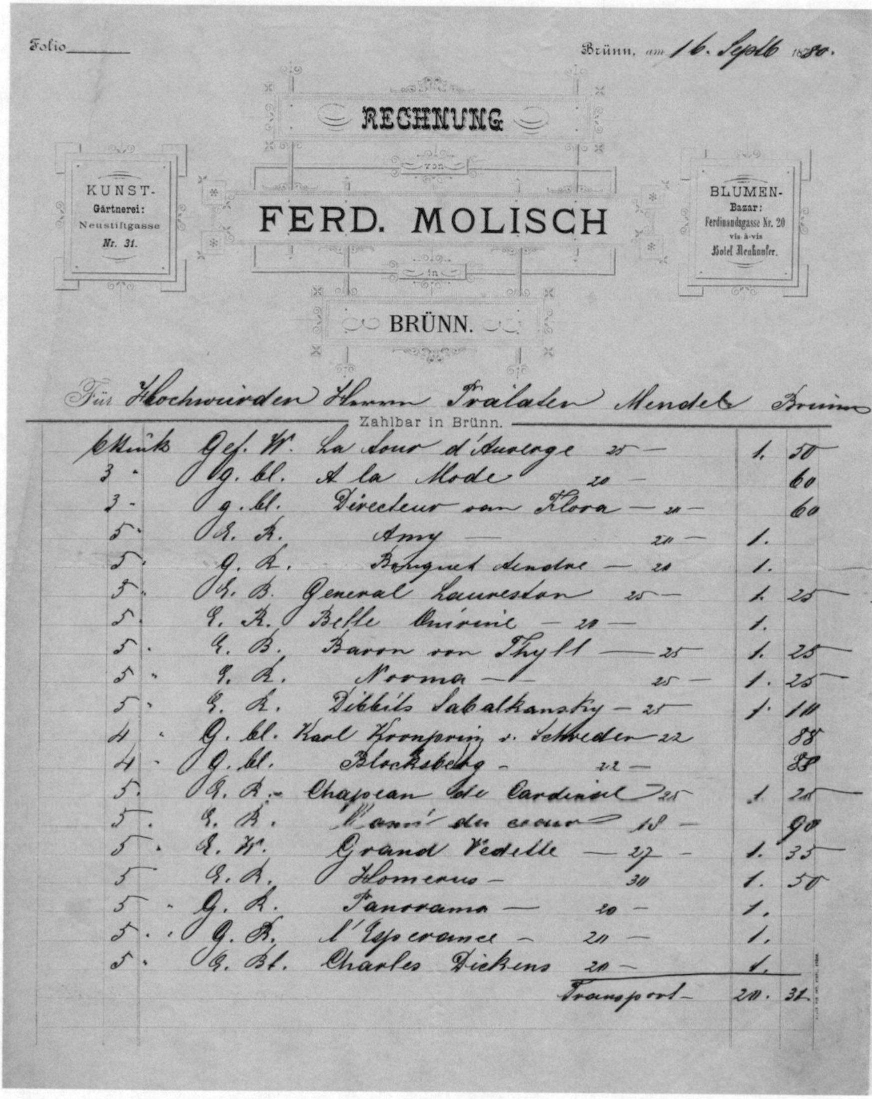

Folio______ Brünn, am 16. Septb 1880.

RECHNUNG von FERD. MOLISCH in BRÜNN.

KUNST-Gärtnerei: Neustiftgasse Nr. 31.

BLUMEN-Bazar: Ferdinandsgasse Nr. 20 vis-à-vis Hotel Neuhauser.

Für Hochwürden Herrn Prälaten Mendel Brünn

Zahlbar in Brünn.

Stück	Gef. W.	La tour d'Auvergne	25	1.	50
3	g. bl.	A la Mode	20		60
3	g. bl.	Directeur von Flora	20		60
5	E. R.	Amy	20	1.	
5	g. R.	Bouquet Agnelle	20	1.	
5	E. B.	General Laurestan	25	1.	25
5	E. R.	Belle Amirine	20	1.	
5	E. B.	Baron von Thyll	25	1.	25
5	E. R.	Norma	25	1.	25
5	E. R.	Dibbits Sabalkansky	25	1.	10
4	g. bl.	Karl Kronprinz v. Schweden	22		88
4	g. bl.	Blocksberg	22		88
5	E. R.	Chapeau de Cardinal	25	1.	25
5	E. R.	L'ami du coeur	18		90
5	E. W.	Grand Vedette	27	1.	35
5	E. R.	Homerus	30	1.	50
5	g. R.	Panorama	20	1.	
5	g. B.	L'Esperance	20	1.	
5	E. Bl.	Charles Dickens	20	1.	
		Transport		20.	31

D733 First page of the account for tulip bulbs by Ferd. Molisch Brno to Prelate Mendel. Dated in Brno, 16 September 1880

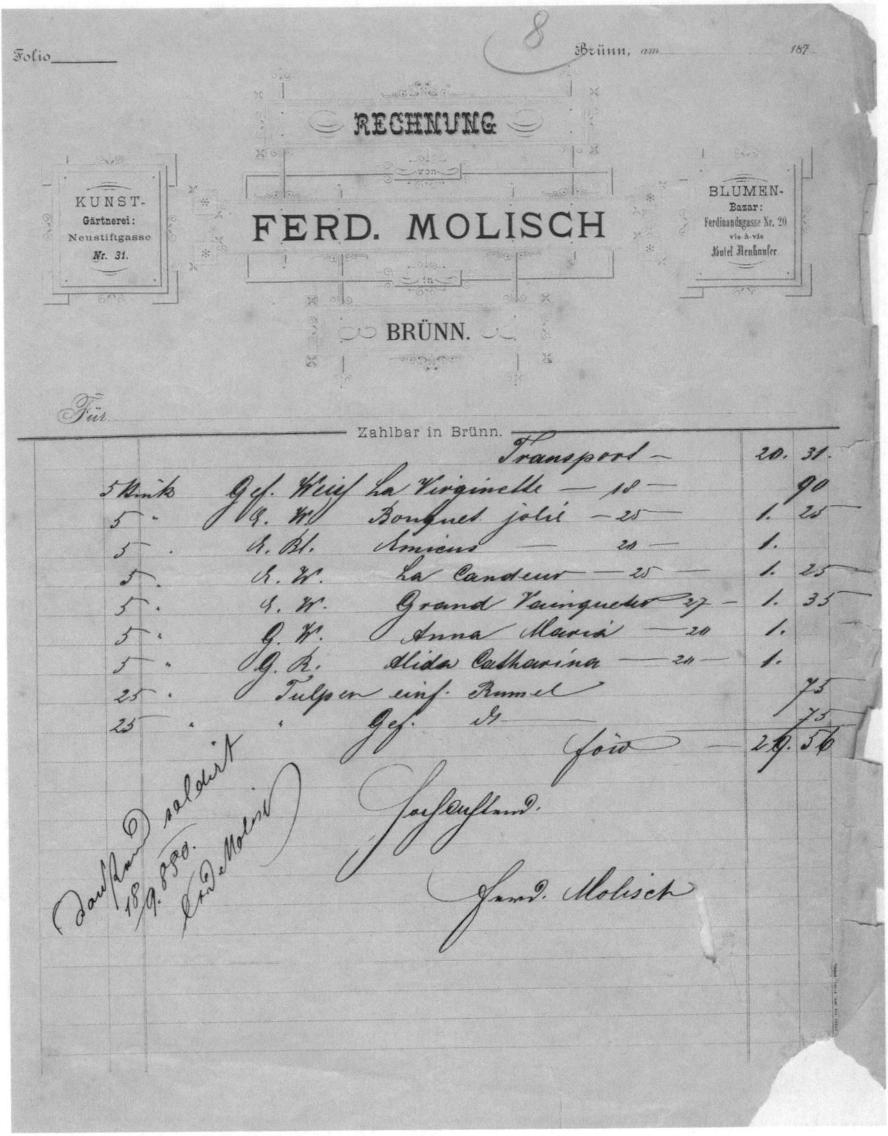
Folio

8

Brünn, am 187

RECHNUNG

von

FERD. MOLISCH

in

BRÜNN.

KUNST-
Gärtnerei:
Neustiftgasse
Nr. 31.

BLUMEN-
Bazar:
Ferdinandsgasse Nr. 20
vis-à-vis
Hotel Neuhauser.

Für

Zahlbar in Brünn.

	Transport	20.	31
5 Stück	Gef. Weiß La Virginelle — 18 —		90
5 "	G. W. Bouquet jolie — 25 —	1.	25
5 "	R. Bl. Amicus — 20 —	1.	
5 "	R. W. La Candeur — 25 —	1.	25
5 "	G. W. Grand Vainqueur 27 —	1.	35
5 "	G. W. Anna Maria — 20	1.	
5 "	G. R. Alida Catharina — 20 —	1.	
25 "	Tulpen einf. Rummel		75
25 "	" Gef. D.		75
	fein	29.	56

Hochachtend

Ferd. Molisch

Dankend saldirt 18/9. 880. Ferd. Molisch

D733 Second page of the account for tulip bulbs delivered by Ferd. Molisch to Prelate Mendel, Dated in Brno, 16 September 1880

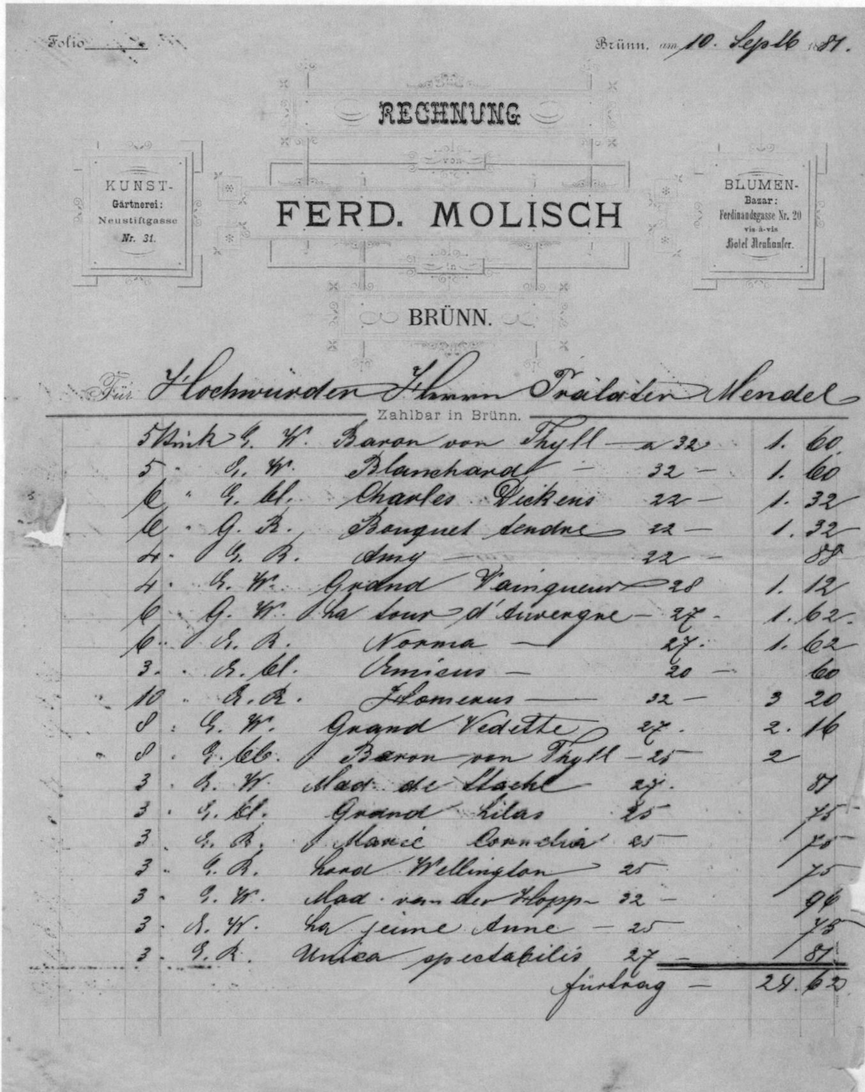

Folio

Brünn, am 10. Septb 1881.

RECHNUNG

von

FERD. MOLISCH

BRÜNN.

KUNST-Gärtnerei: Neustiftgasse Nr. 31.

BLUMEN-Bazar: Ferdinandsgasse Nr. 20 vis-à-vis Hotel Neuhauser.

Für Hochwürden Herrn Prälaten Mendel

Zahlbar in Brünn.

5 Stück	G. W.	Baron von Thyll	à 32	1.	60
5	G. W.	Blanchard	32	1.	60
6	G. Cl.	Charles Dickens	22	1.	32
6	G. R.	Bouquet tendre	22	1.	32
4	G. R.	Amy	22		88
4	G. W.	Grand Vainqueur	28	1.	12
6	G. W.	La tour d'Auvergne	27	1.	62
6	G. R.	Norma	27	1.	62
3	G. Cl.	Ermicus	20		60
10	G. R.	Flomeaus	32	3	20
8	G. W.	Grand Vedette	27	2.	16
8	G. Cl.	Baron von Thyll	25	2	
3	G. W.	Mad. de Stachl	27		81
3	G. Cl.	Grand Lilas	25		75
3	G. R.	Marie Cornelia	25		75
3	G. R.	Lord Wellington	25		75
3	G. W.	Mad. van der Hoop	32		96
3	G. W.	La jeune Anne	25		75
3	G. R.	Unica spectabilis	27		81
		fürtrag		24.	62

D734 First page of the account for tulip bulbs of Ferd. Molisch to Prelate Mendel, Dated in Brno, 10 September 1881

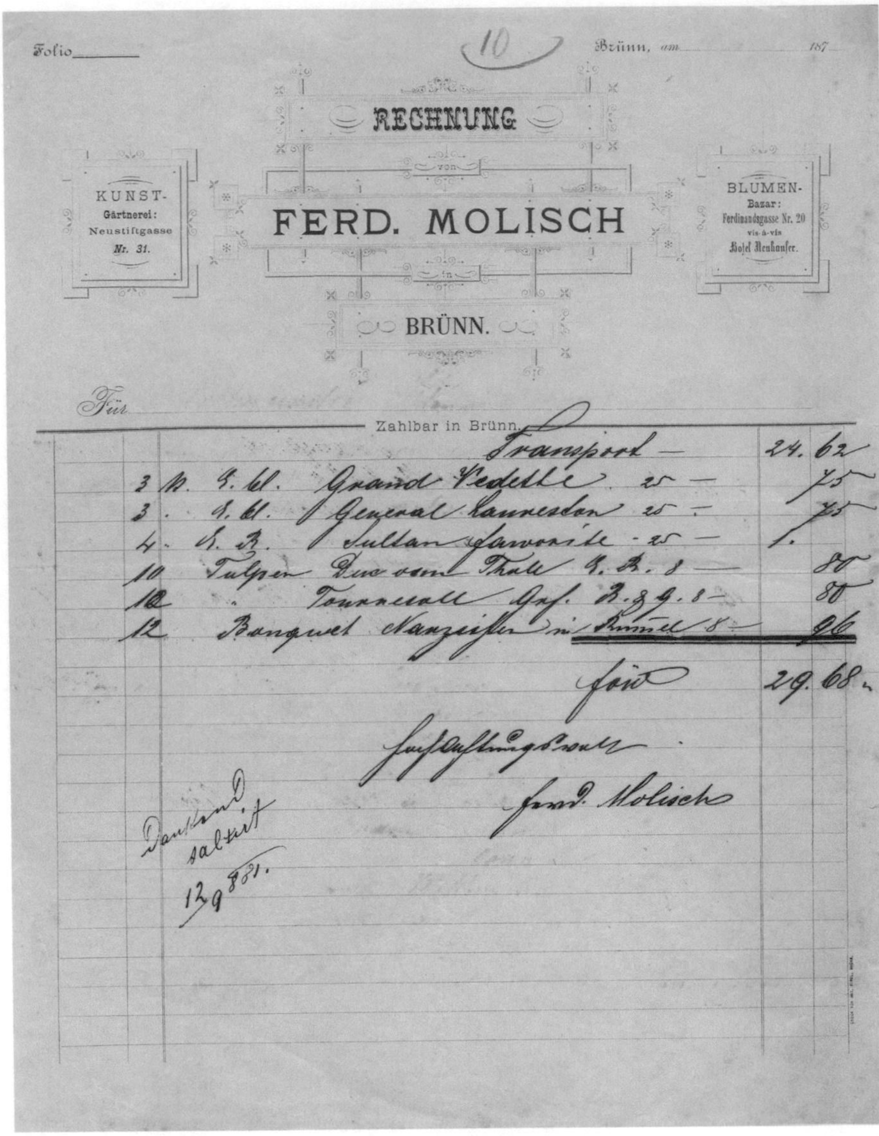

Folio

(10)

Brünn, am 187

RECHNUNG

von

FERD. MOLISCH

in

BRÜNN.

KUNST-Gärtnerei: Neustiftgasse Nr. 31.

BLUMEN-Bazar: Ferdinandsgasse Nr. 20 vis-à-vis Hotel Neuhauser.

Für

Zahlbar in Brünn.

	Transport —	24.	62
3 St.	G. Cl. Grand Vedette 25 —		75
3	G. Cl. General Cannesdon 25 —		75
4	G. B. Sultan Favorite 25 —	1.	
10	Tulpen Duc van Thall G. B. 8 —		80
12	" Tournesoll Gef. B. G. G. 8 —		80
12	Bouquet Narzissen in Bündel 8 —		96
	Sum	29.	68

Hochachtungsvoll

Ferd. Molisch

Dankend saldirt 12/9 881.

D734 The second page of the account for the bulbs of tulips and narcissuses of Ferd. Molisch to Prelate Mendel. Dated in Brno, 10 September 1881. Paid on 12 September 1881

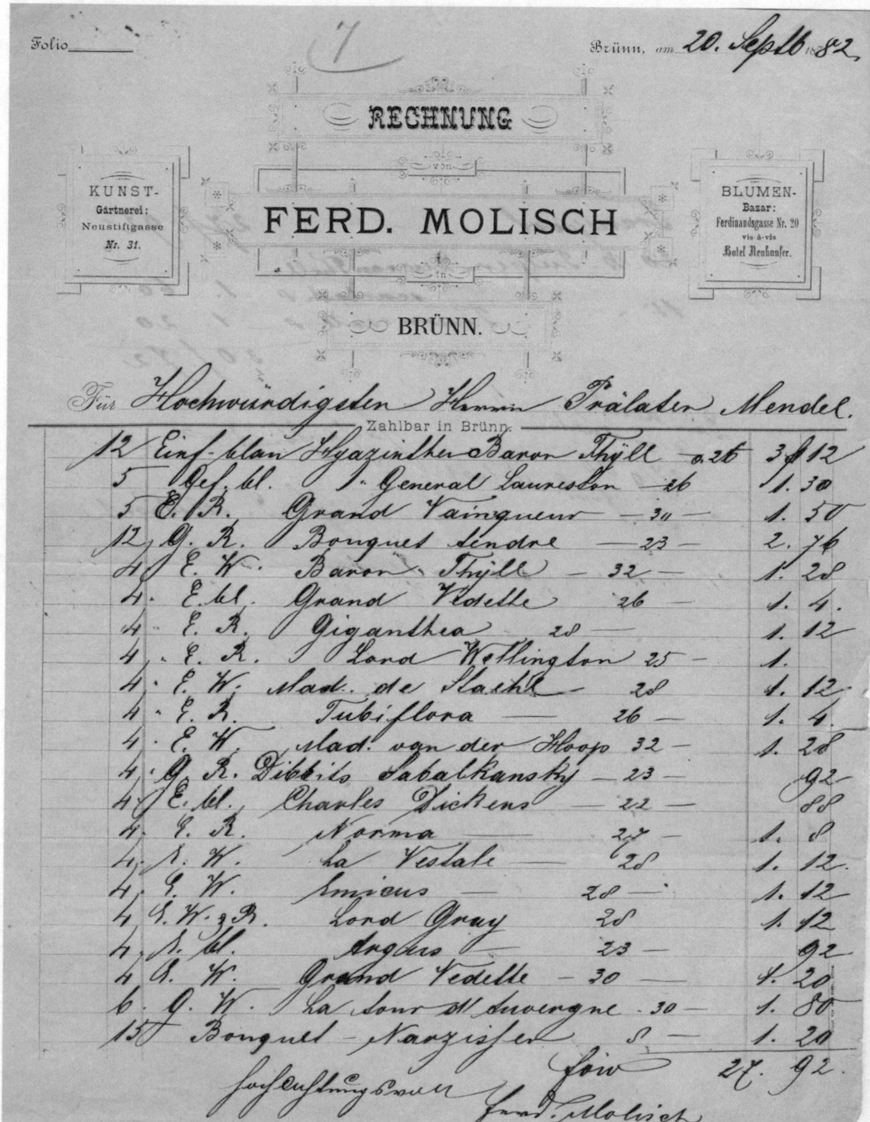
Folio

Brünn, am 20. Septb. 1882

RECHNUNG von FERD. MOLISCH in BRÜNN.

KUNST-Gärtnerei: Neustiftgasse Nr. 31.

BLUMEN-Bazar: Ferdinandsgasse Nr. 20 vis-à-vis Hotel Neuhauser.

Für Hochwürdigsten Herrn Prälaten Mendel.

Zahlbar in Brünn.

12	Einf. blau Hyazinthen	Baron Thyll	à 26	3 f 12
5	Gef. bl.	General Laureston	26	1. 30
5	E. R.	Grand Vainqueur	30	1. 50
12	G. R.	Bouquet Tendre	23	2. 76
4	E. W.	Baron Thyll	32	1. 28
4	E. bl.	Grand Vedette	26	1. 4.
4	E. R.	Gigantea	28	1. 12
4	E. R.	Lord Wellington	25	1.
4	E. W.	Mad. de Staël	28	1. 12
4	E. R.	Tubiflora	26	1. 4.
4	E. W.	Mad. van der Hoop	32	1. 28
4	G. R.	Diebitsch Sabalkansky	23	92
4	E. bl.	Charles Dickens	22	88
4	E. R.	Norma	27	1. 8
4	E. W.	La Vestale	28	1. 12
4	G. W.	Amicus	28	1. 12
4	G. W. g. R.	Lord Gray	28	1. 12
4	E. bl.	Argus	23	92
4	E. W.	Grand Vedette	30	1. 20
6	G. W.	La tour d'Auvergne	30	1. 80
15		Bouquet Narzissen	8	1. 20
		fio		27. 92

Hochachtungsvoll
Ferd. Molisch

D735 The first page of the account for the bulbs of tulips and narcissuses of Ferd. Molisch to Prelate Mendel. Dated in Brno, 20 September 1882

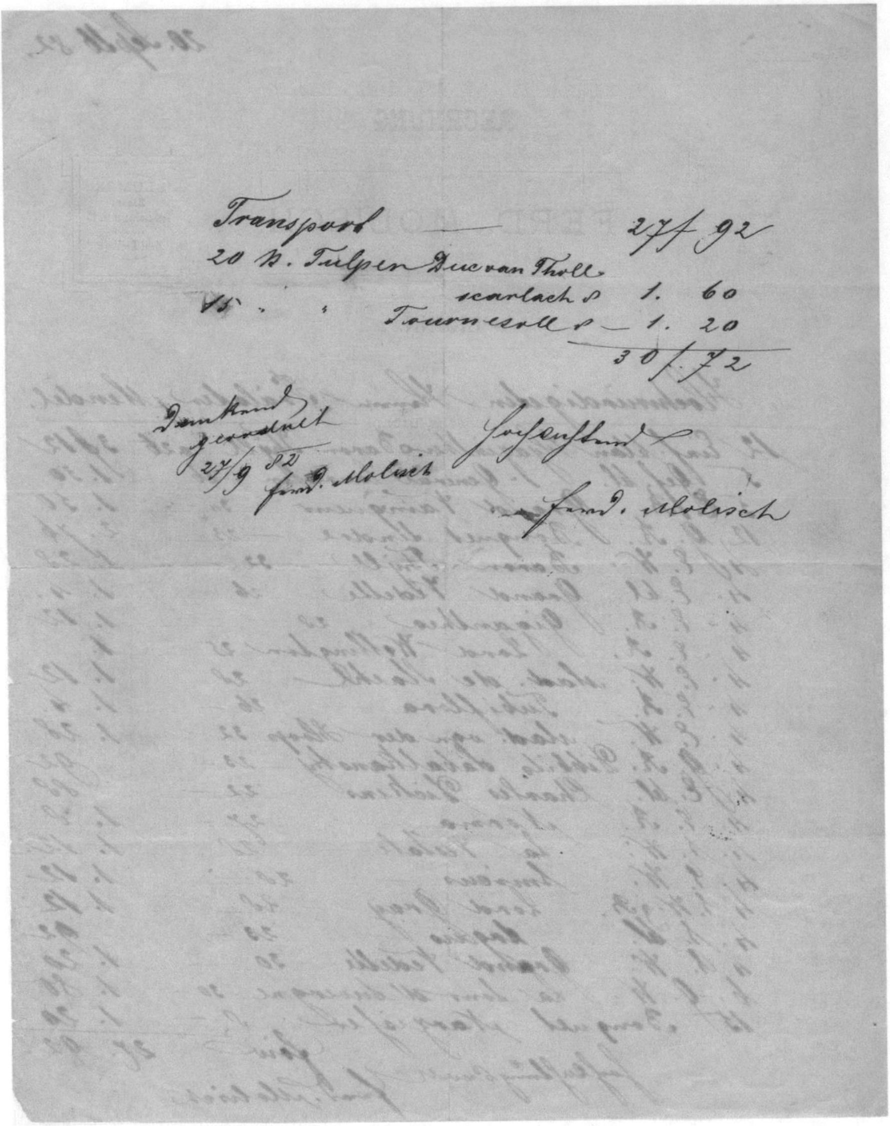

Transport 27f 92
20 St. Tulpen Duc van Thol
scarlach à 1. 60
15 " " Tournesoll à 1. 20
30f. 72

Dankend geordnet
27/9 82
Ferd. Molisch

Hochachtend
Ferd. Molisch

D735 The second page of the account for the bulbs of tulips and narcisses by Ferd. Molisch to Prelate Mendel. Dated in Brno, 20 September 1882

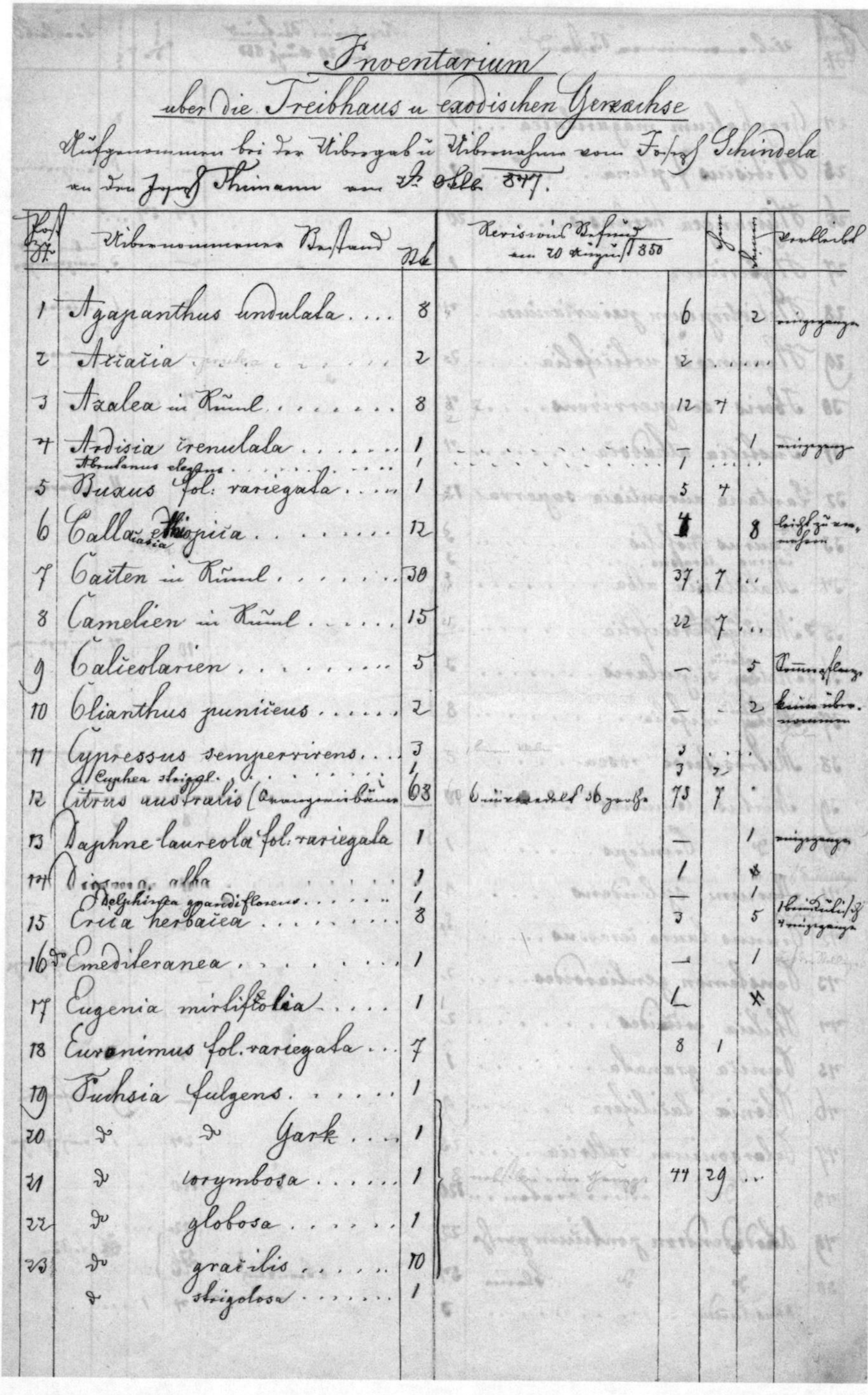

Inventarium

über die Treibhaus u. exotischen Gewächse

Aufgenommen bei der Übergabe u. Übernahme vom Joseph Schindela an den Joseph Thimann am 2. Oktbr. 1847.

Post Nr.	Übernommener Bestand	Nr.	Revisions Befund am 20 August 1850			Verbleib
1	Agapanthus undulata	8		6	2	eingegangen
2	Acacia	2		2		
3	Azalea in Kübel	8		12 7		
4	Ardisia crenulata	1			1	eingegangen
	Abutilon elegans	1		1		
5	Buxus fol. variegata	1		5 7		
6	Calla aethiopica	12		4	8	
7	Cacteen in Kübel	30		37 7		
8	Camelien in Kübel	15		22 7		
9	Calceolarien	5		—	5	
10	Clianthus puniceus	2		—	2	
11	Cypressus sempervirens	3		3		
	Cuphea strigil.	1		3 7		
12	Citrus australis (Orangenbäume)	63		73 7		
13	Daphne laureola fol. variegata	1		—	1	eingegangen
14	Diosma alba	1		1		
	Delphinium grandiflorum	1		—	1	
15	Erica herbacea	8		3	5	
16	do mediteranea	1		—	1	
17	Eugenia mirtifolia	1				
18	Euonimus fol. variegata	7		8 1		
19	Fuchsia fulgens	1				
20	do do Gark	1				
21	do corymbosa	1		44 29		
22	do globosa	1				
23	do gracilis	10				
	do strigolosa	1				

D362 The first page of the inventory of plants grown in the greenhouse, exotic plants and seeds and bulbs taken over by Joseph Thimann from Joseph Schindela in the Augustinian monastery on 02 October 1847. Revision was made on 20 August 1850

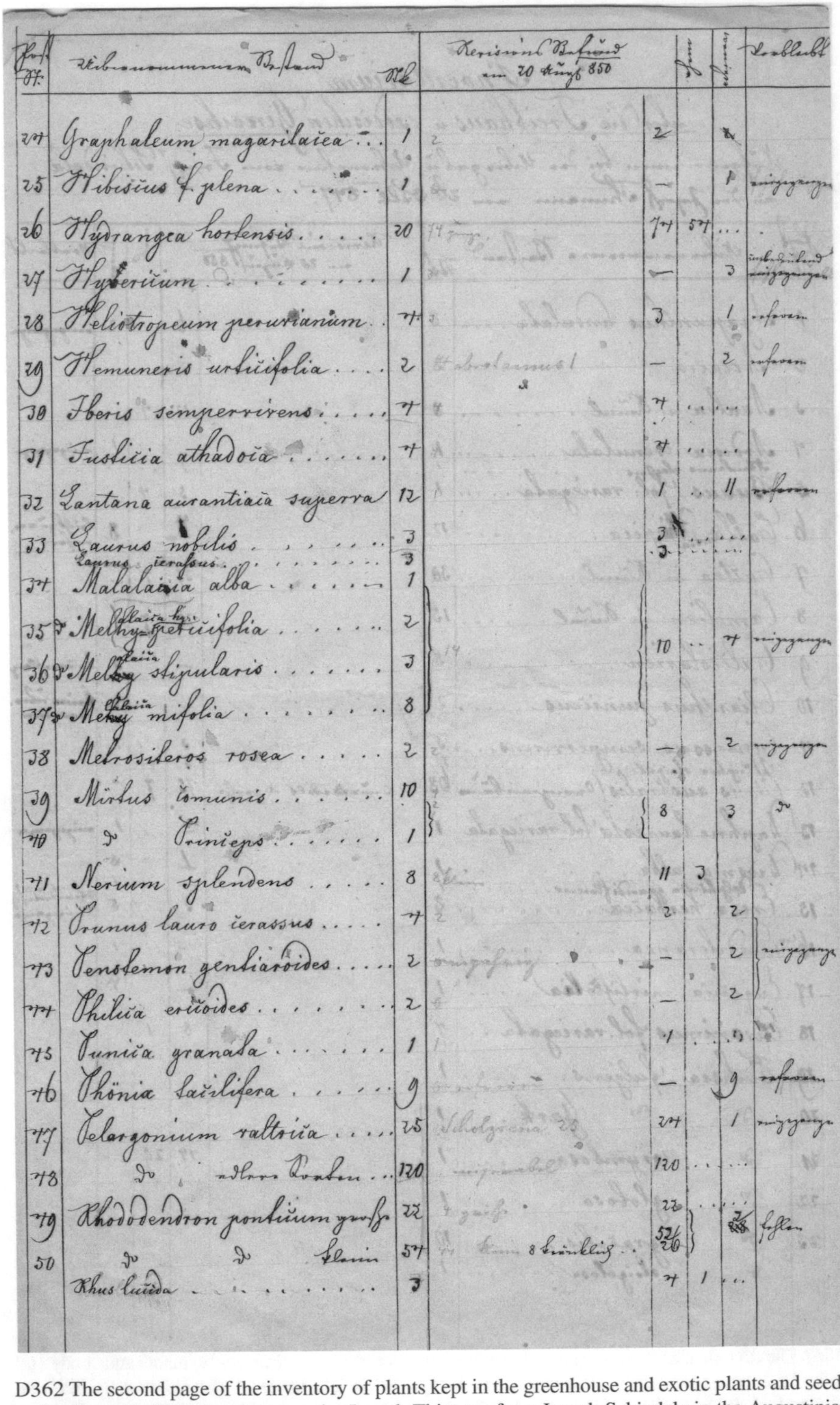

No.	Name		Revisions-Befund am 20 Aug 850				
24	Graphaleum magaritacea	1	2	2			
25	Hibiscus f. plena	1		—		1	
26	Hydrangea hortensis	20		74	54		
27	Hypericum	1		—		3	
28	Heliotropeum peruvianum	4	3	3		1	
29	Hemuneris urticifolia	2		—		2	
30	Iberis sempervirens	4	4	4			
31	Justicia athadoca	4		4			
32	Lantana aurantiaca superra	12	1	1		11	
33	Laurus nobilis	3		3			
	Laurus cerassus	3		3			
34	Malalacia alba	1					
35	Melhyperiicifolia	2					
36	Melaleuca stipularis	3	14	10		4	
37	Melaleuca mifolia	8					
38	Metrosideros rosea	2		—		2	
39	Mirtus comunis	10	2	8		3	
40	do Princeps	1					
41	Nerium splendens	8		11	3		
42	Prunus lauro cerassus	4	2	2		2	
43	Penstemon gentianoides	2		—		2	
44	Philica ericoides	2		—		2	
45	Punica granata	1	1	1			
46	Phönia lacilifera	9		—		9	
47	Pelargonium vattrica	25	Scholziana 25	24		1	
48	do andere Sorten	120		120			
49	Rhododendron ponticum groß	22		22			
50	do do klein	54		52½ / 26			
	Rhus lucida	3		4	1		

D362 The second page of the inventory of plants kept in the greenhouse and exotic plants and seeds and bulbs and offshoots taken over by Joseph Thimann from Joseph Schindela in the Augustinian monastery on 02 October 1847

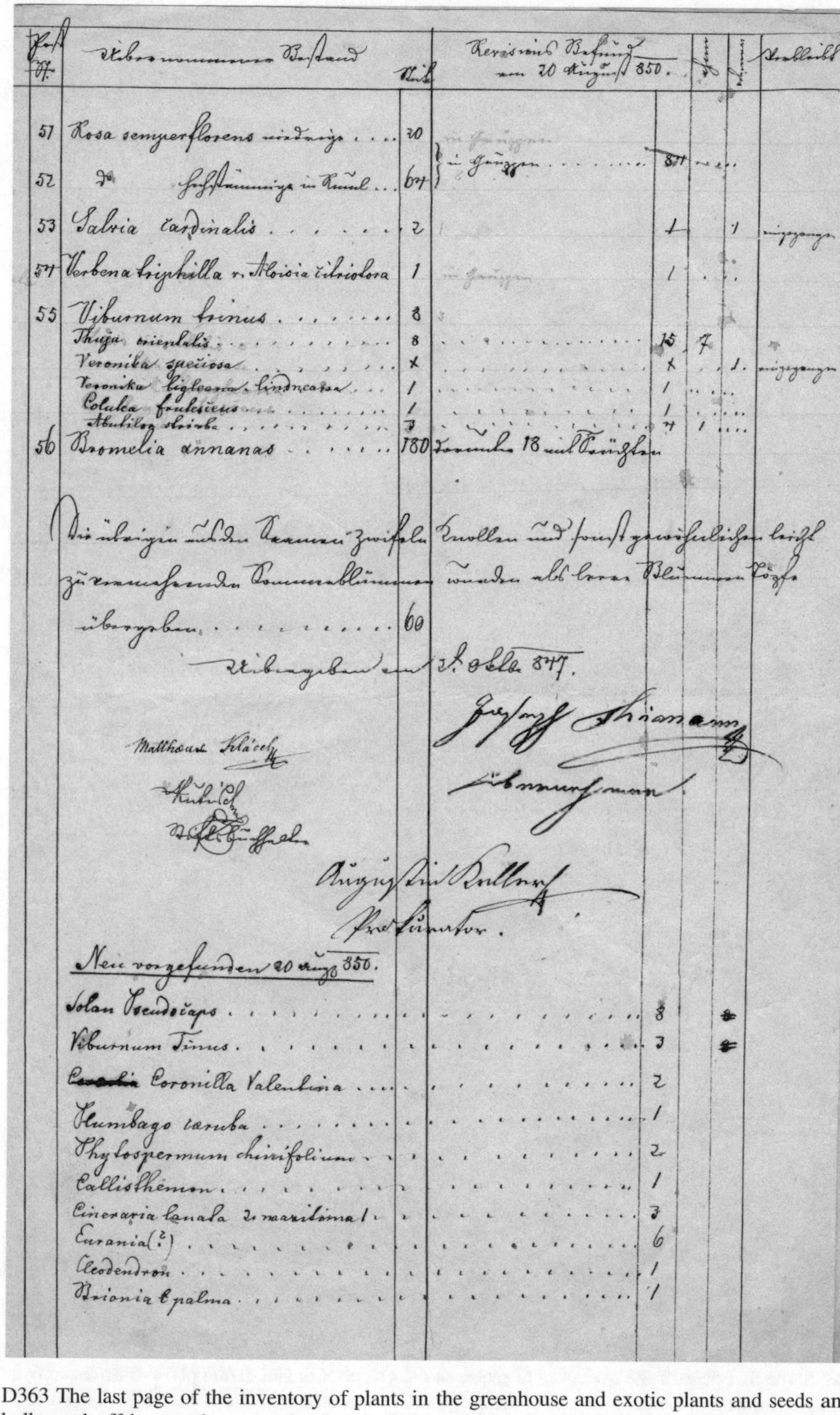

51	Rosa semperflorens ...	30	
52	do ...	64	80
53	Salvia cardinalis ...	2	1 1
54	Verbena triphilla v. Aloisia citriodora	1	1
55	Viburnum tinus ...	8	
	Thuja orientalis ...	8	15 7
	Veronica speciosa ...	x	x 1
	Veronica ...	1	1
	Colutea frutescens ...	1	1
	Abutilon striata ...	3	4 1
56	Bromelia ananas ...	180	18

... 60

... 847.

Matthäus Klácel

Neu vorgefunden 20 Aug 850.

Solan Pseudocaps ...	3
Viburnum Tinus ...	3
Coronilla Valentina ...	2
Plumbago caruba ...	1
Phylospermum chinifolium ...	2
Callisthemon ...	1
Cineraria lanata 2 maritima 1 ...	3
Eurania(?) ...	6
Cleodendron ...	1
Brionia t palma ...	1

D363 The last page of the inventory of plants in the greenhouse and exotic plants and seeds and bulbs and offshoots taken over by Joseph Thimann from Joseph Schindela in the Augustinian monastery on 02 October 1847. Revision was made on 20 August 1850, with ten newly found items added. Signed by Joseph Thimann, Mattheus Klacel, Kubitschek and Augustin Keller

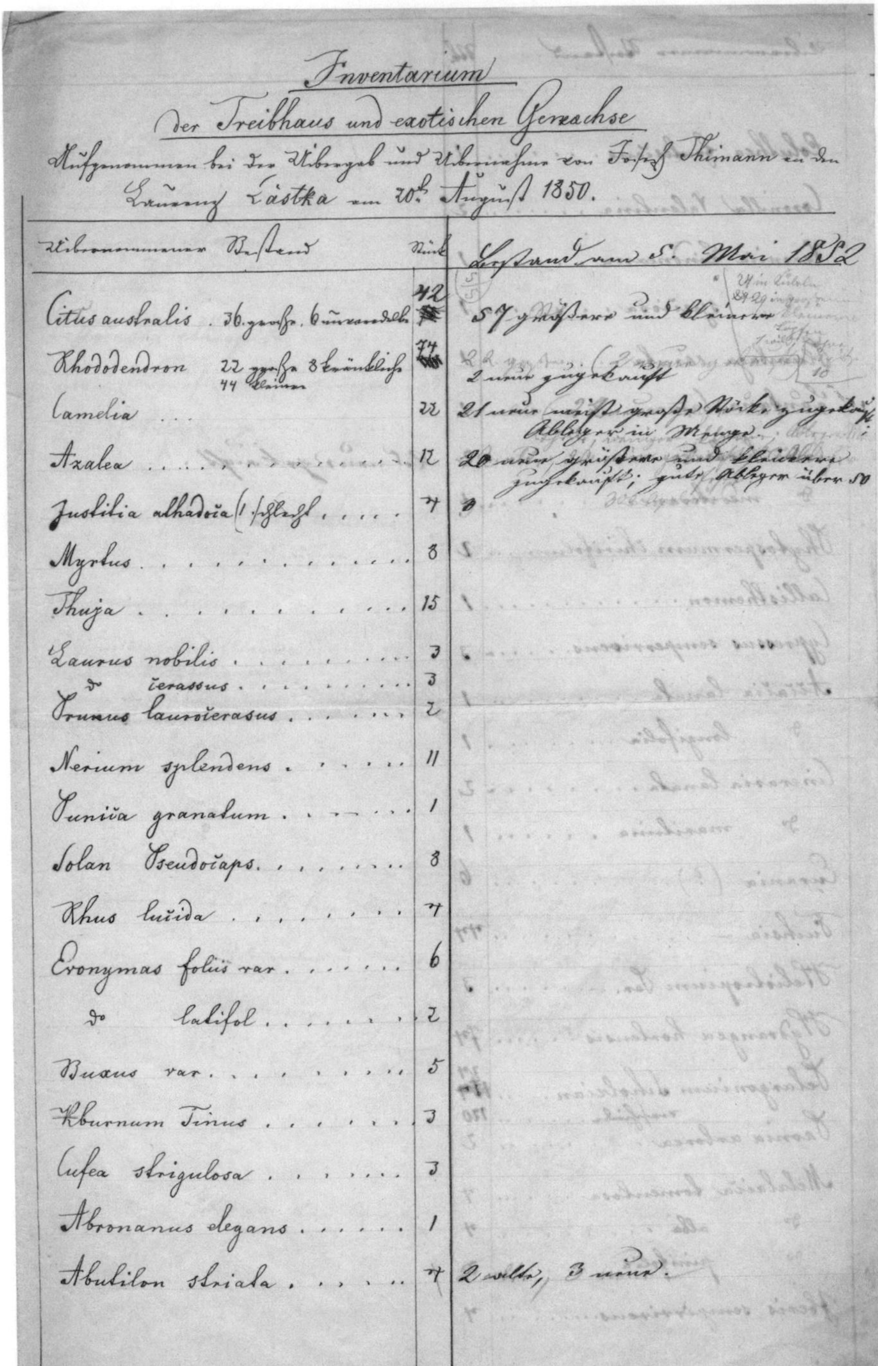

Inventarium

der Treibhaus und exotischen Gewächse

Aufgenommen bei der Übergabe und Übernahme von Joseph Thimann an den Laurenz Čästka am 20. August 1850.

Übernommener Bestand	Stück	Bestand am 5. Mai 1852
Citrus australis . 36 große, 6 [illegible]	42	57 größere und kleinere
Rhododendron 22 große 8 [illegible] 44 kleine	74	2 neue zugekauft
Camelia . . .	22	21 neue meist große Stück zugekauft, Ableger in Menge
Azalea	12	20 neue größere und kleinere zugekauft; [illegible] Ableger über 50
Justitia adhadoia (1 schlecht)	4	0
Myrtus	8	2
Thuja	15	1
Laurus nobilis	3	
do cerassus	3	
Prunus laurocerasus	2	
Nerium splendens	11	
Punica granatum	1	
Solan Pseudocaps.	8	
Rhus lucida	4	
Evonymus foliis var.	6	
do latifol	2	
Buxus var.	5	
Viburnum Tinus	3	
Cufea strigulosa	3	
Abronanus elegans	1	
Abutilon striata	4	2 volle, 3 neue

D288 The first page of the inventory of plants in the greenhouse and exotic plants transferred from Joseph Thimann to Laurenz Castka in the Augustinian monastery in Old Brno on 20 August 1850

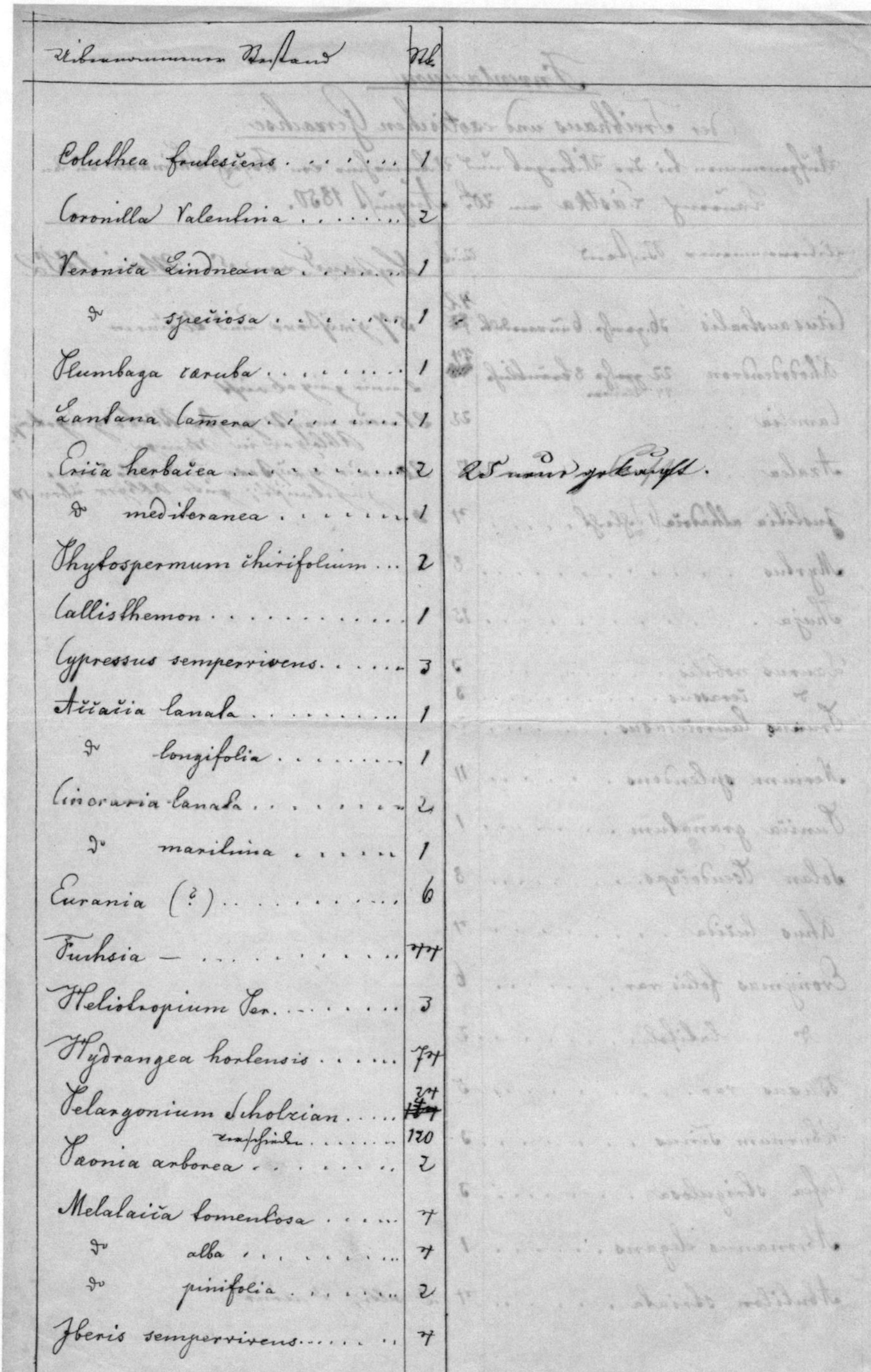

Übernommener Bestand	Nr.	
Coluthea frutescens	1	
Coronilla Valentina	2	
Veronica Lindneana	1	
do speciosa	1	
Plumbago caruba	1	
Lantana camera	1	
Erica herbacea	2	25 neu gepfropft
do mediteranea	1	
Phytospermum chirifolium	2	
Callisthemon	1	
Cypressus sempervirens	3	
Acacia lanata	1	
do longifolia	1	
Cineraria lanata	2	
do maritima	1	
Eurania (?)	6	
Fuchsia —	44	
Heliotropium Per.	3	
Hydrangea hortensis	74	
Pelargonium Scholzian	24 (~~134~~)	
Sämlinge	120	
Paeonia arborea	2	
Melalaica tomentosa	4	
do alba	4	
do pinifolia	2	
Iberis sempervirens	4	

D288 The second page of the inventory of plants in the greenhouse and exotic plants transferred from Joseph Thimann to Laurenz Castka in the Augustinian monastery in Old Brno on 20 August 1850

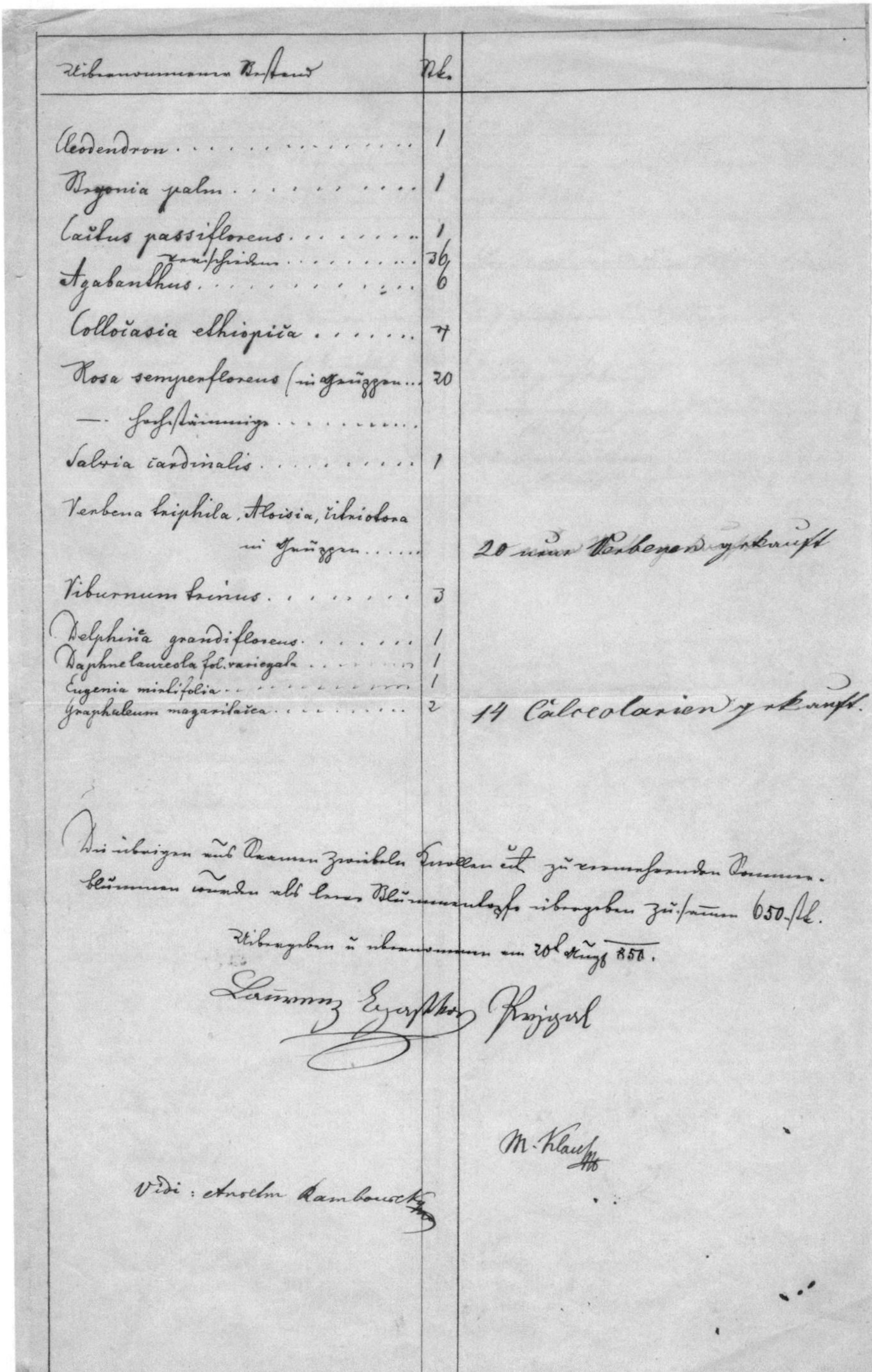

Übernommener Bestand	Nr.	
Clerodendron	1	
Begonia palm.	1	
Cactus passiflorens	1	
verschiedene	36	
Agapanthus	6	
Colocasia ethiopica	7	
Rosa semperflorens (in Töpfen ...	20	
— hochstämmige		
Salvia cardinalis	1	
Verbena triphila, Aloisia, citriodora in Töpfen		20 neue Verbenen gekauft
Viburnum tinus	3	
Delphinia grandiflorens	1	
Daphne laureola fol. variegata	1	
Eugenia mirtifolia	1	
Gnaphalium margaritacea	2	14 Calceolarien gekauft.

Die übrigen aus Samen Zwiebeln Knollen etc. zu vermehrenden Sommer-blumen wurden als [illegible] Blumen[illegible] übergeben zusammen 650 fl.

Übergeben u. übernommen am 20t Aug 850.

Laurenz Czastka [illegible]

M. Klacel

Vidi: Anselm Rambousek

D288 The last page of plants bred in the greenhouse and exotic plants transferred from Joseph Thimann to Laurenz Castka on 20 August 1850. Signed Laurenz Czastka, M. Klacel and Anselm Rambousek

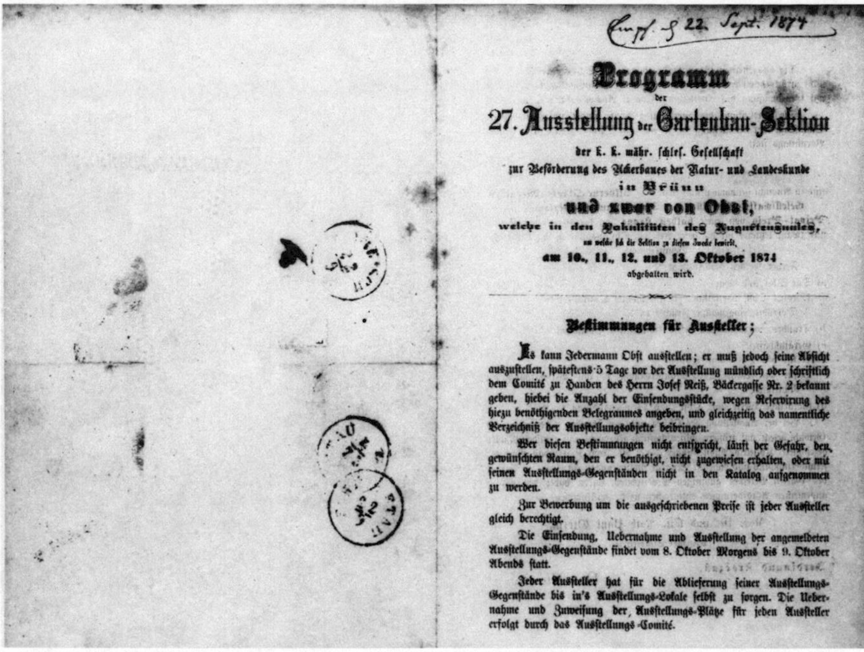

Empf. am 22. Sept. 1874

Programm

der

27. Ausstellung der Gartenbau-Sektion

der k. k. mähr. schles. Gesellschaft
zur Beförderung des Ackerbaues der Natur- und Landeskunde
in Brünn

und zwar von Obst,

welche in den Lokalitäten des Augartensaales,
um welche sich die Sektion zu diesem Zwecke bewirbt,

am 10., 11., 12. und 13. Oktober 1874

abgehalten wird.

Bestimmungen für Aussteller;

Es kann Jedermann Obst ausstellen; er muß jedoch seine Absicht auszustellen, spätestens 5 Tage vor der Ausstellung mündlich oder schriftlich dem Comité zu Handen des Herrn Josef Reiß, Bäckergasse Nr. 2 bekannt geben, hiebei die Anzahl der Einsendungsstücke, wegen Reservirung des hiezu benöthigenden Belegraumes angeben, und gleichzeitig das namentliche Verzeichniß der Ausstellungsobjekte beibringen.

Wer diesen Bestimmungen nicht entspricht, läuft der Gefahr, den gewünschten Raum, den er benöthigt, nicht zugewiesen erhalten, oder mit seinen Ausstellungs-Gegenständen nicht in den Katalog aufgenommen zu werden.

Zur Bewerbung um die ausgeschriebenen Preise ist jeder Aussteller gleich berechtigt.

Die Einsendung, Uebernahme und Ausstellung der angemeldeten Ausstellungs-Gegenstände findet vom 8. Oktober Morgens bis 9. Oktober Abends statt.

Jeder Aussteller hat für die Ablieferung seiner Ausstellungs-Gegenstände bis in's Ausstellungs-Lokale selbst zu sorgen. Die Uebernahme und Zuweisung der Ausstellungs-Plätze für jeden Aussteller erfolgt durch das Ausstellungs-Comité.

D479 The first page of the programme of the 27th exhibit, namely of fruits, organized by the Horticulture Section of the Imperial-Royal Moravia Silesian Society for the Improvement of Agriculture and Knowledge of Nature and the Land in Brünn/Brno, held in Augarten/Lužánky in Brünn/Brno on 10, 11, 12 and 13 October 1874. A written note (Empf. am 22. Sept. 1874) on the first page states, "received on 22 September 1874"

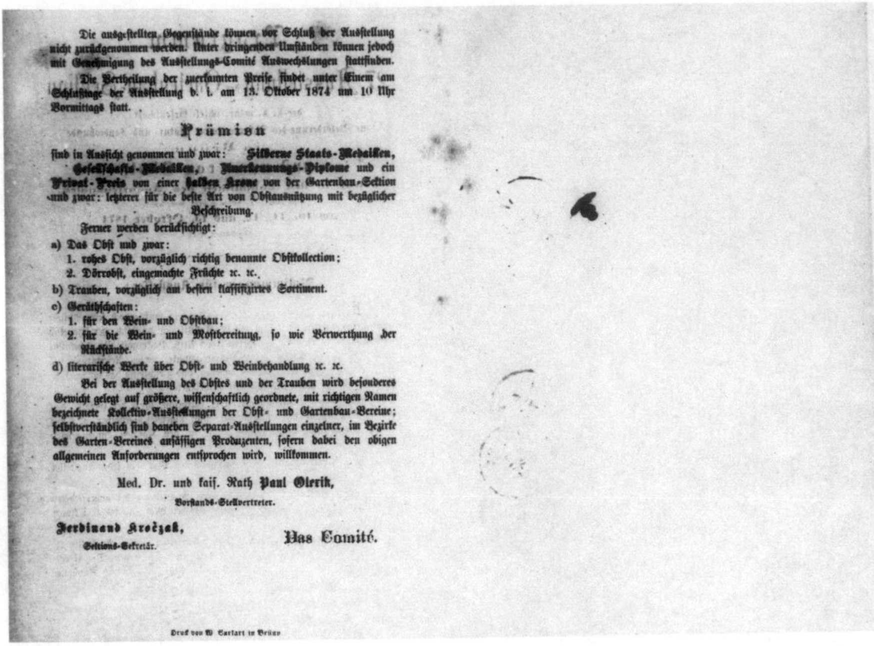

Die ausgestellten Gegenstände können vor Schluß der Ausstellung nicht zurückgenommen werden. Unter dringenden Umständen können jedoch mit Genehmigung des Ausstellungs-Comité Auswechslungen stattfinden.

Die Vertheilung der zuerkannten Preise findet unter Einem am Schlußtage der Ausstellung d. i. am 13. Oktober 1874 um 10 Uhr Vormittags statt.

Prämien

sind in Aussicht genommen und zwar: **Silberne Staats-Medaillen, Gesellschafts-Medaillen, Anerkennungs-Diplome** und ein **Privat-Preis** von einer **halben Krone** von der Gartenbau-Sektion und zwar: letzterer für die beste Art von Obstausnützung mit bezüglicher Beschreibung.

Ferner werden berücksichtigt:

a) Das Obst und zwar:
   1. rohes Obst, vorzüglich richtig benannte Obstkollection;
   2. Dörrobst, eingemachte Früchte ꝛc. ꝛc.

b) Trauben, vorzüglich am besten klassifizirtes Sortiment.

c) Geräthschaften:
   1. für den Wein- und Obstbau;
   2. für die Wein- und Mostbereitung, so wie Verwerthung der Rückstände.

d) literarische Werke über Obst- und Weinbehandlung ꝛc. ꝛc.

Bei der Ausstellung des Obstes und der Trauben wird besonderes Gewicht gelegt auf größere, wissenschaftlich geordnete, mit richtigen Namen bezeichnete Kollektiv-Ausstellungen der Obst- und Gartenbau-Vereine; selbstverständlich sind daneben Separat-Ausstellungen einzelner, im Bezirke des Garten-Vereines ansässigen Produzenten, sofern dabei den obigen allgemeinen Anforderungen entsprochen wird, willkommen.

Med. Dr. und kais. Rath **Paul Olexik**,
Vorstands-Stellvertreter.

Ferdinand Kročzak,
Sektions-Sekretär.

Das Comité.

Druck von W. Burkart in Brünn

D479 The second page of the programme of the 27th exhibition organized by the Horticulture Section of the Agricultural Society, namely of fruits and grapes, to be held from 10 October to 13 October 1874, gives the prizes for collections presented by horticultural societies and individual fruit tree growers. Signed Paul Olexik (vice-chairman) and Ferdinand Kročzak (Secretary of the Section) on behalf of the committee

Mendel's abbatial emblem painted in the library ceiling shows complementarity of antagonisms. The equation of Alpha = Omega may be the simplest expression of hybrid's development where antagonistic elements are in equilibrium. Further attributes of Mendel's emblem are a hat (identical to that of the archbishop), the mitre and the crosier. The white lily is the symbol of chastity; the plough represents Mendel's peasant origin, and the cross in the laurel wreath shows the victory of faith; the flaming heart of St. Augustine symbolizes Mendel as a member of the Order of St. Augustine; shaking hands with a special position of fingers (used commonly by freemasons) may express humanity and friendship

Mendel's younger emblem with the pectoral cross and the Comthur Cross of Franz Joseph

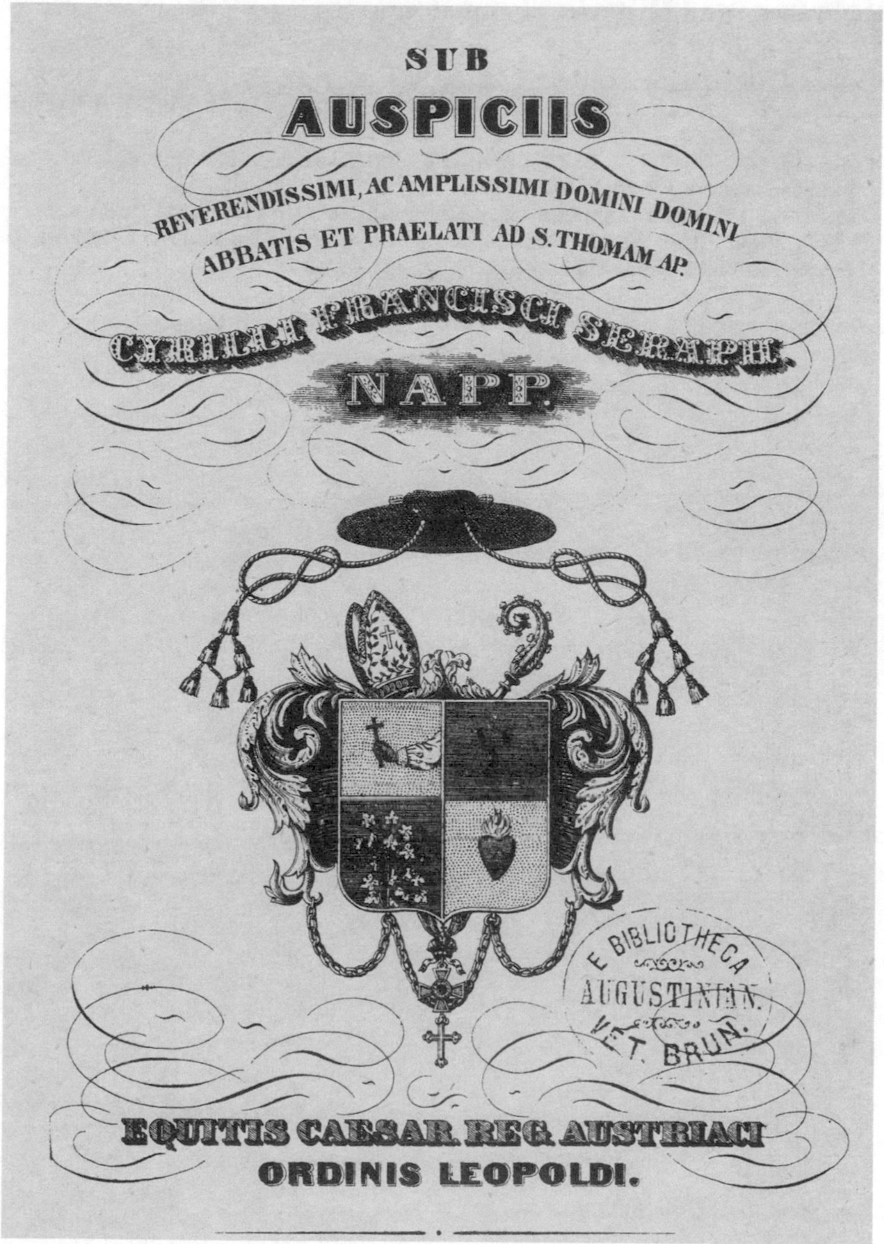

The emblem of Napp shows the Holy Trinity, the flaming heart of St. Augustine, the vine shrub and clergyman's arm holding the cross

References and Historical Printed Sources

Matalová A (1984) Response to Mendelś death in 1884. FM 19:217–221
Stern C, Sherwood ER (eds) (1966) The origin of genetics. A Mendel Source Book. W. H. Freeman and co, San Francisco
van Dijk PJ, Weissing FJ, Noel Ellis TH (2018) How Mendel's interest in inheritance grew out of plant improvement. Genetics 210:347–355
Vávra M (1984) Mendel's cooperation with the Fuchsia breeder J. N. Tvrdý. FM 19:251–256
Vybral V (1968) Die leitende Funktion des Abtes Gregor Mendel in der Mährischen Hypothekenbank und ihr politischer Hintergrund. FM 3:21–34

Chapter 18
Temporospatial Milestones of His Life

A brief life flow of Johann/Gregor Mendel (he signed either Johann Mendel or Gregor Mendel, never Gregor Johann or Johann Gregor Mendel) temporospatial context is essentially connected with Brno. Mendel lived here from 1843 to 1884.

Mendel attended the elementary school in Hynčice for 2 years (1831–1833), the third form at the District Main School (Kreishauptschule) in Lipník (1833–34). As a talented pupil, Johann continued his studies at the 6-year *Gymnasium* in Opava (1834–1840). Heading a teacher's career, Johann enrolled at the Institute of Philosophy in Olomouc (1840–1843). Due to harsh circumstances described in his CV, Johann had to change his plans (and name). One year of noviciate (1843–1844) was followed by 4 years of theological studies in Brno (1844–1848). Psychosomatic indisposition interfered with his chaplain duties (1848–1849) so much that Mendel was brought back to his original intent to be a teacher. His substitute teaching at the *Gymnasium* in Znojmo (1849–1850) and at the Technical Institute in Brno (1851) proved his pedagogic skills, autodidactic ability, industry and perseverance. Despite his failed attempt at the Viennese university in 1850 to get the certificate of proficiency in teaching at Gymnasia Mendel continued in his teaching activities as a substitute. In 1851, Mendel joined the Agricultural Society in Brno which was a learned Society substituting the role of Academy in Moravia and Silesia. After his university training in Vienna (1851–1853) Mendel was prepared for his experiments on plant hybrids. He was teaching at the *Realschule* (1855–1868) and preparing the initial conditions of his eight-year experimental trial with peas (1856–1864). In 1861, he participated in founding the progressive *Naturforschender Verein* that organized lectures on his discovery and published his discovery paper in its academic journal Verhandlungen des naturforschenden Vereines in Brno in 1866.

In 1868 Mendel was elected abbot. His ceiling paintings of 1875 in the Augustinian monastery review his successful activities in agriculture, meteorology, apiculture, pomiculture, horticulture, viniculture and plant hybridization. The paintings in the Library Hall symbolize the golden era of science in the Augustinian monastery under abbots Napp and Mendel.

A. Matalová, E. Matalová, *Gregor Mendel - The Scientist*, Springer Biographies,
https://doi.org/10.1007/978-3-030-98923-1_18

The last 10 years of Mendel's life (1875–1884) were marked heavily with his vain struggle against the increased contributions to the religious fund. The abbatial duties ceased his scientific research that he liked so much. Discussions of scientific questions with his learned colleagues from the *Realschule*, Agricultural Society and *Naturforschender Verein* who came to visit him in his prelate quarters and his loving relation to his two sisters and their families supported Mendel psychically in his resistance *(nimmer sich beugen)* to the increased taxes.

Mendel liked chess, lotteries and humour. In his last letter he invited Alois Schindler to see him in the monastery: *In der Hoffnung Dich recht bald in der bewussten Kerker zu sehen, zeichnet sich Dein immer treuer Vetter Gregor.* (*Hoping to see you very soon in the well-known prison here, I am always yours faithful ally Gregor.)*

Mendel's nephew Alois Schindler lived in the Cow Country abounding in fertile fields, flourishing meadows and woody hills, the country of Mendel's childhood.

Mendel died on 06 January 1884. His burial took place on 09 January.

Requiem in the Church of the Assumption of the Virgin in Old Brno was conducted by the famous music composer Leoš Janáček (1854–1928).

A view from Hynčice to Pohoř Hill. Photo Milan Hofer, 1964

IM4 The brick-built house in Hynčice where Mendel was born on 20 or 22 July 1822

IM 5 The parish church in Vražné where Mendel was christened with the name Johann

IM7 Old elementary school in Hynčice that Mendel attended from 1831 to 1833. His teachers were Thomas Makitta and Johann Schreiber

IM 8 The main district school in Lipník upon Bečva river. Mendel attended only the last form of the 3-year college in the school year 1833/34 and absolved with best results

IM 9 The *Gymnasium* (secondary grammar school) in Opava which Mendel attended from 1834 to 1840. In spite of interfering illness, Mendel finished the *Gymnasium* with best results in 6 years

IM11 Philosophical Institute in Olomouc which Mendel attended from 1840 to 1843. Mendel was a repentant of the first year. He interrupted his studies at the end of the first semester because of exhaustion. In August 1841, his father sold the family farm (fields, animals, gardens and buildings) to Alois Sturm, husband of his older sister Veronika to enable Mendel to finish his philosophical studies in Olomouc

IM13 The house in Olomouc on the Lower Square where Mendel lived during his studies at the Philosophical Institute

IM15 Brno from the time when Mendel came there in 1843. Monochrome lithograph by Adolf Kunicke c. 1838. Moravian Library in Brno

IM72 Old Brno from Červený kopec (Red Hill) with the Old Brno Augustinian church on the left. The bridge across the river Svratka leads to Vienna. Two-toned lithograph by the French artist Nicolas Maria Joseph Chapuy (1790–1858), c. 1852. The Brno Metropolitan Archives

IM16 The church of the Assumption of the Virgin and the eastern wing of the Augustinian monastery in Old Brno that has been the seat of the Augustinians since 1783. Two-toned lithograph signed Winkler. Reproduction by Artistische Anstalt von Reiffenstein and Rösch, Vienna

The Theological Institute in Brno which Mendel attended from 1844 to 1848

The Philosophical Institute in Brno where Mendel took courses in agriculture, horticulture, pomiculture and viniculture and passed examinations therefrom in 1846

IM29 The Dominical church of St. Michael in Brno in which Mendel was ordained priest on 06 August 1847

St. Anna Hospital situated close to the Augustinian monastery was in the pastoral and soul care of Mendel as Kooperator in the Old Brno Parish

IM36 The entry to the *Gymnasium* in Znojmo where Mendel was a substitute teacher in the school year 1849/50

MM The courtyard of the Moravian Museum was the seat of the Agricultural Society that established the museum in 1817. Mendel worked in the Agricultural Society for 33 years. In the background the Brno Cathedral. (The Mendelianum of the Moravian Museum was relocated from the Augustinian monastery to that place in 1999 after the restitution of the Augustinian monastery buildings to the church)

IM38 The old university in Vienna which Mendel attended as an extraordinary student (ausserordentlicher Hörer) from 1851 to 1853

IM39 The house in Vienna where Mendel lived during his studies at the Viennese university from 1851 to 1853. Photo from the first decade of the twentieth century

Weiling (1993) The provisional location of Doppler's Physical Institute in a private house in Erdberg, Vienna

IM43 Theresianum in Vienna where Redtenbacher's chemical laboratory was situated

IM50 The rear garden of the Augustinian monastery in Old Brno with the hothouse and a garden house. Its appearance about 1900

IM51 The small garden in the Augustinian monastery under the windows of the Augustinian refectory on the ground floor and the Library Hall on the first floor. Augustinian Aurelius Thaler founded the garden as an alpinum

The heated orangery for growing and overwintering exotic plants

IM92 The beehouse of Gregor Mendel built in 1871 in the monastic orchard protected from the north by the walls surrounding the monastic area. A cellar for overwintering the bees is excavated in the slope behind the beehouse

IM45 The technically oriented high school (*Realschule*) where Mendel taught physics and natural history from 1854 to 1868 when he was elected abbot of the Augustinian monastery and had to give up his beloved position of the teacher. Here, Mendel reported about his Experiments on Plant Hybrids in 1865 for the *Naturforschender Verein* that published his discovery paper in 1866 in its academic journal *Verhandlungen des naturforschenden Vereines in Brünn*

IM83 The Urban House where the office of the *Naturforschender Verein* was relocated from the *Realschule* in 1870. Director Auspitz was made the chief inspector of Moravian studies and left the *Oberrealschule*. The new director did not prolong the tusculum of the *Naturforschender Verein* in the *Realschule*. Mendel was made responsible for its relocation

IM46 In this urban building Mendel substituted the lectures of Professor J. Helcelet in 1851. From 1854 Mendel was teaching there physics and natural history at the Realschule that was provisionally accommodated in that private house. In 1859, the Realschule was relocated to the new building in Janská Street

IM84 The new building of the Imperial Royal Technical Learning Institute (K. K. Technische Lehr Anstalt) where the lecture meetings of the Natuforschender Verein were held. Mendel often presided over the lectures. He reported here on the tornado in 1870

IM97 As vice-director of the Moravian Mortgage Bank Mendel worked in the Kounic Palace from 1876 to 1878 where the bank had its provisional seat

IM101 The Regional House where the Moravian Mortgage Bank was situated from 1878. There Mendel worked as its vice-director and from 1881 as its Director. In 1883, Mendel resigned from position of the director of the Moravian Mortgage Bank in Brno for health reasons. On 07 May 1883, Mendel presided over the meeting of the Directorate of the Bank for the last time

IM87 The pavilion in Lužánky in Brno where exhibitions of fruits and vegetables were organized by the Agricultural Society in spring and autumn. From 1859 until his death Mendel took part in them as organizer, exhibitor, member of the jury and author of the prize-winning grants from plant hybridization

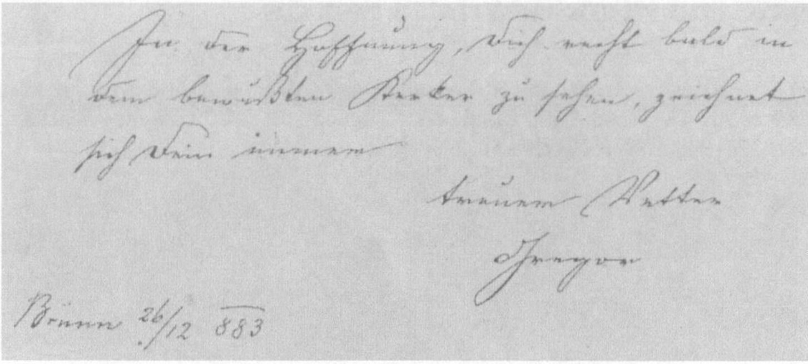

In der Hoffnung, Dich recht bald in dem bewußten Rector zu sehen, zeichnet sich Dein immer

treuer Vetter

Gregor

Brünn 26/12 883

The closing part of Mendel's letter to Alois Schindler dated in Brno, 26 December 1883, signed by Gregor

IM104 The Augustinian vault at the Brno Central Cemetery where Mendel is buried. Mendel's horizontal tomb stone reads in Latin: R. R. D.—GREGORIUS JO. MENDEL—ABBAS—OB. D. 6. JAN. 1884—R. I. P

References and Historical Printed Sources

Weiling F (1993/1994) Johann Gregor Mendel. Der Mensch und Forscher. Forscher in der Kontroverse. J. G. Mendel im Urteil der Zeitgenossen. Medizinische Genetik 5:35–51, 208–222, 274–289, 379–393, 6:35–51, 241–255.